T0185016

Structural Elements in Particle Physics and Statistical Mechanics

NATO ADVANCED STUDY INSTITUTES SERIES

A series of edited volumes comprising multifaceted studies of contemporary scientific issues by some of the best scientific minds in the world, assembled in cooperation with NATO Scientific Affairs Division.

Series B: Physics

Recent Volumes in this Series

Volume 76 – Metal Hydrides
edited by Gust Bambakidis

Volume 77 – Nonlinear Phenomena at Phase Transitions and Instabilities
edited by T. Riste

Volume 78 – Excitations in Disordered Systems
edited by M. F. Thorpe

Volume 79 – Artificial Particle Beams in Space Plasma Studies
edited by Bjørn Grandal

Volume 80 – Quantum Electrodynamics of Strong Fields
edited by W. Greiner

Volume 81 – Electron Correlations in Solids, Molecules, and Atoms
edited by Jozef T. Devreese and Fons Brosens

Volume 82 – Structural Elements in Particle Physics and Statistical Mechanics
edited by J. Honerkamp, K. Pohlmeyer, and H. Römer

Volume 83 – Physics of Ion-Ion and Electron-Ion Collisions
edited by F. Brouillard and J. W. McGowan

Volume 84 – Physical Processes in Laser-Materials Interactions
edited by M. Bertolotti

Volume 85 – Fundamental Interactions: *Cargèse 1981*
edited by Maurice Lévy, Jean-Louis Basdevant, David Speiser, Jacques Weyers, Maurice Jacob, and Raymond Gastmans

This series is published by an international board of publishers in conjunction with NATO Scientific Affairs Division

A Life Sciences B Physics	Plenum Publishing Corporation London and New York
C Mathematical and Physical Sciences	D. Reidel Publishing Company Dordrecht, The Netherlands and Hingham, Massachusetts, USA
D Behavioral and Social Sciences E Applied Sciences	Martinus Nijhoff Publishers The Hague, The Netherlands

Structural Elements in Particle Physics and Statistical Mechanics

Edited by

J. Honerkamp

K. Pohlmeyer

and H. Römer

Albert Ludwigs University
Freiburg, Federal Republic of Germany

PLENUM PRESS • NEW YORK AND LONDON
Published in cooperation with NATO Scientific Affairs Division

Library of Congress Cataloging in Publication Data

NATO Advanced Summer Institute on Theoretical Physics (1981: Freiburg im Breisgau, Germany)
Structural elements in particle physics and statistical mechanics.

(NATO advanced study institutes series. Series B, Physics; v. 82)
"Published in cooperation with NATO Scientific Affairs Division."
"Proceedings of the NATO Advanced Summer Institute on Theoretical Physics... held August 31-September 11, 1981, in Freiburg"—Verso t.p.
Includes bibliographical references and index.
1. Particles (Nuclear physics)—Congresses. 2. Statistical mechanics—Congresses. I. Honerkamp, J. II. Pohlmeyer, K. III. Römer, H. IV. North Atlantic Treaty Organization. Division of Scientific Affairs. V. Title. VI. Series.
QC793.N39 1981 539.7 82-10181
ISBN-13: 978-1-4613-3511-5 e-ISBN-13: 978-1-4613-3509-2
DOI: 10.1007/978-1-4613-3509-2

Proceedings of the NATO Advanced Summer Institute on Theoretical Physics 1981, held August 31-September 11, 1981, in Freiburg, Federal Republic of Germany

Softcover reprint of the hardcover 1st edition 1983
A Division of Plenum Publishing Corporation
233 Spring Street, New York, N.Y. 10013

PREFACE

The NATO ADVANCED SUMMER INSTITUTE ON THEORETICAL PHYSICS 1981 was held in Freiburg, Germany from August 31st until September 11th 1981. It was the twelfth in a series of Summer Institutes organized by German Universities.

Its main objective was a thorough comparison of structures and methods of two different branches of Theoretical Physics, namely Elementary Particle Physics and Statistical Mechanics, and the idea was to exhibit the structural similarities, to trace them until their origins, to compare solution and approximation schemes and to report on those new results and methods in either of the two branches which are indicative of an intimate connection between them. Thus stimulation of a deeper understanding and development of new Methods could be hoped for in both fields.

The contributions to the Summer Institute - lectures and seminars - are contained in this volume.

One group of them gives concise up-to-date information on basic topics in Statistical Mechanics and Phase Transitions, Dynamical Systems, Solvable Lattice Models and Lattice Gauge Theories.

A second group is devoted to special topics which illustrate the interrelationship between Statistical Mechanics and Elementary Particle Physics, like topological quantum numbers on a lattice, model studies on the confinement problem, etc.

Supplementary information on experimental implications and on neighbouring fields is provided in a third group.

We are grateful that the Institute was sponsored by the NATO Advanced Study Institute program and that additional support was given by the Bundesministerium für Wissenschaft und Technologie and the Ministerium für Wissenschaft und Kunst in Baden-Wurttemberg.

It is a pleasure to thank the lecturers and seminar speakers for their carefully prepared contributions.

Special thanks are due to Dr. M. Forger, the Scientific Secretary, to Dr. K. Fredenhagen, who did much redactionary work for these proceedings, Mrs. E. Rupp, our Conference Secretary and to Miss H. Kranz for typing large parts of the manuscripts, and all the others of the Physics Faculty of the University of Freiburg who were involved in the organization of this meeting.

Freiburg
February 1982

J. Honerkamp
K. Pohlmeyer
H. Römer

CONTENTS

CONNECTIVITY: A PRIMER IN PHASE TRANSITIONS AND CRITICAL PHENOMENA FOR STUDENTS OF PARTICLE PHYSICS

H. Eugene Stanley*

Center for Polymer Studies** and Department of Physics
Boston University, Boston, Massachusetts 02215 USA

When the organizers of this Advanced Study Institute on "Structural Elements in Particle Physics and Statistical Mechanics" asked me to present 3 hours of introductory material on critical phenomena, I accepted happily because of the opportunity clearly afforded to broaden my understanding of the important cross-disciplinary overlaps between particle physics and critical phenomena. Now that I find myself standing in front of such an esteemed group of scholars, I am having misgivings. There is hardly anything that I can say here that some of you could not say better!

Moreover, the field of phase transitions and critical phenomena has become so vast that I truthfully do not know just where to begin. Therefore I shall begin at the beginning--I shall assume that some of you, at least, have no prior knowledge of my field and I shall attempt to describe some of the basic methods and results. My presentation shall focus on the overall theme of connectivity, since this is the topic of much of my own research and since most of the concepts of phase transitions can be easily illustrated with the paradigm of connectivity--the percolation problem.

---

*John Simon Guggenheim Memorial Fellow, 1980-1981.
**Supported in part by grants from ARO, ONR, and NSF.

One of the first seminars I ever heard on the Ising model was in the early 1960's from my statistical mechanics professor Roy Glauber. He described a generalization of the Ising model (which itself has no dynamics) to what has come to be called the Glauber model. At the beginning of his seminar, Glauber actually apologized for having an "Ising disease", for at that time workers on the Ising model were thought to be "hooked" on the apparent simplicity of this model. Nowadays we appreciate the fact that it was not the researchers who had the Ising disease, but rather the critics who invented the term! Indeed, it would be difficult to imagine how the recent revolution in phase transitions and critical phenomena understanding could have occurred were it not for the firm foundation of understanding of the "simple" Ising model. By the end of this conference, the non-practitioner of percolation my be wondering if perhaps I and my colleagues have a mutant strain of the Ising disease. You may be right! However it is my hope that our studies of the simple percolation model (and its generalizations) will eventually provide the underpinnings of true progress on systems for which the essential physics would appear to be connectivity.

1. PERCOLATION--FROM SQUARE ONE

Imagine we have a fence that is infinite in spatial extent, and imagine that a randomly-chosen fraction p of the links of the fence are conducting and the remaining fraction $q = 1 - p$ are insulating. A finite section of this infinite fence is shown in Fig. 1. In Fig. 1a, $p=0.2$ and we see that the conducting links form finite clusters. In Fig. 1b, the conducting fraction p has doubled, yet the system still consists of only finite clusters--albeit larger than in Fig. 1a. In Fig. 1c, $p=0.6$ and the system is different globally: in addition to the finite clusters, there exists a single cluster that spans the entire system. There exists a critical value of p, termed the percolation threshold p_c, such that in the limit of an infinite system, the probability of finding an infinite cluster jumps from zero to one at p_c (Fig. 2). Thus the <u>macroscopic</u> connectivity properties of the system change discontinuously as a <u>microscopic</u> parameter p increases infinitesimally from $p_c - \delta$ to $p_c + \delta$.

There are many systems in nature for which the essential physics is connectivity, and one finds that percolation and its various modifications provide adequate descriptions of the rich range of phenomena observed. One striking example is the gelation threshold that occurs, e.g., when monomers with more than two reactive groups are allowed to form chemical bonds (Fig. 3). As the reaction proceeds, there occurs a critical "extent of reaction" α_c above which we find a single "gel" molecule infinite in spatial extent. The system changes from the sol phase below α_c to the gel phase above α_c (Fig. 4a).

A counterpart of bond percolation is site percolation, illustrated schematically in Fig. 4b. Suppose we have a ferromagnet made of a random mixture of two species of atoms,

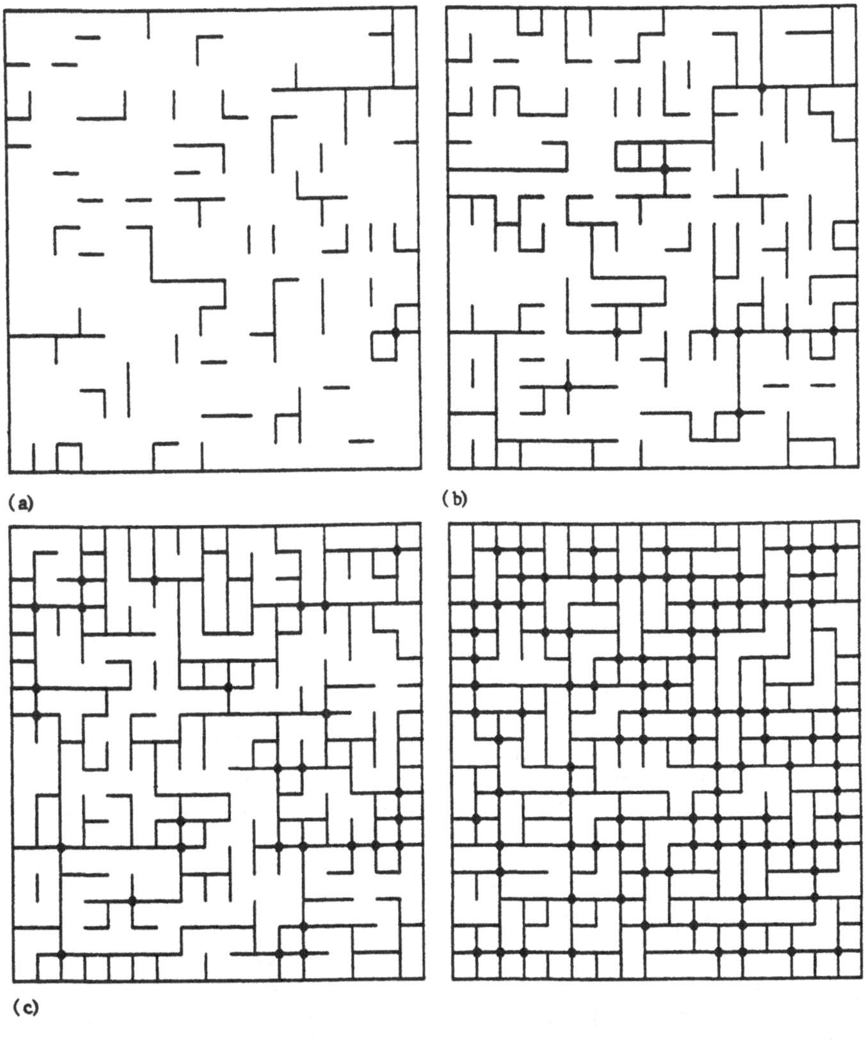

FIG. 1: The phenomenon of bond percolation: a finite section (16x16) of an infinite "fence", in which a fraction p of the links are conducting while the remaining fraction q = 1 - p are insulating. Four choices of the parameter p are shown, (a) p = 0.2, (b) p = 0.4, (c) p = 0.6, and (d) p = 0.8. After Stanley et al. (1980).

A and B. Species A (present in concentration p) has a magnetic moment and interacts with any species A atoms on neighboring lattice sites, while species B (present in concentration $q = 1 - p$) has no magnetic moment. There exists a site percolation threshold p_c such that below p_c an infinite cluster cannot exist and hence magnetic correlations cannot propagate across the lattice; the critical temperature T_c is zero for $p < p_c$. Above p_c an infinite cluster does exist,

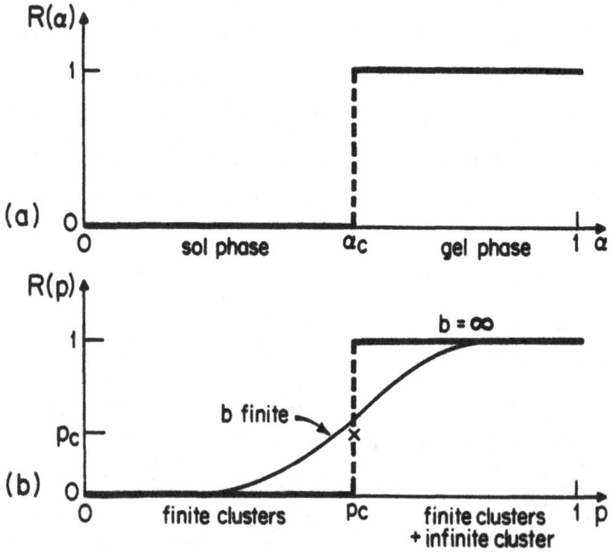

FIG. 2: Illustration of the analogy between (a) the gelation threshold and (b) the percolation threshold. In (a) the function $R(\alpha)$ denotes the probability that in an infinite system, there is at least one molecule infinite in spatial extent. In (b), $R(p)$ denotes the probability that there is at least one cluster infinite in spatial extent. The step function is for an infinite lattice, while the curve denotes the same probability function for a finite lattice of edge b (for which "infinite cluster" is interpreted to mean "spanning cluster"). The rounding apparent for the finite system becomes successively less pronounced as the system grows in size. For the infinite system, $R(p_c) = p_c$, so that p_c is a fixed point and the system is scale invariant. Thus in the renormalization group approach, percolation phenomena naturally divide into three regimes ($p<p_c$, $p=p_c$, and $p>p_c$). After Stanley et al. (1981).

and $T_c \neq 0$. The point Q with coordinates $(p = p_c, T=0)$ in a p-T phase diagram is a very "queer" multicritical point and has been the object of much experimental and theoretical investigation (Fig. 5).

It turns out that solvent effects on gelation are best described by a hybrid model, site-bond percolation, in which the sites of a lattice are occupied by species A (monomers) with probability p_{site} and by species B (solvent molecules) with probability $1 - p_{site}$, while bonds are intact with probability p_{bond} (Fig. 4c). It is also necessary to include lattice-gas correlation among the sites (Coniglio et al. 1979).

Apart from the applications to systems in nature, one of the principal motivations for studying connectivity phenomena is the fact that phenomena near the percolation threshold are remarkably parallel to ordinary "thermal" critical phenomena occurring near the critical point (Fig. 6). Since percolation is a purely

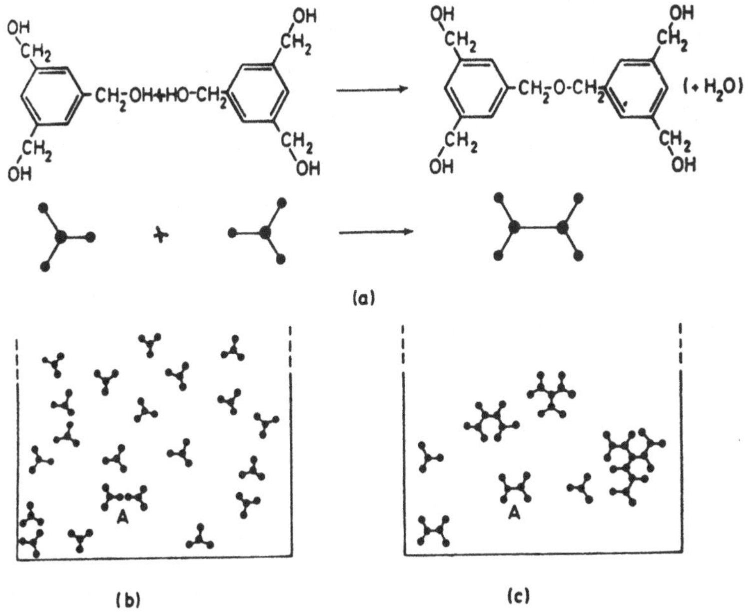

FIG. 3: Illustration of the simplest gelation phenomenon, polyfunctional condensation of f-functional monomers. The f-functional monomer shown in (a) is trimethoyl benzene; it has three "functional" groups that can react to form ether linkages. If f were two, then the most complex structures possible would be chains and rings; however, since f > 2 here, branched networks form. In (b) and (c) beakers are shown at successive stages of reaction. This figure is from Gordon and Ross-Murphy (1975).

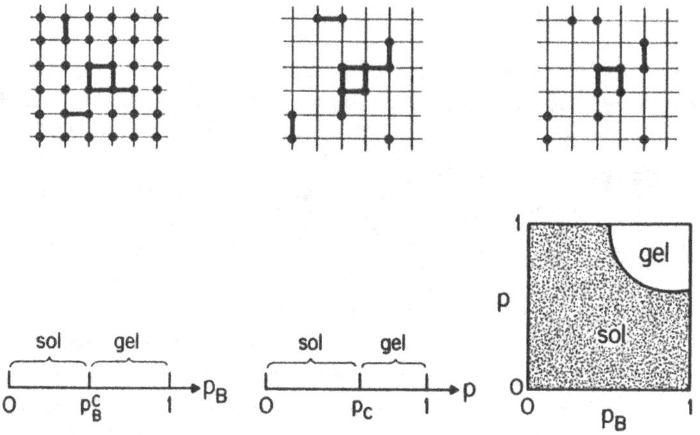

FIG. 4: Schematic illustration of (a) bond percolation, (b) site percolation, and (c) "site-bond" percolation, and their associated phase diagrams. The original Flory theory of polyfunctional condensation is an example of bond percolation, while most models of random magnets involve site percolation. Site-bond percolation is a hybrid model. When correlation is included among the sites, the resulting gelation model is capable of incorporating solvent effects and phase separation. From Stanley et al., (1982).

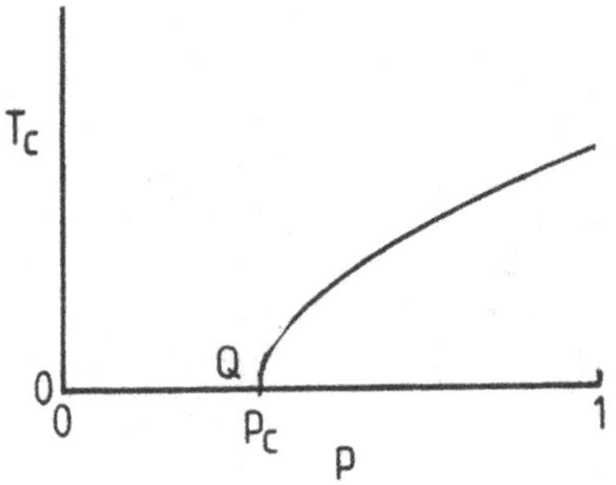

FIG. 5: Multicritical point of a dilute ferromagnet. Both connectivity and thermal fluctuations are singular at point Q and both the pair connectedness length ξ_p and the ordinary correlation length ξ_T diverge (see the recent review, Wu 1982).

geometrical model embodying the full richness of critical phenomena, percolation phenomena are sometimes referred to as "geometric" phase transitions.

In these lectures, I'd like to describe briefly three of the principal approaches used to study connectivity phenomena, (a)

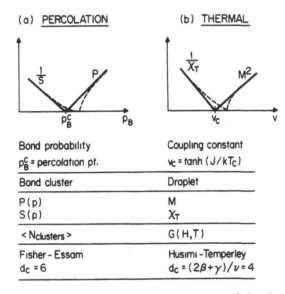

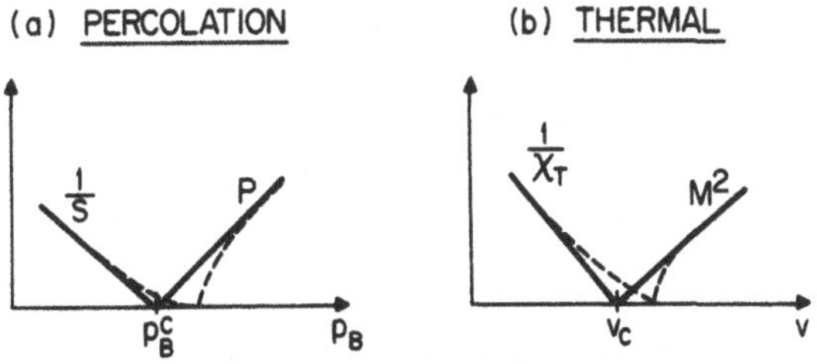

Bond probability p_B^c = percolation pt.	Coupling constant $v_c = \tanh(J/kT_c)$
Bond cluster	Droplet
P(p) S(p)	M χ_T
<$N_{clusters}$>	G(H,T)
Fisher - Essam $d_c = 6$	Husimi - Temperley $d_c = (2\beta+\gamma)/\nu = 4$

FIG. 6: Illustration of the analogies between (a) bond percolation, and (b) an ordinary thermal phase transition (e.g., an Ising or lattice-gas model). Adapted from Stanley and Teixeira (1980).

exact enumeration procedures, (b) Monte Carlo computer simulation, and (c) renormalization group. Then I'll conclude with brief mention of some of the tantalizing relations between connectivity phenomena and various models of thermal critical phenomena such as the Ising model, the Potts model, and a "polychromatic" generalization of the Potts model.

2. EXACT ENUMERATION APPROACHES

2.1 The functions P(s,p) and P(b,p)

Consider, for the sake of specificity, random-site percolation on a square lattice. What is the probability $P(s,p)$ that a randomly-selected site is a member of an s-site cluster? We can answer this question exactly for s small. For $s = 1 - 3$, we find (Fig. 7a)

$$P(s=1\ p) = pq^4 \tag{2.1a}$$

$$P(s=2,p) = 4p^2q^6 \tag{2.1b}$$

$$P(s=3,p) = 3p^3[4q^7 + 2q^8]. \tag{2.1c}$$

For bond percolation, it is customary to calculate $P(b,p)$, the probability that a randomly-chosen bond belongs to a b-bond cluster (Sykes *et al*. 1981 and references therein). Thus (Fig. 7b)

$$P(b=1,p) = pq^6 \tag{2.2a}$$

$$P(b=2,p) = 6p^2q^8 \tag{2.2b}$$

$$P(b=3,p) = 3p^3[2q^9 + 9q^{10}]. \tag{2.2c}$$

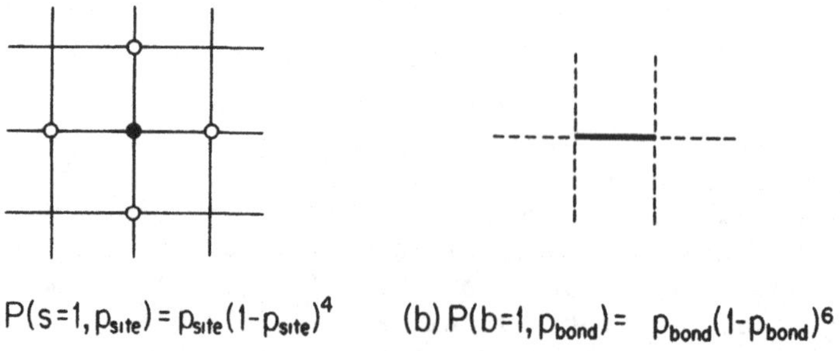

FIG. 7: (a) Derivation for the the square lattice of Eq. (2.1a), the probability that the origin belongs to a one-site cluster [site percolation, site counting]. (b) Derivation of Eq. (2.2a) [bond percolation, bond counting].

2.2 Lattice animals

The general form of $P(s,p)$ for site percolation and $P(b,p)$ for bond percolation is

$$P(L,p) = Lp^L D(L,q) , \tag{2.3a}$$

where the factor p^L arises from the L occupied elements (sites or bonds) and the "perimeter polynomial"

$$D(L,q) = \sum_t g_{Lt} q^t , \tag{2.3b}$$

arises from the unoccupied perimeter that bounds the cluster. Here g_{Lt} is the number of clusters or "lattice animals" with perimeter t. Since $D(L,q)$ gives the number of distinct lattice animals of L elements, grouped according to the number of perimeter elements,

$$A_L = D(L,q=1), \tag{2.4}$$

gives the total number of general L-element lattice animals regardless of perimeter.

The generating function for general lattice animals,

$$G_A(K) = 1 + \sum_{L\geq 1} A_L K^L , \tag{2.5a}$$

displays, for both bond and site percolation, a singularity of the form

$$G_A(K) \sim |K - K_c|^{\theta-1} . \tag{2.5b}$$

The critical parameter K_c is usually denoted $1/\lambda$, where λ is called the growth parameter because asymptotically

$$A_L / A_{L-1} \to \lambda . \tag{2.6}$$

Thus in an enumeration of the number of the animals of size L, one finds λ times as many animals as at the previous order.

2.3 Exact results: linear chain and Bethe lattice

Except for very special cases (such as a d=1 linear chain lattice and a "d=oo" Bethe lattice), we cannot evaluate P(s,p) and P(b,p) for s larger than s_{max}, and b larger than b_{max}. Here S_{max} and b_{max} depend on the complexity of the lattice and on the University computer budget (for the square lattice, $s_{max} = 19$ and $b_{max} = 13$).

For the linear chain, we see that for site percolation (Fig. 8a)

$$P(s,p) = sp^s q^2 \qquad (2.7)$$

if we measure site size, while for bond percolation (Fig. 8b)

$$P(b,p) = bp^b q^2 \qquad (2.8)$$

if we measure bond size (cf. Table 1). Thus

$$A_L = 1 \qquad (2.9)$$

for both site and bond lattice animals. Hence the generating function (2.5a) becomes

$$G_A = (1 - K)^{-1}, \qquad (2.10)$$

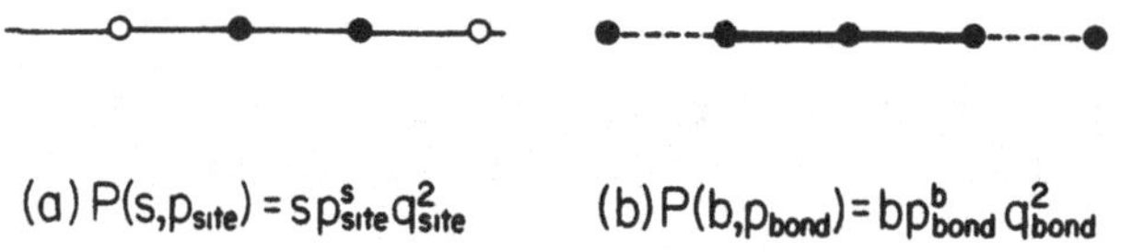

FIG. 8: (a) Derivation for the linear chain of (2.7), the probability that the origin belongs to a s-site cluster [site percolation, site counting]. (b) Derivation of Eq. (2.8) [bond percolation, bond counting]

TABLE 1

Exact results for P(s,p) and P(b,p) for site and bond counting for the d=1 site and bond percolation problems (Reynolds *et al.* 1977). The symbol [KF] indicates that the Kasteleyn-Fortuin correspondence for bond percolation counts cluster size by their *site* content. The symbol [EE] indicates that exact enumeration procedures for site percolation utilize site counting, and for bond percolation utilize bond counting. For site percolation, p denotes site probability, while for bond percolation it denotes bond probability; $q = 1 - p$.

	SITE PERCOLATION	BOND PERCOLATION
SITE COUNTING	$P(s,p) = sp^s q^2$ [EE]	$P(s,p) = sp^{s-1} q^2$ [KF]
BOND COUNTING	$P(b,p) = bp^{b-1} q^2$	$P(b,p) = bp^b q^2$ [EE]

so that the growth parameter is given by $\lambda=1$, and the critical exponent $\theta-1$ defined in (2.5b) is given by

$$\theta - 1 = -1. \qquad (2.11)$$

The A_L can *also* be evaluated exactly for the Bethe lattice, giving the result (Fisher and Essam 1961)

$$\theta - 1 = 3/2. \qquad (2.12)$$

It is believed that θ takes on the value 5/2 for all systems with $d > 8$ (Isaacson and Lubensky 1980). For $1 < d < 8$, the A_L have been evaluated exactly for small values of L, and one finds that θ varies smoothly with d (Gaunt 1980). Parisi and Sourlas (1981) have recently argued that

$$\theta(d) - 1 = \sigma(d') + 1. \qquad (2.13)$$

Here $\sigma(d') + 1$ is the exponent characterizing the singularity (Lee and Yang 1952) in the complex H plane of the Gibbs potential $G(H,T_o)$ for an Ising model of dimensionality $d' = d - 2$ at any temperature $T_o > T_c$ (Fig. 9).

2.4 Critical exponents

From the basic functions $P(s,p)$ one can calculate the analogs for percolation of the thermodynamic functions that are singular at normal critical points. For example, the role of the Gibbs potential $G(T)$ is played by $G(p)$, the mean number of clusters per site regardless of size. Clearly

$$G(p) = \sum_{s \geq 0} n(s,p), \qquad (2.14)$$

where

$$n(s,p) = s^{-1} P(s,p) \qquad (2.15)$$

is the number of s-site clusters per lattice site. Near p_c, one finds that G_{sing}, the singular part of $G(p)$, varies as

$$G_{sing} \sim |p - p_c|^{2-\alpha}, \qquad (2.16)$$

in complete analogy with the behavior of the singular part of the Gibbs potential $G(T)$ near the critical point T . For the

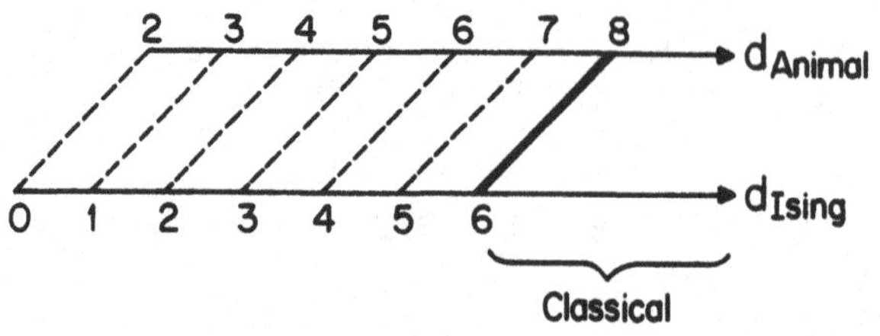

FIG. 9: Schematic illustration of the Parisi-Sourlas (1981) mapping between the lattice animal problem ($d_c=8$) and the Lee-Yang singularity problem in two lower spatial dimensions ($d_c=6$).

linear chain, $n(s,p) = p^s q^2$ for $s \geq 0$ [cf. (2.7)]. Hence from (2.14),

$$G(p) = q. \tag{2.17}$$

That the mean number of clusters per site is the fraction of empty sites is obvious since for a finite system the number of clusters is equal to the number of empty sites. Hence $2 - \alpha = 1$.

The mean size of a finite cluster is also given in terms of the functions $P(s,p)$

$$S(p) = \langle s \rangle = \sum_{s \geq 0} sP(s,p) / \sum_{s \geq 0} P(s,p). \tag{2.18}$$

As $p \to p_c$, one finds

$$S(p) \sim |p - p_c|^{-\gamma}, \tag{2.19}$$

in analogy with the behavior of the isothermal susceptibility for a magnet near its critical point. For the linear chain, (2.7) leads to (Reynolds et al., 1977).

$$S(p) = (1 + p)/(1 - p), \tag{2.20}$$

so that $\gamma = 1$.

The probability of the origin belonging to a finite cluster of <u>any</u> size is given by the denominator of (2.18),

$$P_F(p) = \sum_{s \geq 0} P(s,p), \tag{2.21}$$

and hence the probability of the origin belonging to an infinite cluster is

$$P_\infty(p) = p - P_F(p). \tag{2.22}$$

As $p \to p_c$, one finds

$$P_\infty(p) \sim |p - p_c|^{\beta}, \tag{2.23}$$

in analogy with the behavior of the spontaneous magnetization M for a magnet near the critical point. For a linear chain, substituting (2.7) into (2.21) results in a series that converges for all $p < 1$, and gives $P_F(p) = p$. Hence by (2.22) $P_{oo}(p) = 0$ except for $p = p_c = 1$. Thus $\beta = 0$ for $d=1$.

The analog in percolation of the mean field or "classical" theory is the Bethe lattice, which predicts correct values of the critical exponents for $d > d_c$; in percolation $d_c = 6$ and the values of the classical exponents are summarized, along with their definitions, in Table 2.

The "art" involved in the exact enumeration procedures is to devise tricks that permit one to extrapolate from the information contained in the functions $P(s,p)$ for $s \leq s_{max}$ to obtain estimates of critical properties. One standard procedure is to notice from the general form of the function $P(s,p)$ that if we set $q = 1 - p$, we have an exact expansion thorugh order s_{max}. This "low-density expansion", were it not truncated, would have a singularity on the positive real axis at p_c (Fig. 10a) which might be located using methods such as Padé approximants.

Thus, e.g., by substituting (2.1) into (2.18), we obtain to order two the low-density expansion of the mean cluster size for site percolation on a square lattice,

$$S(p) = 1 + 4p + 12p^2 + \ldots \tag{2.24}$$

Note that when we expand $P_F(p)$ about $p=0$, we obtain

$$P_F(p) = p \tag{2.25}$$

to all orders, as may be verified by substituting (2.1) into (2.21) and setting $q = 1 - p$.

We can also set $p = 1 - q$ in the functions $P(s,p)$ and thereby obtain a family of "high-density expansions" in the variable q (Fig. 10b). Thus, for example, on substituting (2.1) into (2.16), we obtain through order four

$$P_F(p) = q^4 + \ldots \tag{2.26}$$

which is obvious since we must create four empty sites in order to obtain a one-site cluster in a nearly-full lattice.

Table 2. Comparison between exponents for connectivity and thermal critical phenomena.

		PERCOLATION ($d_c=6$)	THERMAL ($d_c=4$)
$2-\alpha$:		$G_{sing}(p) \sim \lvert p - p_c \rvert^{2-\alpha}$	$G_{sing}(T) \sim \lvert T - T_c \rvert^{2-\alpha}$
	$d=1$	1	1
	$d \geq d_c$	3	2
β:		$P_{\infty}(p) \sim \lvert p - p_c \rvert^{\beta}$	$M(T) \sim \lvert T - T_c \rvert^{\beta}$
	$d=1$	0	0
	$d \geq d_c$	1	1/2
γ:		$S(p) \sim \lvert p - p_c \rvert^{-\gamma}$	$\chi(T) \sim \lvert T - T_c \rvert^{-\gamma}$
	$d=1$	1	1
	$d \geq d_c$	1	1
$\alpha + 2\beta + \gamma$:			
	$d=1$	2	2
	$d \geq d_c$	2	2

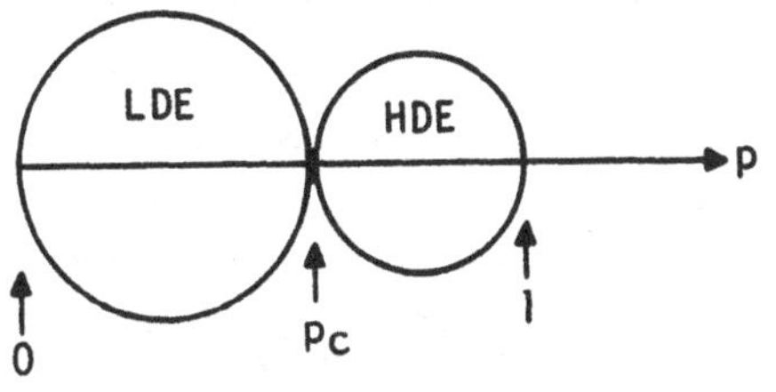

FIG. 10: Schematic illustration of the idealized circles of convergence in the complex p plane for low-density expansions (LDE) and high-density expansions (HDE).

3. COMPUTER SIMULATION APPROACHES

3.1 Methods

Probably no problem is more tantalizingly simple to the computer enthusiast than percolation. Two approaches spring immediately to mind, and each has been developed fairly extensively in recent years.

In the first approach, we start by placing a site at the origin and then use a random number generator to assign a random number to each of the z nearest neighbors of this site. If the random number is less than p, we join the neighbor to the site at the origin, while if the number is larger than p we do not. For p near zero, this process dies out after very few steps and we are left with a tiny cluster. We then repeat this process, say, 10^4 times to obtain good statistics. Among the advantages of this approach is the lack of any "boundary". A drawback is that it is difficult to obtain good statistics close to p_c.

A second approach is to begin with a finite section of a lattice with b^d sites. We assign a random number to every one of the sites, and then proceed to design an algorithm that recognizes clusters efficiently. This procedure gives the distribution $n(s,p)$ defined in (2.15), which means that to generate one 10,000-site cluster near p_c requires generating 10,000 times as many one-site clusters as required in the first approach. A second disadvantage is that the predictions are hampered by "boundary effects," though this apparent "cloud" has a silver lining one can study a sequence of different "box sizes" b and then use the theory of finite-size scaling to extract the exponent ν defined by

$$\xi \sim |p - p_c|^{-\nu} . \tag{3.1}$$

Here the connectedness length ξ characterizes the exponential decay with r (near but not at p_c) of the pair connectedness function $C(\underset{\sim}{r})$, which in turn gives the probability that a site at position $\underset{\sim}{r}$ is occupied <u>and</u> connected to a site at the origin (Fig. 11).

Monte Carlo methods are complementary to exact enumeration procedures in some respects. For example, in the exact procedures, we obtain exact expressions for the cluster distribution function $P(s,p)$ for $s \leq s_{max}$, while in the direct computer simulation procedures we obtain highly approximate numerical extimates of $P(s,p)$ with s_{max} typically 2-3 orders of magnitude larger.

3.2 Scaling

The interpretation of the numerical results is so greatly aided by finite-size scaling that we shall take a short "detour" to explain this procedure (this detour was prompted by a question from the audience) In order to explain finite-size scaling, it is necessary to recognize the empirical fact that the functions $P(s,p)$ appear (Stauffer 1981, Nakanishi and Stanley 1980) to be generalized homogeneous functions in the variables s and $\epsilon \equiv (p_c - p)/p_c$. That is, asymptotically close to the "critical point" $(s=\infty, \epsilon=0)$ $P(s,\epsilon)$ obeys the functional equation (Hankey and Stanley 1972)

$$P(\lambda^{a_s} s, \lambda^{a_\epsilon} \epsilon) = \lambda P(s,\epsilon). \tag{3.2}$$

Equation (3.2) has many implications, among which are a family of "scaling laws" relating the critical-point exponents in percolation. For example, in Sec. 2 we defined the exponents $2-\alpha$, β, and γ that describe the singularities in $G(p)$, $P_\infty(p)$, and $S(p)$ respectively. These exponents can be related by the Rushbrooke scaling law (Stanley 1982b)

$$\alpha + 2\beta + \gamma = 2. \tag{3.3}$$

To see this, we first note that the critical exponent of any function is simply the ratio of the scaling power of the function to the scaling power of the variable describing the path of approach to the critical point. Thus, e.g., $P_F(p)$ obeys the functional equation

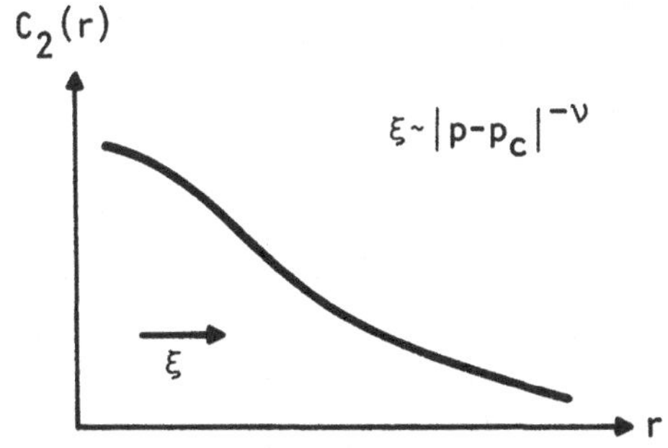

FIG. 11: Schematic illustration of the dependence on site separation $\tilde{r}$ of the pair-connectedness function $C_2(\tilde{r})$ (the probability that the origin is connected to a site at $\tilde{r}$). The decay with r is characterized by a length ξ termed the pair connectedness length, which is singular at the percolation threshold with exponent ν.

$$P_F(\lambda^{a_\epsilon}\epsilon) = \lambda^{1+a_s} P_F(\epsilon). \tag{3.4}$$

The scaling power is $1 + a_s$ since the summation in (1.16) is on the variable s. Since (3.4) is a functional equation valid for all positive λ, we can choose λ as we wish. Thus if we substitute

$$\lambda = |\epsilon|^{-1/a_\epsilon} \tag{3.5}$$

into (3.4), we obtain

$$P_F(\epsilon) \sim |\epsilon|^{\beta} \quad \text{with} \quad \beta = (1 + a_s)/a_\epsilon . \tag{3.6}$$

We attribute the same exponent to P_F and P_∞ since they are related by (2.22) and the function p is itself not singular at p_c.

The exponent γ defined in (2.19) can also be evaluated in terms of the two scaling powers. From the definition (2.18), it follows that $S(\epsilon)$ satisfies the functional equation

$$S(\lambda^{a_\epsilon}\epsilon) = \lambda^{1+2a_s} S(\epsilon), \tag{3.7}$$

where the "scaling power" $1 + 2a_s$ arises from the fact that the summation in (2.18) is on the variable s, and the summand itself is sP_s. Substituting (3.5) into (3.7), we find

$$S(\epsilon) = |\epsilon|^{-\gamma} \quad \text{with} \quad -\gamma = (1 + 2a_s)/a_\epsilon . \tag{3.8}$$

Finally, we can evaluate the exponent $2-\alpha$ defined in (2.16) in terms of a_s and a_ϵ. From (2.14) it follows that the singular part of $G(p)$ obeys the functional equation

$$G_{sing}(\lambda^{a_\epsilon}\epsilon) = G_{sing}(\epsilon), \tag{3.9}$$

since the scaling power $-a_s$ arising from the definition of $n(s,p)$ in (2.15) is cancelled by the scaling power a_s arising from the summation in (2.9). Hence on substituting (3.5) into (3.9) we find

$$G_{sing} \sim |p - p_c|^{2-\alpha} \quad \text{with} \quad 2 - \alpha = 1/a_\epsilon . \tag{3.10}$$

The scaling law (3.3) can now be obtained by eliminating the two unknown scaling powers a_s and a_ϵ from the three equations (3.6), (3.8), and (3.10).

Now we can introduce finite-size scaling. First we recall the functional equation (3.7) obeyed by the mean size function $S(\epsilon)$ for an infinite system. Suppose now that the system is confined to a box of edge b, where b is large but not infinite. The essential Ansatz behind finite-size scaling is that $S(\epsilon,b)$ is a generalized homogeneous function in the variables ϵ and b. Thus finite-size scaling makes the assumption that (Brézin 1981 preprint and references therein)

$$S(\lambda^{a_\epsilon}\epsilon, \lambda^{a_b} b) = \lambda^{a_S} S(\epsilon, b), \tag{3.11}$$

where $a_S = 1 + 2a_s$. Moreover, the new scaling power a_b is assumed to be numerically identical to the scaling power a_ξ characterizing the pair connectedness length, ξ .

A typical prediction of finite-size scaling is the following. Let us make computer simulations of the system directly at p_c, so that $\epsilon = 0$. Choose λ so that the second argument in (3.11) becomes unity. Thus

$$S(\epsilon = 0, b) \sim b^x \tag{3.12a}$$

where the exponent x is given by

$$x = a_S/a_b = (a_S/a_\epsilon)(a_\epsilon/a_\xi) = \gamma/\nu . \tag{3.12b}$$

Thus a plot of $S(\epsilon = 0, b)$ on double logarithmic paper should result, for large b, in a straight line of slope γ/ν (Fig. 12). In this fashion, we can obtain β/ν, $(2-\alpha)/\nu$, and so forth.

Generally speaking, Monte Carlo methods have not led to higher accuracy in estimating critical exponents than, say, exact enumeration procedures. However they have aided our understanding of many features of the percolation problem, as well as giving a graphic illustration of what a "typical" million-site cluster really looks like . . . and how one should describe it (Fig. 13).

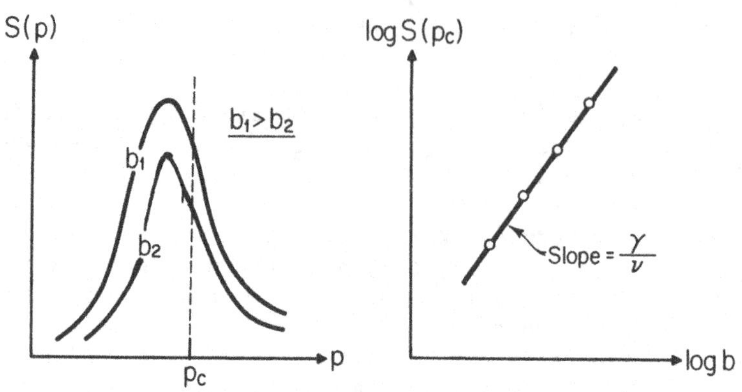

FIG. 12: Schematic illustration of finite-size scaling for S(p), the mean size of a finite cluster, in a sample of linear dimension b. An actual plot that is remarkably linear for b larger than 20 is given in Fig. 3 of Gawlinski and Stanley (1981).

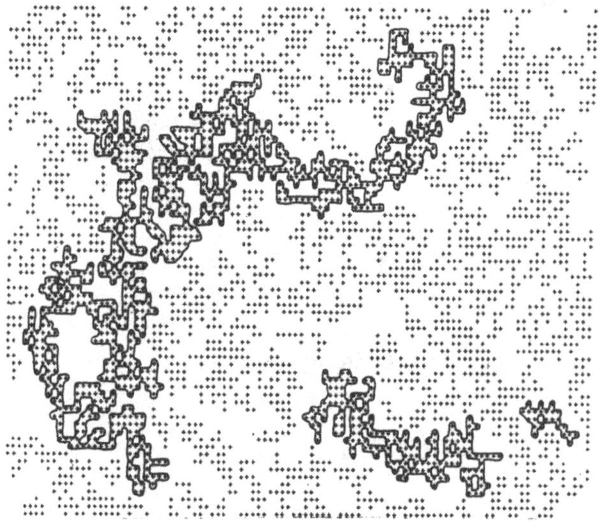

Fig. 13: Computer-generated picture of the diluted square lattice for p=0.53(p_c≃0.59). Here the plus symbol represents the presence of a magnetic site. All sites belonging to clusters of less than 5 sites have been eliminated as an aid to recognizing the shape of the larger clusters, which dominate the scattering. After Stanley et al. (1976).

3.3 Incipient infinite cluster

As an example, we conclude this section by mentioning one particularly fascinating aspect of the "percolation disease" that shows signs of yielding to solution, in large part due to Monte Carlo studies. This is the question of how one describes the incipient infinite cluster that appears as the percolation threshold is approached from below. Consider again the site-dilute random magnet. When the fraction p of magnetic sites is very small, the system consists of small disconnected clusters of magnetically correlated sites. As $p \rightarrow p_c^-$, the mean cluster size increases until at $p = p_c$ a single cluster spans the entire lattice. The percolation threshold p_c is a critical point. By studying the propagation of magnetic correlations through the incipient infinite cluster that appears in the dilute magnet at its percolation threshold, one can obtain information about order propagation near this critical point--a classic unsolved problem (Zernike 1940).

Since the incipient infinite cluster dominates the behavior of the system it is important to be able to describe its structure. If a cluster is considered as a network of wires carrying electrical current between two parallel bus bars, it can be decomposed into a conducting "backbone" and many "dangling ends" that do not contribute to the electrical conductivity (and hence order propagation) between the ends (Fig. 14). Describing the topology of this backbone is a formidable problem.

The backbone bonds may be divided into two classes, conveniently visualized as "red" and "blue" (Stanley 1977) Red (blue) bonds are singly connected (multiply-connected); removing a single red (blue) bond breaks (does not break) the connection between two parallel bus bars touching the most eastern and most

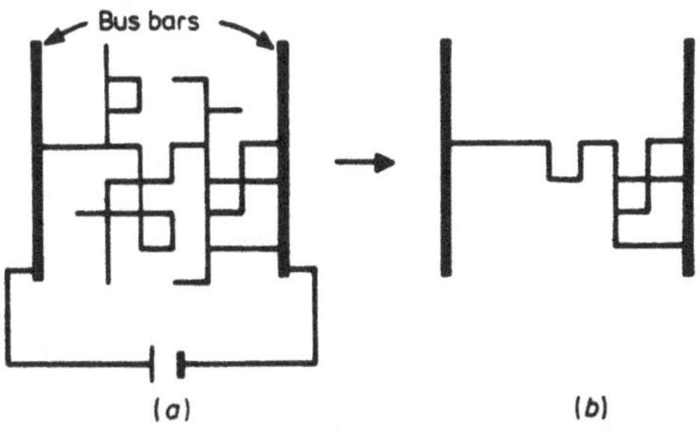

Fig. 14: Schematic illustration of the backbone of a cluster: bus bars are attached to the cluster extremeties and a potential difference is applied in (a). Current flows in the backbone bonds shown in (b), but not to the "dangling ends" which are attached to the backbone at only one vertex. After Shlifer et al. (1979).

western bonds. An example of this backbone decomposition is shown in Fig. 15, where "red" bonds are shown as heavy lines and "blue" bonds as light lines (the dangling ends are shown dashed).

The distinction between red and blue bonds seemed important in connection with the dilute Ising magnet--e.g., propagation of order along a string of red bonds would be analogous to propagation in a simple one-dimensional system, while propagation of order in a "blue blob" is more like propagation in the d-dimensional magnet. For this reason, I posed the "red-blue" problem in a talk given at the annual Canadian Undergraduate Physics Association in Toronto in November 1977, and an undergraduate member of the audience, Rob Pike, came up afterward to announce that he thought he could solve the problem.

Pike indeed succeeded in formulating a computer algorithm that partitions bonds into three separate classes: red, blue, and yellow (the dangling ends). One worry that clouded the Pike project was the possibility that as $p \to p_c^-$, the mean number of red bonds S_{red} would approach zero; this would occur, e.g., by the joining up of yellow dangling ends:

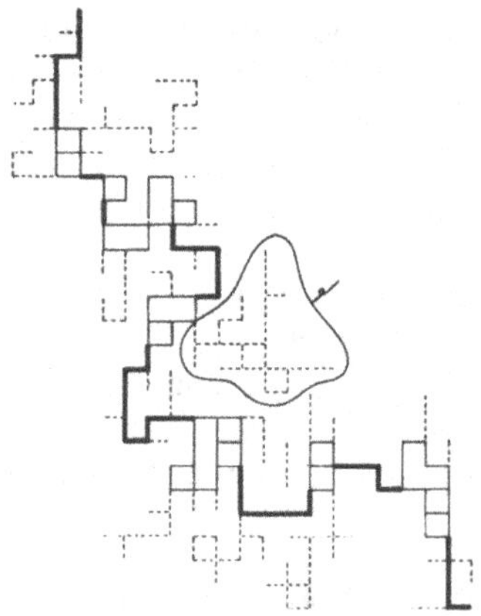

Fig. 15: Computer-generated 'incipient infinite cluster' for the bond percolation problem on a 25 x 20 square lattice with $p=p_c=0.5$. Note the qualitative feature of 'ramification' or 'stringiness' of this cluster, corresponding to the fact that the effective cluster dimensionality d_p is less than d (for d=2, $d_p \simeq 1.8$). If we think of this cluster as representing a dilute magnetic system in which only a fraction p of the exchange integrals are non-zero, then it is clear that magnetic correlations will spread from one end of the lattice to another along a path that is dominated by the 'backbone bonds' (shown as full lines) rather than by the 'dangling ends' (shown as broken lines). The backbone bonds that are singly-connected are shown as bold lines while those that are multiply-connected are shown as light lines. After Stanley (1977).

Fortunately, Pike's Monte Carlo data showed clearly that all three functions S_{red}, S_{blue}, and S_{yellow} diverge. The corresponding critical exponents are

$$\gamma_{red} \simeq 1.0, \qquad \gamma_{blue} \simeq 1.7, \qquad \gamma_{yellow} \simeq 2.4. \tag{3.13}$$

In connection with his analysis of the dilute ferromagnet, Coniglio (1981) has very recently proved <u>rigorously</u> that $\gamma_{red} = 1$.

From (3.13) and the result that the connectedness length $\xi \sim |p - p_c|^{-\nu}$ with $\nu = 4/3$ for both the full cluster <u>and</u> the backbone, we can evaluate the fractal dimensions (Stanley 1977; Mandelbrot 1977) D^+ and D_B^+ of the incipient infinite cluster and of the incipient infinite backbone (cf. Fig. 16). From the results (Stanley 1982a)

$$D^+ = (\gamma + d\nu)/2 \qquad D_B^+ = (\gamma_B + d\nu)/2 \quad , \tag{3.14a}$$

we find

$$D^+ \simeq 91/48 \simeq 1.9 \qquad D_B^+ \simeq 157/96 \simeq 1.6, \tag{3.14b}$$

where we have used the extended den Nijs conjecture (Stauffer 1981) $\gamma = 43/18 = 2.4$ and the conjecture $\gamma_{blue} = (\gamma_{red} + \gamma_{yellow})/2$.

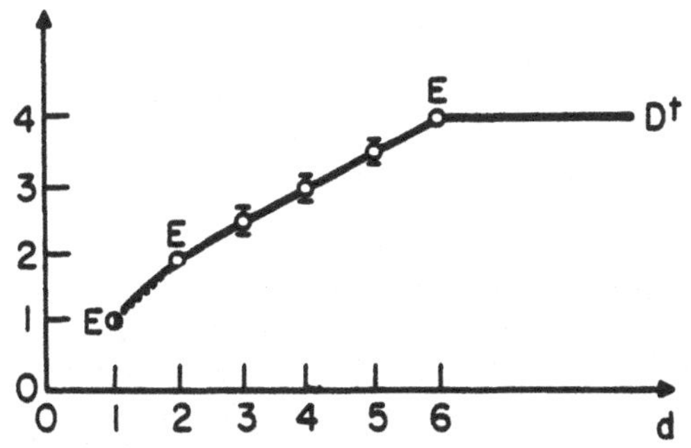

<u>FIG. 16</u>: Schematic illustration of the dependence upon d of the fractal dimension D^+. The behavior near d=6 is based on the predictions of renormalization group calculations. The points for d=3,4,5 are based on results from series expansions and computer simulations. The letter E denote "exact". After Stanley (1982a).

3.4 Some open questions

The utility of the red-blue decomposition of the backbone to the classic problem of order propagation in the dilute ferromagnet has recently been demonstrated--cf. Fig. 17 (Coniglio 1981). However there remain interesting questions relating to cluster topology.

One important open question is the relation to cluster properties of the exponent t describing the approach to zero as $p \rightarrow p_c^+$ of the electrical conductivity, $\sigma_{el} \sim (p - p_c)^t$. The beautiful experiments of Deutcher (1981) and other should serve as a strong stimulus for a program designed to put the exponent t on a firmer conceptual foundation.

Still another question concerns the new exponent γ_{min} defined through $L_{min} \sim |p_c - p|^{-\gamma_{min}}$, where L_{min} is the shortest "cow path length" through the cluster backbone (from one bus bar to the other). Pike and Stanley (1981) find $\gamma_{min} = 1.35 \pm 0.02$ when averaging over all clusters and $\gamma_{min} = 1.49 \pm 0.02$ when averaging over only the largest clusters.

Another open question concerns the nature of the "universality classes" for percolation (roughly speaking, two systems are said to be in the same universality class if they have the same critical exponents). Suppose, for example, that the elements are not constrained to the sites or bonds of a lattice. Do the exponents change? The answer to this question appears to

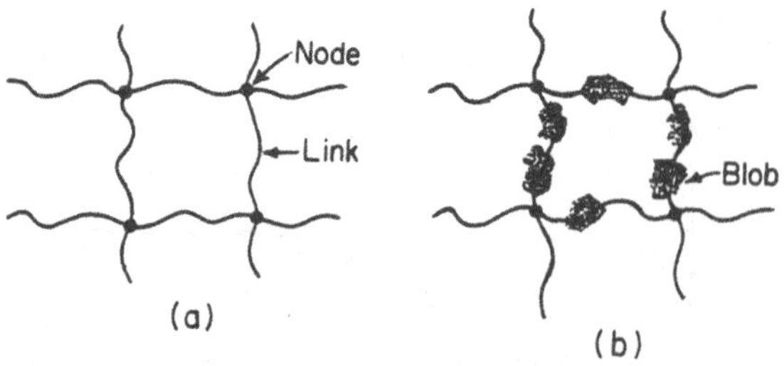

FIG. 17: Schematic illustration of (a) the conventional "nodes and links" picture of the infinite cluster just above the percolation threshold (Stauffer 1981), and (b) the "nodes, links, and blobs" generalization conjectured in Stanley (1977) and strongly supported by exact results (Coniglio 1981) and Monte Carlo calculations (Pike and Stanley 1981). The generalization consists of replacing the original "link" by a combination of singly-connected "red bonds" and multiply-connected "blue bonds" (blobs), as indicated perhaps more clearly in the schematic illustration of the incipient infinite cluster shown in Fig. 14. After Stanley (1982a).

be "no", at least for a simple $d=2$ system of overlapping discs (Fig. 18; see Gawlinski and Stanley 1981 and references therein). Preliminary computer simulation analysis of a three-dimensional system of interacting particles also suggests that the basic functions $P(s,p)$ are themselves unchanged (Fig. 19).

4. RENORMALIZATION GROUP APPROACHES

Percolation has been actively studied using both momentum-space and position-space renormalization group approaches. The former leads to the prediction that the upper marginal dimension $d_c=6$. Hence the utility of expansions in the variable (d_c-d) for systems of the dimension $d=2$ and 3 is limited.

A variety of position-space renormalization group approaches have been developed recently, and some hold promise of providing accuracy comparable to or even exceeding that provided by the exact enumeration approaches. These have been recently reviewed elsewhere (Stanley et al. 1982), so I'll limit my remarks here to a brief description of one particular avenue. (Reynolds et al., 1980).

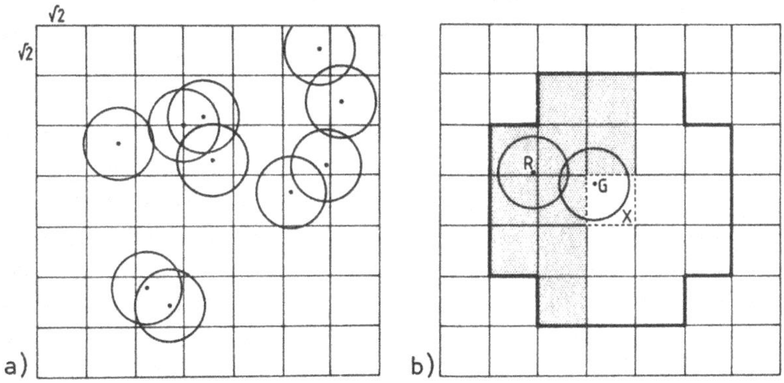

FIG. 18. (a) Typical configuration of ten discs situated in a $\sqrt{2}L \times \sqrt{2}L$ area with $L=7$ at a percolation threshold x_c. The ten discs shown form three clusters, two four-disc clusters and one two-disc cluster. Note that the diagonal of the covering mesh equals the disc diameter, so that the two discs can overlap if and only if their centres are either in the same cell or in two different mesh cells that are no further than the fourth-nearest neighbours in separation. (b) Schematic illustration of the algorithm used. Two discs are shown in this example. Disc G, coloured 'green', is in a cell where the pointer is. Disc R, coloured 'red', is in a cell which is the fourth-nearest neighbour of cell x. The algorithm calculates the centre-to-centre distance d between disc G and disc R. Since $d < 2$, disc G and disc R belong to the same cluster. After Gawlinski and Stanley (1981).

As a simple example, partition a square lattice into cells of edge b. The cells play the role of "renormalized sites" (Fig. 20). If the sites are occupied with probability p, then the cells may be defined to be occupied with probability

$$p' = R(p), \tag{4.1}$$

where

$$R(p) = \sum_{\text{state}} \pi_{\text{state}} \; f_{\text{state}} \tag{4.2}$$

is the probability that the configuration spans. The summation is over all 2^N states of the system, where $N = b^2$,

$$\pi_{\text{state}} = p^s q^{N-s} \tag{4.3a}$$

is the probability of a particular state with s occupied sites and

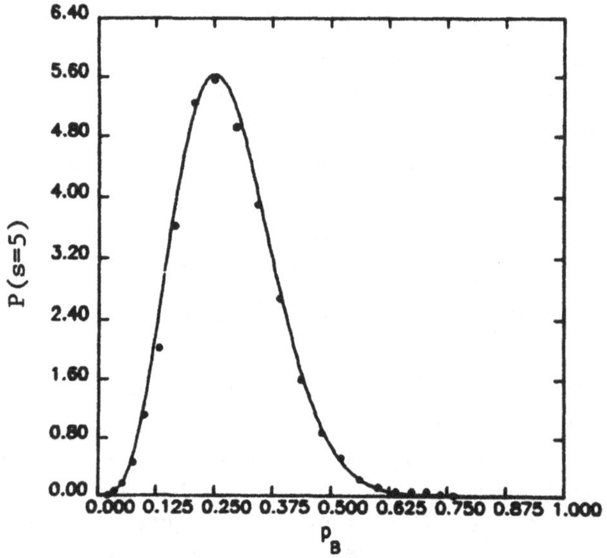

FIG. 19: Illustration of the function P(s=5,p) for bond percolation with site counting on an ice lattice. The solid curve is given by the exact expression $P(s=5,p) = 455p^4(1-p)^{12}$, while the points represent the connectivity analysis (Geiger and Stanley, unpublished) of the Rahman-Stillinger molecular dynamics tapes for a ST2 model interaction potential for liquid water. The agreement between exact calculations for a lattice and Monte Carlo calculations for an interacting continuum system is striking; no adjustable parameters have been used. Adapted from Stanley *et al.* (1981).

$$f_{state} = 1 \quad \text{(if the state "spans")}$$

$$f_{state} = 0 \quad \text{(otherwise)}. \tag{4.3b}$$

The simplest example is $b=2$; in this case there are 2^4 states, some of which are shown in Fig. 21. If "spanning" is defined from East-to-West, then (4.2) becomes simply

$$R(p) = p^4 + 4p^3q + 2p^2q^2. \tag{4.4}$$

The function $R(p)$ has been evaluated for $b=3$ and 4 exactly, and by Monte Carlo methods for b up to 500. As b increases,

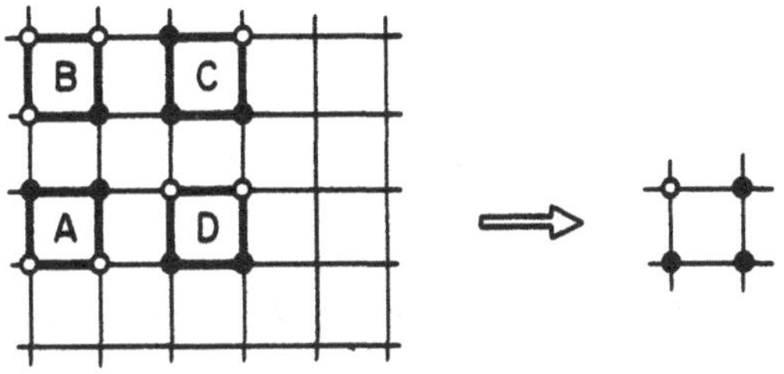

FIG. 20: Rescaling a lattice by forming cells out of groups of sites. Four-site cells on the square lattice are shown, with dots representing occupied sites. In this example, cells C and D are both "occupied" (each cell can be traversed), and therefore connected on the cell level. On the site level, however, we cannot connect them. On the other hand, cells A,B, and C are connected on the site level, though only the next-neighbor cells, A and C, are occupied. These incorrect connections force one to either introduce nearest-neighbor, further-neighbor, and multi-site probabilities, or to go to larger cells where the "interfacing" error plays a smaller relative role. After Reynolds et al. (1977).

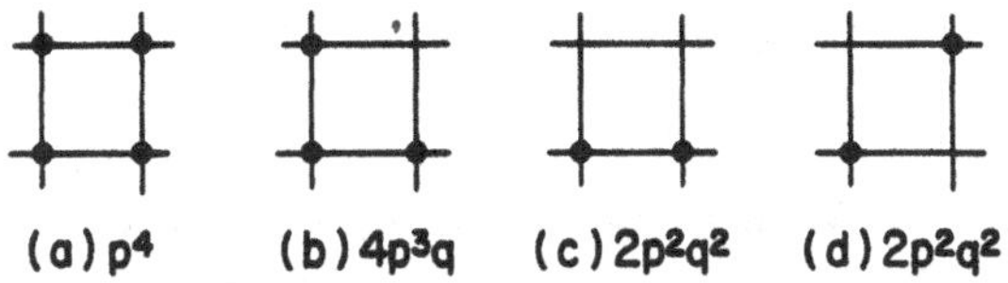

FIG. 21: Some of the 2^4 configurations that arise in the position-space renormalization group for site percolation for the square lattice using a $b = 2$ cell. Configurations (a) - (c) span from East to West, while configuration (d) does not.

$R(p)$ becomes sharper and approaches the step function sketched in Fig. 2b

Equation (4.1) serves as a simple though highly approximate position-space renormalization group transformation with fixed points given by

$$p^* = R(p^*). \tag{4.5}$$

When we substitute (4.5) into (3.1), we find two trivial fixed point at $p^* = 0,1$ and also a critical fixed point at

$$p^*(b=2) = 0.62. \tag{4.6a}$$

This estimate should be compared with those provided by the exact enumeration approaches,

$$p_c = 0.593 \pm 0.002. \tag{4.6b}$$

Although the agreement is not impressive, it is not terribly poor for a "single-shot" technique.

The strength of the position-space renormalization group approach is that one can systematically consider cells for larger and larger values of b. One finds that $p^*(b)$ varies smoothly and predictably with b, following the relation

$$p^*(b) - p_c(\infty) \sim b^{-1/\nu}, \tag{4.7}$$

suggested by finite-size scaling considerations. Extrapolating a sequence of estimates for $p^*(b)$ for a range of b from 2 to 500, we find the estimate (Fig. 22)

$$p^*(b=\infty) = 0.5931 \pm 0.006, \tag{4.8}$$

which is believed to be possibly even more accurate than the estimate (4.6b) from exact enumeration procedures.

To calculate the connectedness length exponent from the renormalization transformation (4.1) we note that all lengths in the rescaled system have been reduced by a factor of b from the lengths in the original system. Hence the connectedness length transforms as

$$\xi' = b^{-1}\xi \quad . \tag{4.9}$$

Substituting (3.1) into (4.9), we find

$$|p' - p^*|^{-\nu} = b^{-1}|p - p^*|^{-\nu}. \tag{4.10}$$

Since the transformation R(p) is analytic near p^*, when we subtract (4.5) from (4.1), we find

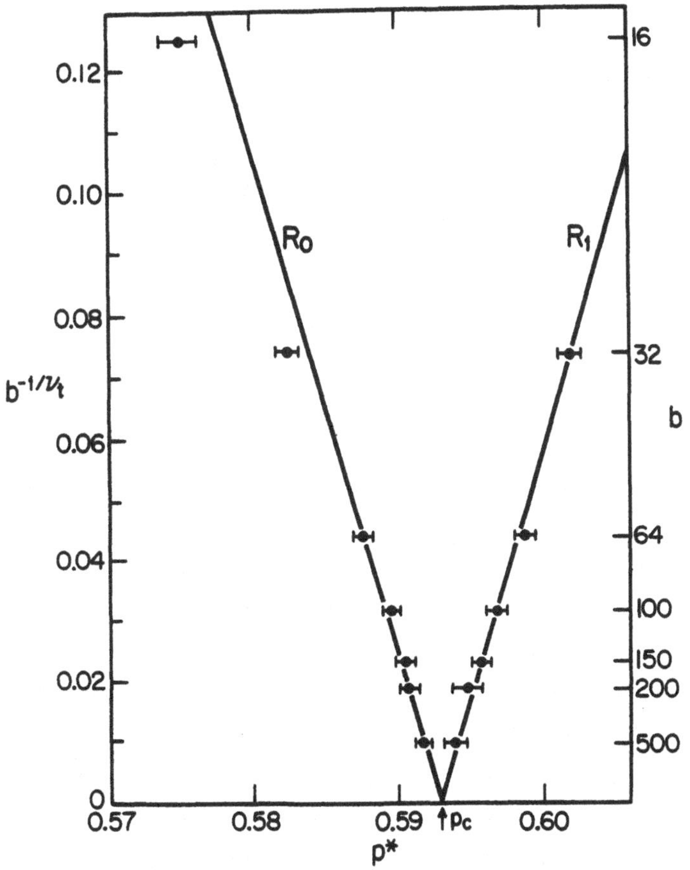

FIG. 22: Extrapolation of the sequence of fixed points $p^*(b)$ obtained from two different connectivity rules, R_0 and R_1 using Eq. (4.7). We have chosen the trial value of ν, $\nu_t = 4/3$, for this plot Both curves approach approximately the same final estimate of p_c as $b \to \infty$. After Reynolds et al. (1980).

$$p' - p^* = R(p) - R(p^*) \cong \lambda(p - p^*), \qquad (4.11)$$

where

$$\lambda = dR/dp|_{p^*}. \qquad (4\ 12)$$

In order for (4.10) and (4.11) to be consistent, we must have $\lambda^{-\nu} = b^{-1}$, or

$$\nu(b) = \ln b/\ln \lambda\ . \qquad (4.13)$$

Again, the error in $\nu(b)$ will decrease as b increases in a <u>predictable</u> fashion

$$\nu(b) - \nu(\infty) \sim A_1(\ln b)^{-1} + A_2(\ln b)^{-2}. \qquad (4.14)$$

Figure 23 shows a sequence of estimates from $b=5$ to $b=500$. A least squares fit to these estimates results in the $b=\infty$ intercept

$$\nu(b=\infty) = 1.33 \pm 0.01, \qquad (4.15)$$

which is in good agreement with the den Nijs conjecture $\nu=4/3$ for $d=2$.

5. BOND PERCOLATION AND THE "MONOCHROMATIC" POTTS MODEL

5.1 The Potts Model

I'd like to conclude these lectures with a brief discussion of the relation of bond percolation to a model of great recent interest, the Potts model, and the relation of the lattice animal problem to a particular generalization of the pure Potts model. These connections were first noted by Kasteleyn and Fortuin (1969) and by Harris and Lubensky (1981) respectively.

The Potts model is a natural generalization of the ordinary lattice-gas or Lenz-Ising model in which each spin variable s_i can exist in more (or less!) than two states (Fig. 24).

I'll begin by showing how bond percolation arises as a limiting case of the Potts model. For concreteness, we shall illustrate several stages of the argumentation with a simple linear chain lattice with period boundary conditions, which has N sites and E=N edges. With each site i, associate a variable $s_i = 1, 2, \ldots, Q$. When two neighboring spins s_i and s_j are in the <u>same</u> state, they have an "attraction" parametrized by the dimensionless coupling constant $K = J/kT$, while if they are in <u>different</u> states there is zero interaction energy (Fig. 25a). Hence the Boltzmann factor for a pair $\langle ij \rangle$ of neighboring sites is given by

$$\exp[K\,\delta(s_i, s_j)] = 1 + v\,\delta(s_i\, s_j), \tag{5.1a}$$

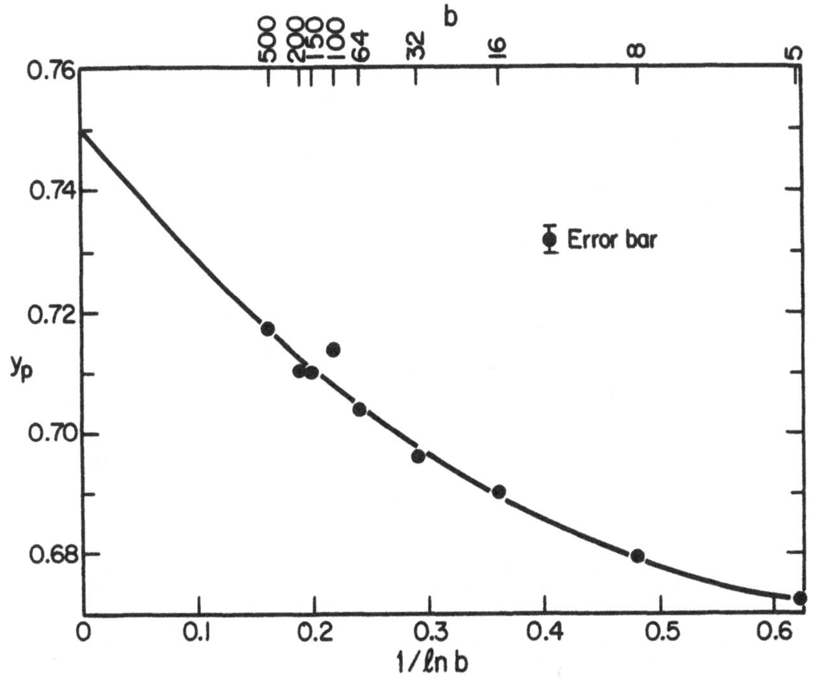

FIG. 23: Quadratic extrapolation against 1/ln b for $y_p = 1/\nu$, using Eq. (4.14). The error bar shown is representative. After Reynolds et al. (1980).

where

$$v = e^K - 1. \tag{5.1b}$$

The Zustandsumme or partition function is the sum over all Q^N states of the system. Each state is weighted by the appropriate Boltzmann factor, which in turn is the product over all nearest-neighbor pairs $\langle ij\rangle$ of the Boltzmann factor (5.1a). Thus

$$Z_N(Q;N) = \sum_{s_1=1}^{Q} \cdots \sum_{s_N=1}^{Q} \prod_{\langle ij\rangle} [(1 + v\,\delta(s_i,s_j))] \tag{5.2}$$

Since there are E edges in the lattice, there are 2^E terms that arise when the product in (5.2) is expanded. Each term becomes in 1:1 correspondence with a "graph" if we simply draw a line on the corresponding edge $\langle ij\rangle$ of the lattice for each factor $v\,\delta(s_i,s_j)$. For example, the term unity corresponds to a graph with no lines, and gives rise to a contribution of order v^0. There are E separate graphs, each with one line, giving rise to contributions of order v. In general, there are $\binom{E}{e}$ graphs giving rise to contributions of order v^e, where $e = e(G)$ is the number of lines or "edges".

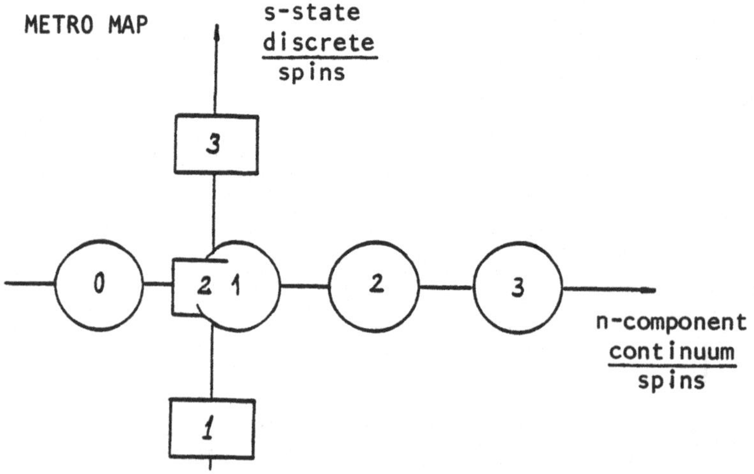

FIG. 24 Schematic illustration of a "metro map" showing how the Ising model has been generalized, first by allowing the two Ising spin states to become s discrete states (first proposed by Potts 1952), and next by allowing the two spin states to be replaced by a continuum of spin states confined to an n-dimensional spin space (first proposed by Stanley 1968). The percolation problem corresponds to the $s=1$ "station" on the North-South metro line. After Stanley (1981).

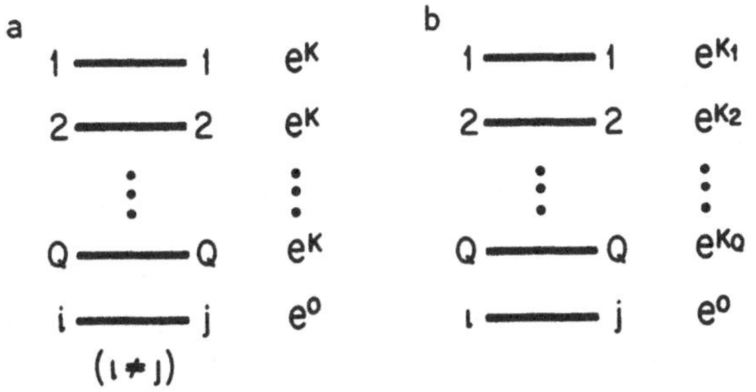

FIG. 25: Boltzmann factors for (a) the ordinary "monochromatic" Potts model, in which neighboring spins in the same spin state are weighted by an additional factor exp(K), and (b) the "polychromatic" Potts model, in which the additional factor $\exp(K_\alpha)$ depends upon which of the Q states the neighboring spins are in.

The N-fold summation in (5.2) can be regarded as a operator that "filters out" a factor Q^n, where $n = n(G)$ is the number of clusters in the configuration.*
Hence the N-fold summation is replaced by a single summation over the 2^E graphs,

$$Z_N(Q;v) = \sum_{2^E \text{ graphs}} Q^n v^e. \tag{5.3}$$

For the closed linear chain n is simply E-e for all values of e except e=E (for which n=1, not 0); the case N = E = 4 is illustrated in Fig. 26 Hence (5.3) may be trivially evaluated, with the result

$$Z_N(Q;v) = (Q + v)^N + (Q - 1)v^N. \tag{5.4}$$

Note that as $T \rightarrow \infty$, v approaches zero so that $Z_N(Q) = Q^N$. Also, for Q=2, (5.4) reproduces the familiar result of Ernst Ising's Ph.D. thesis.

---

*Here we count clusters by their site content. Thus for the configuration with no lines, there are N clusters; for the E configurations with one intact line, there are N-1 clusters. The analog of Eq. (2.8) is $P(s,p) = sp^{s-1}q^2$ (cf. Table 1).

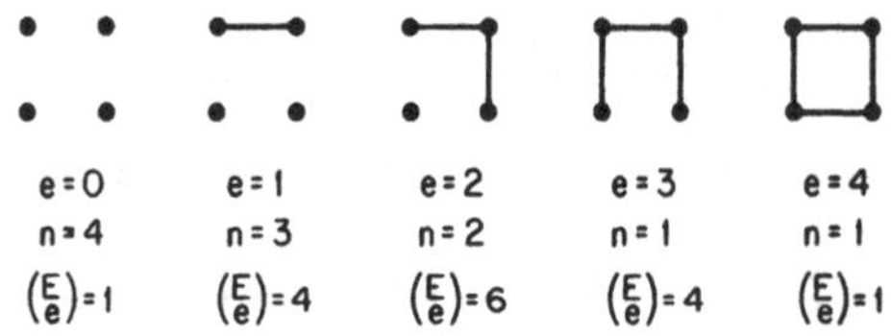

FIG. 26: Illustration of the use of the relation (5.3) to calculate the Potts model partition function for the linear chain with periodic boundary conditions. Here $N = E = 4$ for simplicity, but the argument leading to (5.4) holds for all N.

5.2 Bond percolation

The preceding formulation sounds tantalizingly similar to the bond percolation problem, which we now describe in similar terms. Imagine that each of the E bonds of our lattice is randomly intact with probability p, and define a cluster to be a set of sites joined by intact bonds. Clearly there are 2^E states of the system, and the probability of each state is

$$\pi(\text{state}) = p^e q^{E-e} \tag{5.5}$$

where now $e = e(\text{state})$ is the number of intact bonds in a given state.

In bond percolation the average $\langle A \rangle$ of any quantity $A(\text{state})$ is

$$\langle A \rangle = \sum_{2^E \text{ states}} A(\text{state})\, \pi(\text{state}), \tag{5.6}$$

where the summation is over the 2^E states of the system. Substituting (5.5) into (5.6), we obtain

$$\langle A \rangle = q^E \sum_{2^E \text{ states}} A(\text{state})\, (p/q)^e. \tag{5.7}$$

Motivated by the resemblance between (5.7) and (5.3), let us choose $A(\text{state}) = Q^n$ where $n = n(\text{state})$ is the number of clusters in the state. Thus (5.7) becomes

$$\langle Q^n \rangle = q^E \sum_{2^E \text{ states}} Q^n \, (p/q)^e. \tag{5.8}$$

If we let the "high-temperature variable" v defined in (5.1b) be the ratio of intact to broken bond probabilities (Fig. 27), then (5.8) gives a direct relation between the average of Q^n in bond percolation and the Potts model partition function,

$$\langle Q^n \rangle = q^E \, Z(Q, v=p/q). \tag{5.9}$$

In percolation, the analog of the Gibbs potential is $\langle n \rangle$, the mean number of clusters regardless of size,

$$\langle n \rangle = (\partial/\partial Q)[\ln \langle Q^n \rangle]_{Q=1}. \tag{5 10}$$

Combining (5.9) and (5.10), we finally obtain the Kasteleyn-Fortuin result that $\langle n \rangle$ is the logarithmic derivative of the Potts model partition function, evaluated at $Q=1$,

$$\langle n \rangle = (\partial/\partial Q)[\ln Z_N(Q, v=p/q)]_{Q=1} \tag{5.11}$$

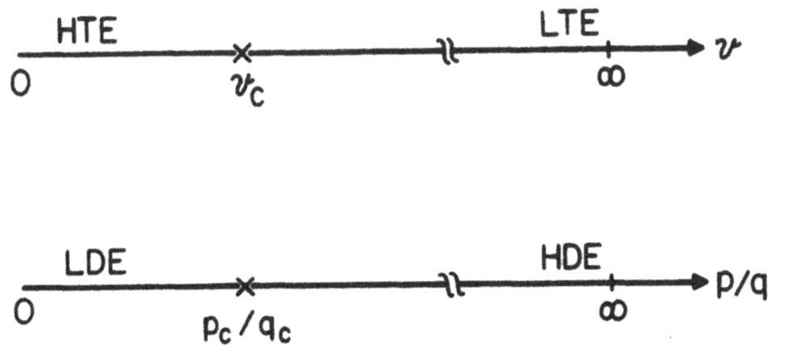

FIG. 27: Illustration of the correspondence between the "monochromatic" Potts model and the random bond percolation problem, showing that the variable v defined in Eq. (5.1b) corresponds to the ratio p/q of intact to broken bonds. In particular, high-temperature expansions (HTE) for the Potts model correspond to low-density expansions (LDE) for bond percolation, while low-temperature expansions (LTE) correspond to high-density expansions (HDE). Similarly, the critical parameter $v_c=Q$ corresponds to the ratio p_c/q_c; for the square lattice, $v_c = Q^{1/2} = 1$.

For the closed linear chain, the left hand side of (5.11) was evaluated directly in Eq. (2.17),

$$\langle n \rangle \propto Nq. \tag{5.12}$$

This result is also obtained by substituting (5.4) into the right-hand side of (5.11). Incidentally, Eq. (5.12) can be obtained by inspection: the number of clusters in the closed linear chain is equal to the number of broken bonds, Nq, since each broken bond gives rise to a new cluster.

6. BOND LATTICE ANIMALS AND THE "POLYCHROMATIC" POTTS MODEL*

Just as bond percolation corresponds to the $Q=1$ limit of the Potts model, so also the lattice animal problem corresonds to the $Q=1$ limit of a generalized Potts model that we shall call the polychromatic Potts model.

In the ordinary or "monochromatic" Potts model, we place a colored bond between two adjacent sites that are in the same spin state s_j. Clearly this can be generalized by placing Q different colors of bonds between neighboring spins that are in the same spin state. Thus when two neighboring spins s_i and s_j are in the same state, they have an interaction parametrized by the dimensionless coupling constant $K_\alpha = J_\alpha /kT$, where $\alpha = 1,2, \ldots ,Q$ denotes the state α (Fig. 25b). Equation (3.1a) for the Boltzmann factor is replaced by

$$\exp[K_\alpha \delta(s_i,s_j,\alpha)] = 1 + v_\alpha \delta(s_i,s_j,\alpha). \tag{6.1a}$$

Here $\delta(s_i,s_j,\alpha) = 1$ if $s_i = s_j = \alpha$ and zero otherwise, and

$$v_\alpha = \exp(K_\alpha) - 1. \tag{6.1b}$$

In analogy to (5.2), the partition function is now

$$Z_N(Q,\{v_\alpha\}) = \sum_{s_1=1}^{Q} \cdots \sum_{s_N=1}^{Q} \prod_{\langle ij\rangle} [1 + \sum_{\alpha=1}^{Q} v \,(s_i,s_j, \,)]. \tag{6.2}$$

Clearly (6.2) reduces to (5.2) if $K_\alpha = K$ for all states

*This section is based on joint work with F.Y. Wu.

To obtain a graphical expansion for the polychromatic Potts model, we proceed exactly as in the monochromatic case by associating a graph of the lattice with each of the 2^E terms obtained by expanding the product in Eq. (6.2). There is no change for the zero-line graph, which gives rise to a contribution Q^N when operated on by the N-fold summation in (6.2). For the E one-line graphs, the summation gives rise to the factor $[v_1 + \ldots + v_Q]Q^{N-2}$. Consider now the $\binom{E}{2}$ two-line graphs. If the two lines form two disconnected one-bond clusters, then we have a contribution $[v_1 + \ldots + v_Q]^2 Q^{N-4}$, while if the two lines form a single two-bond cluster, then there is a contribution $[v_1^2 + \ldots + v_Q^2]Q^{N-3}$. In general, we see that the appropriate generalization of (5.3) is

$$Z_N(Q;\{v_\alpha\}) = \sum_{2^E \text{ graphs}} \prod_{\text{clusters}} [v_1^{b(c)} + \ldots + v_Q^{b(c)}]. \tag{6.3}$$

Here the product is over all clusters c appearing in a given graph, <u>including the one-site clusters</u>, and b(c) is the number of bonds in cluster c.

We can see that (6.3) reduces to (5.3) if $K_\alpha = K$ for all states α. The Q terms in the square brackets of (6.3) become $Qv^{b(c)}$. When we form the product over all clusters of the factor Q, we obtain Q^n, where n is the total number of clusters in the graph. When we form the product over all clusters of $v^{b(c)}$, we obtain v^e since e, the total number of lines in the graph, is simply the summation over all clusters of the number of bonds in that cluster, $e = \sum_{\text{clusters}} b(c)$.

A special case of the polychromatic Potts model was introduced by Wu (1978) and shown by Harris and Lubensky (1981) to reduce to the bond animal generating function. This case is obtained by setting $K_1 = 0$ so that two spins have no interaction if they are both in state $\alpha = 1$. The remaining Q-1 states are treated on an equal footing by choosing $K_\alpha = K$ for $2 \leq \alpha \leq Q$. For this case, then, (6.3) reduces to

$$Z_N(Q;v) = \sum_{2^E \text{ graphs}} Q^P (Q-1)^{n-P} v^e, \tag{6.4}$$

where P denotes the number of one-site clusters (isolated "points").

Motivated by the form of Eq. (5.11), consider the logarithmic derivative of Eq. (6.4) evaluated at Q=1. The presence of the factor $(Q-1)^{n-P}$ in (6.4) means that when we form the derivative of Z_N with respect to Q, only those graphs with $n - P = 1$ make a contribution after setting $Q=1$. Hence we can replace the summation in (6.4) over all 2^E graphs by a summation over all graphs consisting of only a single cluster. Thus if we form the logarithmic derivative evaluated at $Q=1$, and use the fact that $Z_N(Q=1;v) = 1$ from (6.4), then we find

$$(\partial/\partial Q)[\ln Z_N(Q;v)]_{Q=1} = \sum{}' \; v^e, \tag{6.5}$$

where the prime on the summation indicates that it is over only those graphs made up of a single cluster. We can group together all terms in (6.5) with the same number of bonds,

$$(\partial/\partial Q)[\ln Z_N(Q;v)]_{Q=1} = \sum_{b=0}^{\infty} A_b v^b, \tag{6.6}$$

where A_b is the total number of b-bond "lattice animals". Thus (6.6) is the generating function for the bond animal problem.

Just as we illustrated the correspondence between the monochromatic Potts model and the bond percolation problem using the linear chain, so also we can illustrate the correspondence between the polychromatic Potts model and the lattice animal problem on the same lattice. We have already noted above that $A_b = 1$ for an infinite linear chain lattice, so that the bond animal generating function is simply

$$G_A(K) = \sum_{b=0}^{\infty} A_b K^b = (1 - K)^{-1}. \tag{6.7}$$

To calculate the partition function of the Potts model in the $N=\infty$ limit, we can appeal to the transfer matrix method. Thus in the large-N limit, we have

$$Z_N(Q;v) = \text{trace } T^N \sim (\lambda_{max})^N, \tag{6.8}$$

where λ_{max} is the largest eigenvalue of the transfer matrix T. For example, for $Q=3$, we have

$$T = \begin{pmatrix} e^{\circ} & 1 & 1 \\ 1 & e^K & 1 \\ 1 & 1 & e^K \end{pmatrix}. \tag{6.9}$$

For all Q, we find that the largest eigenvalue is given by

$$2\lambda_{max} = (Q + v) + [(Q + v)^2 - 4v]^{1/2}. \qquad (6.10)$$

Forming the logarithmic derivative, we have for large N

$$(\partial/\partial Q)[N^{-1} \ln Z_N(Q;v)]_{Q=1} = (1 - v)^{-1}. \qquad (6.11)$$

The agreement of (6.7) and (6.11) confirms the general result (6.6) for the linear chain.

ACKNOWLEDGEMENTS

Most of the research mentioned in these lectures was carried out in collaboration with others, as indicated by the references cited. In particular, I wish to thank A. Coniglio, W. Klein, H. Nakanishi, S. Redner, P. J. Reynolds, D. Stauffer, and F. Y. Wu for their continuing collaboration on these and other topics.

REFERENCES

This list is not intended to be complete. Certain important articles have been omitted when their results are clearly discussed and placed into perspective in a more recent work. The interested reader will find extensive references to earlier works in the articles cited below.

CONIGLIO, A. (1981) "Geometrical structure and thermal phase transition of the dilute s-state Potts and n-vector model at the percolation threshold" in Disordered Systems and Localization (eds. C. Castellani, C. Di Castro, and L. Peliti). Springer Verlag, Heidelberg, p. 51-55.

CONIGLIO, A., STANLEY, H. E., and KLEIN, W. (1979) "Site-bond correlated percolation problem: A statistical mechanical model of polymer gelation" Phys. Rev. Lett. 42, 518-522.

DEUTSCHER, G. (1981) "Experimental relevance of percolation" in

Disordered Systems and Localization (eds. C. Castellani, C. Di Castro and L. Peliti). Springer Verlag, Heidelberg, p. 26-40.

FISHER, M. E. and ESSAM, J. W. (1961) "Some cluster size and percolation problems" J. Math. Phys. 2, 609-619.

GAUNT, D. S. (1980) "The critical dimension for lattice animals" J. Phys. A 13, L97-L101.

GAWLINSKI, E. T. and STANLEY, H. E. (1981) "Continuum percolation in two dimensions: Monte Carlo tests of scaling and universality for non-interacting discs" J. Phys. A 14, L291-L299.

GORDON, M. and ROSS-MURPHY, S. B. (1975) "The structure and properties of molecular trees and networks" Pure and Appl. Chem. 43, 1-26.

HANKEY, A. and STANLEY, H. E. (1972) "Systematic application of general homogeneous functions to static scaling and universality" Phys. Rev. B6, 3515-3542.

HARRIS, A. B. and LUBENSKY, T. C. (1981) "Generalized percolation" Phys. Rev. B 24, 2656-2670.

ISAACSON, J. and LUBENSKY, T. C. (1980) "Flory exponents for generalized polymer problems" J. Phys. Lett. (Paris) 41, L469.

KASTELEYN, P. W. and FORTUIN, C. M. (1969) "Phase transitions in lattice systems with random local properties" J. Phys. Soc. Japan 26S, 11.

LEE, T. D. and YANG, C. N. (1952) "Statistical theory of equations of state and phase transtitions. II. Lattice gas and Ising model" Phys. Rev. 87, 410-419.

MANDELBROT, B. (1977) Fractals: Form, Chance and Dimension, W. H. Freeman, San Francisco.

NAKANISHI, H. and STANLEY, H. E. (1980) "Scaling studies of percolation phenomena in systems of dimensionality two to seven: Cluster numbers" Phys. Rev. B 22, 2466-2488.

PARISI, G. and SOURLAS, N. (1981) "Critical behavior of branched polymers and the Lee-Yang edge singularity" Phys. Rev. Lett. 46, 871-874.

PIKE, R. and STANLEY, H. E. (1981) "Order propagation near the percolation threshold" J. Phys. A 14, L169-L177.

POTTS, R. B. (1952) "Some generalized order-disorder transformations" Proc. Cambridge Phil. Soc. 48, 106-109.

REYNOLDS, P. J., STANLEY, H. E., and KLEIN, W. (1977) "Ghost fields, pair connectedness, and scaling: Exact results in one-dimensional percolation" J. Phys. A10, L203-L209.

REYNOLDS, P. J., STANLEY, H. E. and KLEIN, W. (1980) "Large-cell Monte Carlo renormalization group for percolation" Phys. Rev. B 21, 1223-1245.

SHLIFER, G., KLEIN, W., REYNOLDS, P. J., and STANLEY, H. E. (1979) "Large-cell renormalization group for the backbone problem in percolation" J. Phys. A 12, L169-L174.

STANLEY, H. E. (1968) "Dependence of critical properties on dimensionality of spins" Phys. Rev. Lett. 20, 589-592.

STANLEY, H. E. (1977) "Cluster shapes at the percolation threshold: An effective cluster dimensionality and its connection with critical-point exponents" J. Phys. A10, L211-220.

STANLEY, H. E. (1981) "New directions in percolation including some possible applications of connectivity concepts to the real world" in Disordered systems and localization (eds. C. Castellani, C. Di Castro, and L. Peliti), Springer, Heidelberg, p. 59-83.

STANLEY, H. E. (1982a) "Geometric analogs of phase transitions: an essay in honor of Laszlo Tisza" In L. Tisza Festschrift (A. Shimony and H. Feshbach, eds.). M.I.T. Press, Cambridge, Mass.

STANLEY, H. E. (1982b) Introduction to Phase Transitions and Critical Phenomena, 2nd edition. Oxford Univ. Press, London and New York.

STANLEY, H. E., BIRGENEAU, R. J., REYNOLDS, P. J. and NICOLL, J. F. (1976) "Thermally-driven phase transitions near the percolation threshold in two dimensions" J. Phys. C9, L553-560.

STANLEY, H. E., CONIGLIO, A., KLEIN, W., NAKANISHI, H., REDNER, S., REYNOLDS, P.J. and SHLIFER, G. (1980) "Critical phenomena: past, present, and future" in Proc. Intl. Symp. on Synergetics (ed., H. Haken), Springer Verlag, Heidelberg, Chap. 1.

STANLEY, H. E., TEIXEIRA, J., GEIGER, A. and BLUMBERG, R. L. (1981) "Interpretation of the unusual behavior of H_2O and D_2O at low temperature: Are concepts of percolation relevant to the puzzle of liquid water?" Physica 106A, 260-277.

STANLEY, H. E., REYNOLDS, P. J., REDNER, S. and FAMILY, F. (1982) "Position-space renormalization group for models of linear polymers, branched polymers and gels" in Real-Space Renormalization (eds. T. W. Burkhardt and J. M. J. van Leeuwen). Springer, Heidelberg, Chap. 7.

STANLEY, H. E. and TEIXEIRA, J. (1980) "Interpretation of the unusual behavior of H_2O and D_2O: Tests of a percolation model" J. Chem. Phys. 73, 3404-3424.

STAUFFER, D. (1981) "Scaling properties of percolation clusters." In Disordered systems and localization (eds. C. Castellani, C. Di Castro, and L. Peliti). Springer, Heidelberg, p. 9-25.

SYKES, M. F., GAUNT, D. S., and GLEN, M. (1981) "Perimeter polynomials for bond percolation processes" J. Phys. A 14, 287-292.

WU, F. Y. (1978) "Percolation and the Potts Model" J. Stat. Phys. 18, 115.

WU, F. Y. (1982) "The Potts model" Rev. Mod. Phys. 54, xxx.

ZERNIKE, F. (1940) "The propagation of order in cooperative phenomena" Physica 7, 565.

ALGEBRAIC ASPECTS OF EXACT MODELS*

M. Gaudin

Centre d'Etudes Nucléaires de Saclay
Service de Physique Théorique

Several statistical or dynamical systems modelling a variety of physical phenomena, such as spin chains, two dimensional spin lattices and chemical crystals, or particles in δ function interaction, share the same momentous underlying structures, which are essentially the applicability of Bethe's superposition ansatz for wave functions, the commutativity of transfer matrices, and the existence of an important ternary operator algebra. Their close relationship among each other, and with features like integrability and S matrix factorization discussed in field theoretical context, became familiar in the course of time.

In the following we shall outline algebraically the appearance of these structures and interrelations among them, exemplarily for the eight vertex model (chap.1) and for δ function interacting particles of general spin (chap.2) and spin 1/2 (chap.3).

1. THE EIGHT VERTEX MODEL

1.1. Definition and Equivalences

The eight vertex model describes a system of chemical dipole bonds on a two dimensional quadratic lattice.

It emerges from the six vertex model, when the "ice condition" imposed on the latter is dropped. This condition means electric neutrality at every lattice site, i.e. excludes vertices n°7 and 8(fig.1.1).

*From: M.Gaudin, Saclay Notes CEA-N-1559(1),(2) (1972,1973) and Saclay Note to appear; extracted and translated by K.-H. Rehren

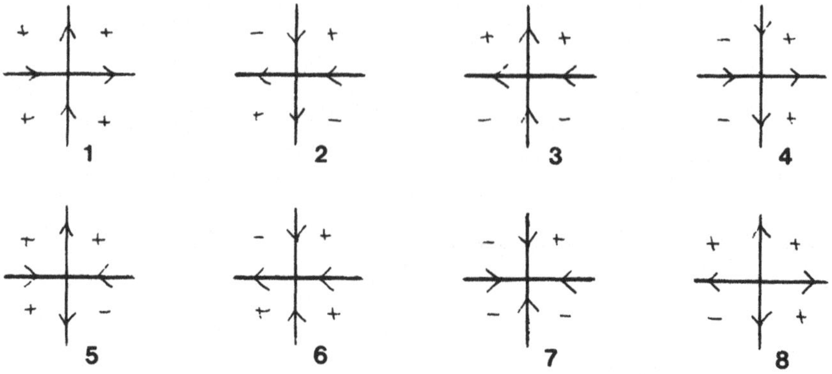

Fig. 1.1. The eight vertices

The thermodynamical problem proceeds by the evaluation of the partition function of a torus-like system of size N x M, as a function of the statistical weights $\omega_j = \exp(-\beta e_j)$ $(j=1,\dots,8)$ associated with every vertex:

$$Z(\omega_1,\dots,\omega_8) = \sum_{\Gamma} \prod_{V} \omega_{i(V)}. \tag{1.1}$$

The product extends over the MN sites V, while the sum runs over all compatible configurations $\Gamma : V \to i(V)$ of vertices. Periodic boundary conditions are chosen for technical conveniency. The free energy is well defined and can be computed as the twofold limit

$$-\beta f = \lim_{M\to\infty} \lim_{N\to\infty} (MN)^{-1} \log Z. \tag{1.2}$$

Whereas the six vertex model is solved in generality, and its thermodynamical features are well explored, the solution to the eight vertex model is only known for special choices of the parameters. There are several equivalencies with Ising type or dimer models. We shall present some examples of these.

- Equivalence with an Ising model

The following prescription is meaningful, because every vertex contains an even number of incoming arrows: given a vertex configuration Γ on the quadratic lattice, assign to each plaquette a "spin" + or - , such that the down and left pointing arrows of Γ separate opposite spins on the neighbouring plaquettes. The construction defines two spin configurations s' and s" on the dual lattice which are symmetric to each other by a global exchange $+ \leftrightarrow -$.

Provided that the statistical weights of the vertex model

satisfy

$$\omega_1\omega_2\omega_3\omega_4 = \omega_5\omega_6\omega_7\omega_8 \tag{1.3}$$

then its partition function is the same as that of the spin model on the dual lattice with nearest neighbour (vertical, horizontal, diagonal, antidiagonal) interaction. The general solution is not known.

- Equivalence of the self conjugate model and an Ising model with four-spin interaction.

The self conjugate eight vertex model is defined to be invariant under the inversion of all four arrows in a vertex. Its parameters are

$$a := \omega_1 = \omega_2 \qquad b = \omega_3 = \omega_4 , \tag{1.4}$$

$$c = (\omega_5\omega_6)^{1/2} \qquad d := (\omega_7\omega_8)^{1/2}. \tag{1.5}$$

It equivalates an Ising model with four-spin interaction between two independent sub-lattices (Kadanoff and Wegner, 1971). Its partition function

$$Z = \sum_{\sigma} \exp \sum_{j=1}^{M}\sum_{k=1}^{N} \left(K_3\sigma_{jk}\sigma_{j+1,k+1} + K_4\sigma_{j+1,k}\sigma_{j,k+1} + K\sigma_{jk}\sigma_{j+1,k+1}\sigma_{j,k+1}\sigma_{j+1,k}\right) \tag{1.6}$$

has been calculated by Baxter (1971).

- Equivalence with a dimer system.

Define the lattice R_1 to be a regular arrangement of tetrahedra such that the joining apices project onto the middle of the links of the original quadratic lattice which carries the arrow system (fig. 1.2). The edges of the tetrahedra carry dimer bonds according to the correspondence recipe fig. 1.3.

The dimer partition function equals that of the self conjugate vertex model if $\omega_1\omega_2 + \omega_3\omega_4 + \omega_5\omega_6 = \omega_7\omega_8$. Baxter studied the case $\omega_5 = \omega_6 = 0$ which can be treated by the Pfaffian method (Kasteleyn 1967). The method can even be extended if $\omega_1\omega_2 + \omega_3\omega_4 - \omega_5\omega_6 = \omega_7\omega_8$.

1.2. Transfer Matrix and Symmetry of the Self Conjugate Model

The self conjugate eight vertex model describes a ferroelectric with vertex energy invariant under inversion of the dipoles which constitute the hydrogene bonds. This assumption means

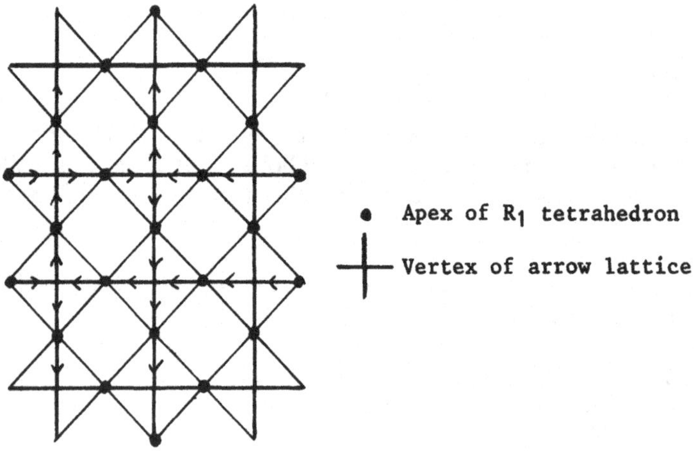

Fig. 1.2. The dimer model

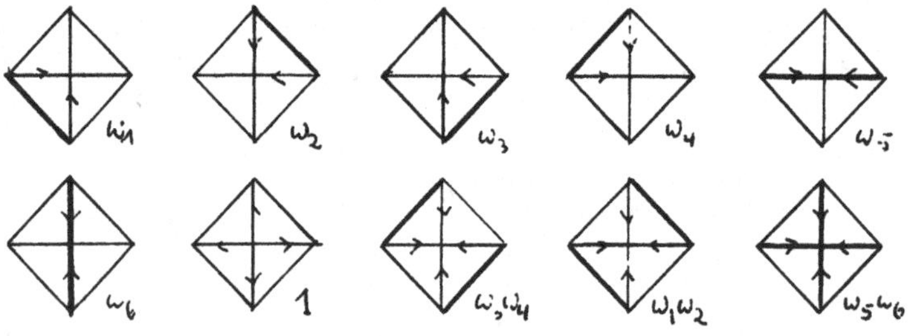

Fig. 1.3. Correspondence between dimers and vertices

vanishing of the electric field, and is reflected in (1.4). The parametrization (1.5) is justified by the fact that $\omega_5, \dots, \omega_8$ enter the partition function only via the products $c^2 = \omega_5\omega_6$, $d^2 = \omega_7\omega_8$.

The partition function Z(eq.1.1) can be written as $Z = \mathrm{Tr}\, T^M$ where the transfer matrix T is a $2^N \times 2^N$ square matrix. Its indices are two configurations of "vertical" arrows along two adjacent "horizontal" rows (fig.1.4). Its matrix elements are written

$$(1.7) \quad \langle \alpha | T | \beta \rangle = \sum_{\{\lambda\}} (\alpha_1 \lambda_1 | R | \beta_1 \lambda_2) \dots (\alpha_N \lambda_N | R | \beta_N \lambda_1)$$

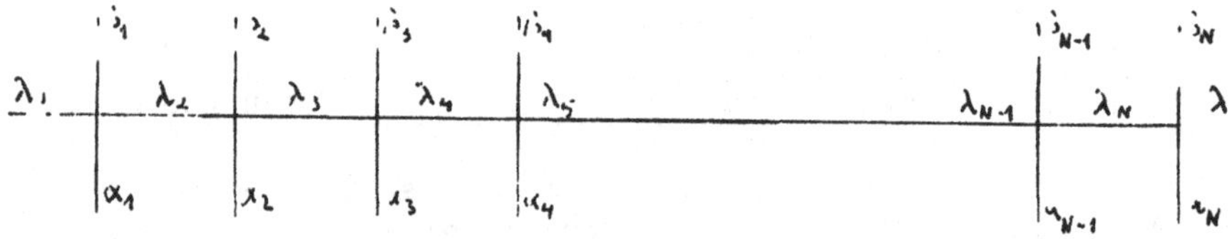

Fig. 1.4. Matrix element of the transfer matrix.

where the 4 x 4 matrix R expanded on tensor products of two independent sets of Pauli matrices $\sigma^i \otimes \tau^j$ $(\sigma^0 = \tau^0 = 1)$, is

$$R = \sum_{\alpha=0}^{3} w_\alpha \sigma^\alpha \otimes \tau^\alpha = (w\,\sigma\,\tau),$$

(1.8) $$w_0 = \frac{a+b}{2}, \quad w_1 = \frac{c+d}{2}, \quad w_2 = \frac{c-d}{2}, \quad w_3 = \frac{a-b}{2}.$$

Introducing independent σ matrices for each site n = 1,...,N and writing $R_n = (w\,\sigma_n\,\tau)$, the transfer matrix is

(1.9) $$T(w) = \operatorname*{tr}_\tau R_1 \ldots R_N = \operatorname*{tr}_\tau (w\sigma_1\tau) \ldots (w\sigma_N\tau).$$

The maximal eigenvalue t of T determines the free energy per site in the thermodynamical limit: $-\beta f = \lim_{N\to\infty} N^{-1} \log t$.

Proposition: Z(w) is invariant under the $2^4 \cdot 4!$ operations which permute the w_α and change signs (hypercubic group).

The operations of this group are to be regarded as duality operations interchanging high and low temperature regions.

1.3. Relation between the XYZ Hamiltonian and the Transfer Matrix

Proposition (Sutherland 1970):

The XYZ Hamiltonian

$$H = \frac{1}{2} \sum_n \sum_{\alpha=1}^{3} J_\alpha \sigma_n^\alpha \otimes \sigma_{n+1}^\alpha = \sum_n H_{n,n+1}$$

commutes with a one parameter family of transfer matrices $T(w(J))$ taken at

(1.10) $$w_\alpha(J) = (J + J_\alpha)^{1/2} \qquad (\alpha = 0, \ldots, 3)$$

with fix J_α obeying $J_0 = J_1 + J_2 + J_3$.

The proof is based on the important identity

(1.11) $$[H_{n,n+1}, R_n R_{n+1}] = \frac{1}{2}\sqrt{P(J)}\left(R_n \frac{dR_{n+1}}{dJ} - \frac{dR_n}{dJ} R_{n+1}\right)$$

with $$P(J) = \prod_{\alpha=0}^{3} (J + J_\alpha).$$

In turn Sutherland's result emerges from Baxter's theorem on the commutativity of the transfer matrices T(J) among themselves (sect.1.4).

Proposition:

Normalize $T_0(\mathfrak{z}) = 2^{-N}\mathfrak{z}^{-\frac{N}{2}}T(\mathfrak{z})$. Then

(1.12) $$H = -\frac{N}{2}\mathfrak{z}_0 + 2T_0^{-1}\frac{dT_0}{d\mathfrak{z}^{-1}}\Big|_{\mathfrak{z}^{-1}=0},$$

$$T_0^{-1}\Big|_{\mathfrak{z}^{-1}=0} = C = \text{cyclic permutation operator.}$$

1.4. One Parameter Family of Commuting Transfer Matrices

Proposition: (Baxter 1971)

(1.13) $$[T(\mathfrak{z}), T(\mathfrak{z}')] = 0 \qquad \forall\, \mathfrak{z}, \mathfrak{z}'.$$

Proof: We write

$$T = T(w) = \operatorname{tr}_{\tau} R_1 \cdot R_N , \qquad T' = T(w') = \operatorname{tr}_{\tau'} R_1' \ldots R_N'$$

with $R_n = (w\sigma_n\tau)$, $R_n' = (w'\sigma_n\tau')$, $[\tau, \tau'] = 0$.
Due to the trace structure of T and T', TT' will equal T'T if there is a similarity transformation

(1.14) $$X^{-1}R_nR_n'X = R_n'R_n \qquad \forall n.$$

The ansatz

(1.15) $$X = \sum_{\alpha=0}^{3} x_\alpha \tau^\alpha \otimes \tau'^\alpha \equiv (x\,\tau\tau')$$

requires to solve the ternary equation

(1.16) $$(w\sigma\tau)(w'\sigma\tau')(x\tau\tau') = (x\tau\tau')(w'\sigma\tau')(w\sigma\tau).$$

It represents 6 independent equations for 4 unknown x's.
The system to be solvable imposes conditions on the w and w' which become

(1.17) $$\frac{w_i^2 - w_j^2}{w_k^2 - w_l^2} = \frac{w_i'^2 - w_j'^2}{w_k'^2 - w_l'^2} = \frac{x_i^2 - x_j^2}{x_k^2 - x_l^2}, \qquad (ijkl \text{ different})$$

and are actually solved by (1.10):

(1.18) $$w_\alpha^2 = \mathfrak{z}_\alpha + \mathfrak{z}, \quad w_\alpha'^2 = \mathfrak{z}_\alpha + \mathfrak{z}'.$$

With $\mathfrak{z}_0 - \mathfrak{z}_1 + \mathfrak{z}_2 + \mathfrak{z}_3$, an arbitrary w fixes uniquely $\mathfrak{z}_\alpha$ and $\mathfrak{z}$, and leaves one free parameter $\mathcal{J}'$ for the w'. The dependence on the $\mathcal{J}_\alpha$ which label different commuting families, exhibits a structure related with elliptic functions:

Consider the quartic Jacobi relation among Θ-functions of module ℓ

$$S\{\Theta_i(-V)\Theta_j(V')\Theta_k(V\ V'-\zeta)\Theta_l(\zeta)\} \quad 0 \tag{1.19}$$

with the permutation operator

$$S = I + (12)(34) + (13)(42) + (14)(23)$$

acting on the indices $\{ijkl\}$ $\{1,2,3,4\}$. Divided by $\prod_i \Theta_i(\zeta)$ it acquires the same structure as the independent components of (1.16). The latter is thus identically fulfilled if

$$\text{(1.20)}\qquad \begin{aligned} w_\alpha &\propto \frac{\Theta_{\alpha+1}(V)}{\Theta_{\alpha+1}(\zeta)}, \qquad w'_\alpha = \frac{\Theta_{\alpha+1}(V')}{\Theta_{\alpha+1}(\zeta)} \\ x_\alpha &\propto \frac{\Theta_{\alpha+1}(\zeta+V'-V)}{\Theta_{\alpha+1}(\zeta)} \end{aligned}$$

with the free parameter V.
This is

$$\text{(1.21)}\qquad \begin{aligned} \frac{w_0}{w_3} &= \frac{\operatorname{sn}(V,\ell)}{\operatorname{sn}(\zeta,\ell)} \qquad \frac{w_1}{w_3} = \frac{\operatorname{cn}(V,\ell)}{\operatorname{cn}(\zeta,\ell)} \\ \frac{w_2}{w_3} &= \frac{\operatorname{dn}(V,\ell)}{\operatorname{dn}(\zeta,\ell)} \end{aligned}$$

Inversely, the parameters V, ℓ and ζ are calculated from w by

$$\text{(1.22)}\qquad \begin{aligned} \ell^2 &= \frac{w_2^2-w_3^2}{w_2^2-w_0^2} : \frac{w_1^2-w_3^2}{w_1^2-w_0^2} \\ \operatorname{sn}\zeta &= \left(\frac{w_1^2-w_3^2}{w_1^2-w_0^2}\right)^{1/2} \\ \operatorname{sn} V &= \frac{w_0}{w_3}\operatorname{sn}\zeta . \end{aligned}$$

The symmetry of the partition function (sect. 1.2) allows us to choose $w_0^2 < w_3^2 < w_2^2 < w_1^2$, guaranteeing $0 < \ell < 1$, ζ and V real.

A reparametrization should be quoted. We pass from $\tau_\ell = K_\ell'/K_\ell$ to $\tau = K'/K := 2\tau_\ell$ belonging to the module $k = (1-\ell)/(1+\ell)$. Define v, η by

$$(1.23) \quad \frac{\eta}{K} = i\frac{\zeta}{K_\ell'}, \quad \frac{v}{K} = i\frac{V}{K_\ell'}, \quad \left(\frac{K'}{K} = 2\frac{K_\ell}{K_\ell'}\right).$$

This yields Baxter's parametrization $(\Theta(\cdot) \equiv \Theta(\cdot|\tau))$

$$a:b:c:d = W_0+W_3 : W_0-W_3 : W_1+W_2 : W_1-W_2 =$$

$$(1.24) \quad = \frac{\Theta_1(v+\eta)\,\Theta_4(v-\eta)}{\Theta_1(2\eta)\,\Theta_4(0)} : \frac{\Theta_1(v-\eta)\,\Theta_4(v+\eta)}{\Theta_1(2\eta)\,\Theta_4(0)} :$$

$$: \frac{\Theta_4(v+\eta)\,\Theta_4(v-\eta)}{\Theta_4(2\eta)\,\Theta_4(0)} : \frac{\Theta_1(v+\eta)\,\Theta_1(v-\eta)}{\Theta_1(2\eta)\,\Theta_4(0)} .$$

1.5. A Representation of the Symmetric Group π_N

The ternary equation (1.16) is the basis for the commutativity of the transfer matrix. If we associate a parameter V_i with each site and define

$$(1.25) \quad \Lambda_{ij} \equiv X_{ij}(V_i - V_j) := \frac{1}{2}\frac{\Theta_1(2\zeta)}{\Theta_1(2\zeta+V_i-V_j)}\cdot\left(w(\zeta+V_i-V_j)\,\sigma_i\,\sigma_j\right)$$

it is displayed as the following important algebra

$$(1.26) \quad \begin{aligned} X_{ij}X_{ji} &= 1\!\!1 \\ X_{ji}X_{jk}X_{ik} &= X_{ik}X_{jk}X_{ji} \\ X_{ij}X_{k\ell} &= X_{k\ell}X_{ij} \qquad (ijk\ell \text{ different}) \end{aligned}$$

The spin representation of the symmetric group π_N is generated by

$$(1.27) \quad P_{ij} = X_{ij}(V_i - V_j = 0) = \frac{1}{2}(1\!\!1 + \vec{\sigma}_i\,\vec{\sigma}_j) = (ij).$$

The operators

(1.28) $Y_{34}^{ij} = X_{34}(V_i - V_j)\, P_{34}$

satisfy the algebra

$$Y_{34}^{ij}\, Y_{34}^{ji} = \mathbb{1}$$
(1.29) $$Y_{12}^{jk}\, Y_{23}^{ik}\, Y_{12}^{ij} = Y_{23}^{ij}\, Y_{12}^{ik}\, Y_{23}^{jk}$$
$$Y_{12}^{ij}\, Y_{34}^{kl} = Y_{34}^{kl}\, Y_{12}^{ij}.$$

(34 stands for (nn + 1).

Consequently, the neighbour transposition operators Y_{34} acting on the group ring as $Y_{34}|ijkl\,\rangle = Y_{34}^{kl}|ijkl\,\rangle = |ijlk..\rangle$ are involutions $[Y_{34}^2 = \mathbb{1}]$, their products are rotations of order 2 or 3 $[(Y_{12}Y_{34})^2 = (Y_{12}Y_{23})^3 = \mathbb{1}]$. This fixes the group generated by the generators $Y_{n,n+1}$ uniquely to be isomorphic with the symmetric group. The representation is isomorphic with the regular one which is continuously approached as $\zeta \to 0$.

1.6. Diagonalization of the Transfer Matrix

In the context of the six vertex model, the diagonalization of the transfer matrix passes by the solution of a functional equation of type

(1.30) $T(\varphi)\,Q(\varphi) = g(\varphi - i\phi)\,Q(\varphi + 2i\phi) + g(\varphi + i\phi)\,Q(\varphi - 2i\phi)$

which has been generalized by Baxter (1971) even for the inhomogeneous ferroelectric models. The idea is to construct a similar relation for the eight vertex model: We thus shall look for a matrix Q(w), a numeric function $\phi(w)$, and a commutative and invertible translation operation t : w → tw such that:

(1.31) $T(w)\,Q(w) = \phi^N(w)\,Q(tw) + \phi^N(tw)\,Q(t^{-1}w),$

(1.32) $[Q(w), Q(tw)] = 0.$

Indeed, such quantities can be constructed by the consistent ansatz

(1.33) $\langle \alpha_1 \ldots |Q|\beta_1 \ldots\rangle = \mathrm{tr}\,(\alpha_1|S|\beta_1)\ldots(\alpha_N|S|\beta_N)$

each $(\alpha|S|\beta)$ being an (infinite) matrix subject to the selection rule

(1.34) $$(\alpha | S_{nm} | \beta) \neq 0 \iff |n - m| = 1.$$

The outcome of this quite tedious construction is essentially the following:

1.7. The Coupled Spectrum Equations

In Baxter's elliptic parametrization (1.24), t is just a simple translation $t : v \to v + 2\eta$ acting on the family parameter v. T(v) and Q(v) can be simultaneously diagonalized in a basis independent of v. The eigenvalues (also denoted Q(v)) are entire functions of v with exactly N/2 zeros inside the rectangle (2K,iK') (eq.(1.23)) of the complex v plane:

(1.35) $$Q(v) = e^{-(i\nu'\pi v/2K)} \prod_{j=1}^{N/2} \Theta_1 \Theta_4 (v - v_j)$$

where the conserved quantum number ν' is (modulo 2) the number of down $\downarrow$ arrows in the eigenvector $|\alpha_1 \dots \alpha_N\rangle$.

The zeros v_j are determined by $\frac{N}{2}$ coupled equations

(1.36) $$\left(\frac{\Theta_1(v_j - \eta \,|\, \frac{\tau}{2})}{\Theta_1(v_j + \eta \,|\, \frac{\tau}{2})} \right)^N = e^{-i\frac{\nu'\pi\eta}{K}} \prod_{i \neq j} \frac{\Theta_1(v_j - v_i - 2\eta \,|\, \frac{\tau}{2})}{\Theta_1(v_j - v_i + 2\eta \,|\, \frac{\tau}{2})}$$

This system admits several predictions about the thermodynamics of the eight vertex model.

2. IDENTICAL PARTICLES WITH DELTA FUNCTION INTERACTION

2.1 The Bethe Ansatz

We consider N particles of the same mass and the same internal symmetry moving on an axis.
The Hamiltonian is

(2.1) $$H = - \sum_{j=1}^{N} \frac{\partial^2}{\partial x_j^2} + 2c \sum_{i<j} \delta(x_i - x_j).$$

$2c$ measures the strength of the attractive or repulsive interaction force.

Since H is totally symmetric solutions can be classified

according to their symmetry type, characterized by a Young tableau τ.

Every continuous solution of the Schrödinger equation, whatever be its asymptotic behavior, will be called elementary solution. The Schrödinger equation can be seen as a wave equation of a particle moving in $\mathbb{R}^N$ among $N(N-1)/2$ infinitely refracting thin laminae (Mc Guire (1964), Gaudin (1971), Derrida (thesis 1974)). Bethe's ansatz allows to construct elementary solutions as finite superpositions of an incoming wave with all geometrically transmitted or reflected waves. Its particularity is just the absence of refracted waves.

Wherever all x_j are distinct the Schrödinger equation is solved by a plane wave. Calling D_Q the sector of $\mathbb{R}^N$

(2.2) $\quad D_Q : x_{Q1} < x_{Q2} < \dots < x_{QN}$

for every permutation $Q \in \pi_N$, the Bethe ansatz is

$$\Psi\big|_{x\in D_Q} = \sum_{P\in\pi_N} \langle Q \| P\rangle\, e^{i(P^{-1}k,\,Q^{-1}x)} \quad , k \in C^N \tag{2.3}$$

supplemented by conditions on the coefficients $\langle Q \| P\rangle$ to guarantee for continuity of the wave function and for the proper discontinuity of its derivative at the sector boundaries. These are exactly

$$\langle Q(34) \| P\rangle = x_{P3,P4}\langle Q \| P\rangle + (1 + x_{P3,P4})\langle Q \| P(34)\rangle \tag{2.4}$$

with $x_{jk} = \frac{ic}{k_j - k_k}$, and (34) standing for any neighbour transposition. That the $(N-1)(N!)^2$ equations (2.4) for $(N!)^2$ unknown $\langle Q\|P\rangle$'s are compatible, is a nontrivial feature: if we choose arbitrary values for $\langle I\|P\rangle$ (I = identity, $P\in\pi_N$), all $\langle Q\|P\rangle$ can be calculated since Q is a product of neighbour transpositions. However, the result must be unequivocal notwithstanding whether we represent (e.g.) (13) = (12)(23)(12) or (13) = (23)(12)(23). This is indeed true: considering (2.4) as the action of linear operators T_{jj+1} on the group ring

$$\langle Q(34) \| P\rangle = \sum_{P'} \langle Q \| P'\rangle\langle P' | T_{34} | P\rangle \tag{2.5}$$

with only nonvanishing matrix elements

$$\begin{aligned} \langle P | T_{34} | P\rangle &= x_{P3,P4} \\ \langle P(34) | T_{34} | P\rangle &= 1 + x_{P3,P4} \end{aligned} \tag{2.6}$$

we verify the algebra

$$T_{34}\, T_{34} = \mathbb{1}$$

(2.7) $$T_{12}\, T_{23}\, T_{12} = T_{23}\, T_{12}\, T_{23}$$

$$[T_{34}, T_{56}] = 0.$$

It is this representation of the symmetric group which guarantees for the compatibility of (2.4).

2.2 Yang's Representation

Besides $T_{j,j+1}$ we shall quote another representation of π_N. Yang defines neighbour transposition operators $Y_{j\,j+1}$ acting on group ring vectors $|\bar{P}\rangle := \sum_Q \langle Q \| P \rangle\, Q^{-1}$:

(2.8) $$Y_{34}\, |\bar{P}\rangle := |\overline{P(34)}\rangle = \frac{(34) - x_{P3,P4}}{1 + x_{P3,P4}}\, |\bar{P}\rangle.$$

Displaying $|\bar{P}\rangle = |P1, .., PN\rangle$ we find

(2.9) $$Y_{34}\, |ijk\ell ...\rangle =: Y_{34}^{k\ell}\, |ijk\ell ...\rangle = |ij\ell k ...\rangle$$

together with the algebra (which in turn justifies the definition (2.8))

$$Y_{12}^{ij}\, Y_{12}^{ji} = \mathbb{1}$$

(2.10) $$Y_{12}^{jk}\, Y_{23}^{ik}\, Y_{12}^{ij} = Y_{23}^{ij}\, Y_{12}^{ik}\, Y_{23}^{jk}$$

$$[Y_{12}^{ij}, Y_{34}^{k\ell}] = 0.$$

Let us now proceed to the problem of N particles confined in a volume L, in view of the thermodynamical limit. We shall impose periodic boundary conditions on the momenta of the ansatz (2.4).

Proposition: The periodic boundary conditions on the system of identical particles with δ function interaction are expressed by the following eigenvalue problem:

(2.11) $$Z_P\, P|\bar{P}\rangle = \exp(i k_{P1} L)\, P|\bar{P}\rangle$$

where $Z_P(k) := X_{P2\,P1} \cdots X_{PN.\,P1}$, $X_{ij} = \dfrac{1 - x_{ij}\,(ij)}{1 + x_{ij}}$

for any i, j.

The X's obey the same algebra as the analogous operators in

(1.26). Therefore we have the similarity relations

(2.12) $$Z_{P(34)} = X_{P3,P4}\, Z_P\, X_{P3,P4}^{-1} \quad ,$$

and (2.11) reduces to only N independent equations ($P = C^j$, C = cyclic permutation)

(2.13) $$(Z_j - \exp ik_jL)\, C^j \,|\, \overline{Cj} \rangle = 0$$

where $Z_j = X_{j+1,j} \cdots X_{Nj} X_{1j} \cdots X_{j-1\,j}$. This system is indeed compatible due to the

Lemma: The operators Z_j commute with each other.

Hence periodic boundary conditions get along well together with the Bethe ansatz.

2.3. The Ternary Algebra and Integrability

There is an important analogy between the operators Z_j of the sect. 2.2, and the transfer matrices of a vertex model, which is expressed in the

Lemma:

(2.14) $$Z_j = - \lim_{k \to k_j} Z(k)$$

where

(2.15) $$Z(k) := \operatorname{tr}_0 X_{10} \cdots X_{N0} \;,\quad X_{j0} = \frac{1 - x_{j0}(ic)}{1 + x_{j0}} \;,\quad x_{j0} = \frac{ic}{k_j - k}$$

with an auxiliary site index o. The operation tr_0 is an appropriate generalization to symmetry types of Young tableaux with n lines of the trace over Pauli matrices τ for n = 2 (cf. chap. 1). We see that the commutativity of the Z_j is due to the

Proposition: (2.16) $$[Z(k), Z(k')] = 0 \qquad \forall k, k'$$

which is proven in exact analogy with (1.13) exploiting vitally the ternary algebra among the X's.

The operators $Z(k)$ can be regarded as a one parameter family of transfer matrices for a vertex model in which there are n possible states for each line. If n = 2, it becomes an inhomogeneous six vertex model.

The commutativity of transfer matrices is closely related with the factorization of a scattering matrix into two-particle

S matrices (Zamolodchikov (1979)). In this picture, the ternary relations act as the associativity conditions on a multiplet of asymptotic field operators $A_\alpha(k)$ which are the o - components of $X_{1o}X_{2o}\cdot X_{No}$. If these conditions are met, then the three particle S matrix can be unequivocally determined as a product of two-particle matrices

$$\begin{aligned} &S_{ij,\lambda\mu}(k_1-k_2)\,S_{\lambda\ell,\alpha\nu}(k_1-k_3)\,S_{\mu\nu,\beta\gamma}(k_2-k_3) \equiv \\ (2.17)\quad &\equiv S_{j\ell,\lambda\mu}(k_2-k_3)\,S_{i\mu,\nu\gamma}(k_1-k_3)\,S_{\nu\lambda,\alpha\beta}(k_1-k_2) = \\ &= S_{ij\ell,\alpha\beta\gamma}(k_1,k_2,k_3). \end{aligned}$$

The ternary relations have more extensive implications than the mere commutativity of transfer matrices. We call <u>transition matrix</u> (from site 1 to site j) the product

(2.18) $$T_j(k) = X_{1o}\,X_{2o}\cdots X_{jo}\,.$$

They satisfy the difference equation

(2.19) $$T_j^{-1} = X_{oj}\,T_{j-1}^{-1}$$

which is the first equation of a Lax pair, to be supplemented by a time evolution equation

(2.20) $$i\frac{d}{dt}T_j^{-1} = M_j\,T_j^{-1}$$

in order to develop a (discrete) inverse scattering method (Faddeev (1979)). The compatibility condition

(2.21) $$i\frac{d}{dt}X_{oj} = M_jX_{oj} - X_{oj}M_{j-1}$$

must match with the Hamiltonian time evolution

(2.22) $$i\frac{d}{dt}X_{oj} = \left[(j-1,j)+(j,j+1)\,,\,X_{oj}\right].$$

That such operators M_j exist, is again guaranteed for precisely by the ternary relations among the X's :

(2.23) $$M_j = \frac{1}{k^2+1}\left\{\mathbb{1} + (j,j+1) + (oj)\cdot(o,j+1) + ik\left[(o,j+1)(oj) - (oj)(o,j+1)\right]\right\}.$$

It implies the existence of integrals of the motion generated by $\mathfrak{X}(k)$, and the integrability of the system.

For each j, the algebra of $T_j(k)$ is an isomorphic realization of that of the field operators A(k). With regard to sect. 3.2, let us quote explicitly the algebra of transition matrices for spin-1/2 fermions (n=2). We place ourselves in the conjugate representation

$$(2.24) \qquad X_{j0} = \frac{1 + x_{j0}\,(j0)}{1 + x_{j0}}$$

and represent $(j0) = \frac{1}{2}(\mathbb{1} + \vec{\sigma}_j \otimes \vec{\tau})$. Calling $T_{\alpha\beta}$ the components of T as a τ matrix, $T_{12} = T_+$, $T_{21} = T_-$, $T_{11} - T_{22} = T_z$, we have the commutator algebra

$$[T_+(k), T_+(k')] = [T_z(k), T_z(k')] = 0$$

$$(2.25) \qquad T_{11}(k)\,T_+(k') = (1 + x_{00'})\,T_+(k')\,T_{11}(k) - x_{00'}\,T_+(k)\,T_{11}(k')$$

$$T_{22}(k)\,T_+(k') = (1 - x_{00'})\,T_+(k')\,T_{22}(k) + x_{00'}\,T_+(k)\,T_{22}(k').$$

3. IDENTICAL PARTICLES WITH DELTA FUNCTION INTERACTION: GENERAL SOLUTION FOR TWO INTERNAL STATES

3.1 The Problem of Spin-1/2 Fermions

We shall continue the line of chap. 2 for the special case of spin-1/2 fermions. The Hamiltonian (2.1) does not act on the internal variables. Hence a wave function totally antisymmetric in its N arguments which carries the total spin $S_z = \frac{N}{2} - M = \bar{M} - \frac{N}{2}$, factorizes into a sum of products of a spin function $\chi_{\tilde{\tau}}$ with an orbital function ψ_τ of conjugate symmetry type $\tau, \tilde{\tau}$ (Hamermesh, Group Theory (1964)). $\tilde{\tau} = [\bar{M}, N-\bar{M}]$ having maximally two lines is determined by S_z. Thus $\tau = [1^{2\bar{M}-N}\, 2^{N-\bar{M}}]$, i.e. ψ_τ is antisymmetric separately in the $\bar{M}$ arguments of the first column, and in the $N-\bar{M}$ arguments of the second one, and cannot be further antisymmetrized (Fock condition). Finally the sum $\Psi = \sum \chi_{\tilde{\tau}} \psi_\tau$ carries over all allowed numberings of the Young tableau τ.

In terms of expansion (2.3) of the orbital wave functions, the belonging to the symmetry type expresses itself as conditions on the amplitudes $\langle Q \| P \rangle$. In Yang's picture (sect. 2.2) they just imply that the group ring vectors $|P\rangle$ belong to the representation τ too.

Since we are already familiar with its conjugate spin - $\frac{1}{2}$ representation $\tilde{\tau}$ from chap. 1, let us replace the algebra to solve by its conjugate:

$$|\bar{P}\rangle \to |\tilde{P}\rangle = \sum_{Q} I(Q)\, Q^{-1} \langle Q \| P\rangle$$

$$(3.1)\qquad (oj) \to (\widetilde{oj}) = -(oj) = -\tfrac{1}{2}(\mathbb{1} + \vec{\sigma}_j \otimes \vec{\tau})$$

(and omit the $\sim$ thereafter).
The spectrum equations become

$$(3.2)\qquad z_j = z(k_j) = \exp ik_jL \qquad (j = 1, \;, N)$$

where z_j denote the eigenvalues of the operators Z_j built from (eq.(2.15))

$$(3.3)\qquad X_{no} = \frac{1 + x_{no}\, \tfrac{1}{2}(\mathbb{1} + \vec{\sigma}_n \otimes \vec{\tau})}{1 + x_{no}}$$

Now Z(k) appears as the transfer matrix for an inhomogeneous six vertex model with weights

$$(3.4)\qquad \begin{aligned} &a_n = 1, \qquad b_n = y_n^{-1}, \qquad c_n = x_n y_n^{-1} \\ &\left(x_n = x_{no} = \frac{ic}{k_n - k}\right., \qquad \left. y_n = 1 + x_n\right) \end{aligned}$$

for the vertices 1,2 resp 3,4 resp. 5,6. Actually, the relative invariants of this inhomogeneous model do not depend on the site:

$$(3.5)\qquad \frac{\mathcal{I}_2}{\mathcal{I}_1} = \frac{\mathcal{I}_3}{\mathcal{I}_1} = \frac{a_n^2 + b_n^2 - c_n^2}{2a_n b_n} = 1$$

This feature allows to apply Bethe's method in Baxter's (1971) generalization.

The transfer matrix commutes with the operator of the total spin component S_z, thus there is an $S_z = M$ eigenvector of Z:

$$(3.6)\qquad |\bar{I}\rangle \equiv |M\rangle = \sum_{(n)} c(n_1, \;, n_M)\, S^-_{n_1} \ldots S^-_{n_M}\, |\tfrac{1}{2}N\rangle.$$

The sum carries over all ordered M-fold indices $1 \leq n_1 < \ldots < n_M \leq N$. Z mixes only states $(n),(m)$ with either $n_i > m_i\ \forall i$ or $n_i \leq m_i\ \forall i$ (fig. 3.1). Redefining in case $(m_1, \;, m_M) \to (m_2, \;, m_M, m_1 + N)$ such that $1 \leq n_1 < \; < n_M \leq N$ and $m_i \leq n_i \leq m_{i+1}$, and imposing cyclic invariance

$$(3.7)\qquad c(m_1, \ldots, m_M) = c(m_2, \ldots, m_M, m_1 + N)$$

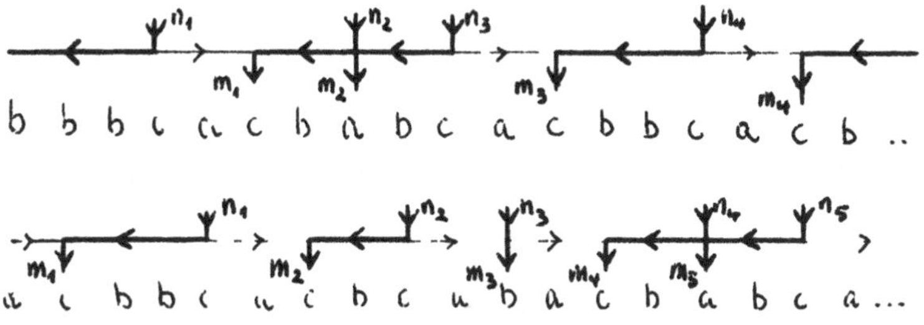

Fig. 3.1. Elements of the six vertex transfer matrix

we can simply write $\mathcal{T}|\bar{I}\rangle = z|\bar{I}\rangle$ as

$$\sum_{(n)} c(n) \prod_{i=1}^{M} D(m_i,n_i)\,\bar{D}(n_i,m_{i+1}) = z\,y\,c(m) \tag{3.8}$$

where

$$D(m,n) = \begin{cases} x_n & m<n \\ x_n^{-1} & m=n \end{cases}$$

$$\bar{D}(n,m) = \begin{cases} y_{n+1} \cdots y_{m-1}\, x_n & n<m \\ y_n x_n^{-1} & n=m \end{cases} \tag{3.9}$$

$$y = y_1 y_2 \cdot y_N\,, \qquad x_0 = x_{N+1} = 1.$$

Yang (1968) has solved (3.8) by

$$c(n) = \sum_{Q\in\Pi_M} B(Q)\,\varphi(v_{Q1},n_1)\,.\;\varphi(v_{QM},n_M) \tag{3.10}$$

with

$$\varphi(v,n) = y'_1 \cdots y'_{n-1}\, x'_n$$

$$x'_n = x_n(v) = \frac{ic}{k_n - v}\,, \qquad y'_n = y_n(v) = 1 + x'_n. \tag{3.11}$$

The coefficients B(Q) have to be chosen of the form

$$B(Q) = \prod_{1\le j<l\le M} \left(1 - \frac{ic}{v_{Qj} - v_{Ql}}\right). \tag{3.12}$$

In order to satisfy the cyclic conditions (3.7), we must determine the M parameters v_a from the momenta k_j, by the coupled spectrum equations

$$\prod_{b(\neq a)} \frac{v_a - v_b - ic}{v_a - v_b + ic} = \prod_{j=1}^{N} \left(1 + \frac{ic}{k_j - v_a}\right). \tag{3.13}$$

Now the eigenvectors (3.6) of Z are determined.

We calculate the eigenvalues $\mathfrak{z}$

$$(3.14) \quad \mathfrak{z}(k) = \prod_{a=1}^{M} \left(1 + \frac{ic}{k - v_a}\right) + y^{-1} \prod_{a=1}^{M} \left(1 - \frac{ic}{k - v_a}\right),$$

and the periodicity (3.2) finally imposes the spectrum condition on k_j

$$(3.15) \quad \mathfrak{z}(k_j) = \prod_{a=1}^{M} \left(1 + \frac{ic}{k_j - v_a}\right) = \exp ik_j L .$$

Remark that (3.13) implies, that $\mathfrak{z}(k)$ has no pole at $k = v_a$.

3.2 The Operator Method

There is another important method to diagonalize the transfer matrix of the fermion model or the eight vertex model. Faddeev (1979) constructs the Bethe wave function from a reference state acted on by a product of field operators. The essential entry is the algebra of transition matrices (eq.(2.25)); the reference state is $|\frac{1}{2}N\rangle$. Then the eigenvectors of $Z(k) = T_{11} + T_{22}$ are

$$(3.16) \quad |M\rangle = T_{+}(v_1) \ldots T_{+}(v_M) \; |\tfrac{1}{2}N\rangle,$$

and the spectrum equations (3.13) - (3.15) are reproduced.

REFERENCES:

R. Baxter : Stud.Appl.Math. 50 (1971),51
Phys.Rev.Lett. 26 (1971),832
B. Derrida: Solution d'un modèle à trois corps: étude de la diffusion (Thèse de 3° cycle, Univ.Paris 1976)
L. D. Faddeev: Quantum completely integrable models of field theory (Leningrad 1979)
M. Gaudin: J.Math.Phys. 12 (1971), 1674 and 1677
M. Hamermesh: Group Theory (Addison-Wesley, 1964)
L. P. Kadanoff, F. J. Wegner: Phys.Rev. B4 (1971),3989
W. P. Kasteleyn: Graph Theory and crystal physics (in: Graph theory and theoretical physics, Harary Ed. A. P. 1967)
J. B. Mc Guire: J.Math.Phys. 5 (1964), 622
B. Sutherland: J.Math.Phys.11 (1970), 3183
C. N. Yang: Phys.Rev.168 (1968), 1920
A. B. Zamolodchikov: Comm.Math.Phys. 69 (1979), 165

ELEMENTARY METHODS FOR STATISTICAL SYSTEMS,

MEAN FIELD, LARGE n, AND DUALITY

Claude Itzykson

DPh-T, CEN Saclay, 91191 Gif-sur-Yvette Cedex, France

I. REGULARIZATION AND RENORMALIZATION

Renormalizable field theories are somehow paradoxical. They are singled out by precise constraints involving geometric and internal invariances to such an extent that they are determined by a few renormalized parameters. Yet any practical calculation involves a regularization scheme in some disguise or another to allow for manageable expressions which break some of these sacrosanct invariances, in particular the geometric ones. This breaking is characterized by some length (a) or momentum scale ($\Lambda \sim 1/a$). To the extent that the regularization does not viciate the content of the continuous theory for momenta much smaller than Λ or lengths much larger than a, there is a fundamental need for a set of renormalization transformations with an ultraviolet fixed point. As the renormalized parameters are kept fixed an increase in the cutoff scale Λ should affect the bare parameters in a way which leads to well defined predictions for physical correlations. Asymptotically free models for which the bare couplings are driven to zero are especially attractive because perturbative methods appear ultimately justified. Ironically such theories are plagued with difficulties concerning their spectrum and physical properties. The prime example of such a situation is afforded by Yang Mills field theories for which command of the strong coupling regime is mandatory to extract meaningful statements about their physical content including the confinement properties. It is therefore fortunate that statistical methods can be adapted to these problems and yield valuable information, even though they strongly rely on the cutoff theory. One may hope in the future to find similar methods directly in the continuum framework.

In this perspective, these lectures are devoted to a survey of some approximation schemes developped in the context of statistical mechanics. This review will of course not be meant to be complete in any way. It is rather designed to convey the flavour of some of the arguments that can be used to neglect under certain circumstances most of the fluctuations. This is a rather unlucky feature but provides a departing point from which more elaboration leads to increased insight into the models.

The confluence point of statistical mechanics and (euclidean) field theory is in the use of discretized path integrals. Discretization means here that continuous spacetime has been replaced by a regular lattice. After specification of the dynamical variables these integrals require three types of ingredients.

(i) An a priori measure on the dynamical variables. This is generally taken for granted and only poorly defined in the continuous case. Discretization allows for precise definitions. As we shall deal mostly with variables taking their values in a group or homogeneous space, the point measure will naturally be dictated by its invariance properties. In the particular case of a compact simple Lie group this will be the unique Haar measure normalized to unity.

(ii) A Boltzman weight factor expressed as an exponential

$$\exp \beta S$$

where the action S (minus the energy in statistical mechanics) is bounded from above. Any local part in S could have been absorbed in the factorized a priori measure. In most cases one requires S to couple only neighbooring degrees of freedom or more generally to imply short range interactions. This is mandatory if we are interested in the final analysis in the description of a local relativistic invariant field theory*.

(iii) A discussion of the relevance of boundary conditions. This ties up both with the question of observables and the possibility of transitions to ordered phases in the system. Also the link with some topological constraints present in a continuous model has not yet been sharply pinpointed.

The list of relevant models can be quite extensive. Also one can probe deeper and deeper the structure of a given system. We shall mostly have in mind models with global of local symmetries. The prototype of the former is the classical Heisenberg model with

---

* *In this context it would be interesting to have the analog of the Osterwalder-Schrader requirements for a short range euclidean lattice theory. These insure in particular the positivity of the underlying Hilbert space. Of course in the end only that part of the Hilbert space pertaining to the vicinity of the largest eigenvalue of some transfer matrix does matter. So these requirements could be weakened.*

O(n) symmetry (sometimes also called non linear σ-model). The action is given by a sum over nearest neighbours of scalar products between unit spins

$$S = \sum_{(ij)} \vec{S}_i \cdot \vec{S}_j$$

and generalizations thereof. The case n=1 is the Ising model.

Gauge invariant models [1] have as a prototype the SU(n) lattice gauge theory where the action involves the trace over plaquettes variables - i.e. ordered products of link variables taking their values in SU(n) along elementary circuits of the lattice

$$S = \sum_p \text{tr } U_p + \text{c.c.}$$

The case when the link variables are just ±1 is an Ising like gauge model. In the sequel we shall limit ourselves to these types, and study in gross generality some of their properties.

A. Variational Principle

Perhaps the most straightforward and rigorous way to implement the mean field approximation, is to use the variational principle of statistical mechanics which instructs us to minimize energy minus entropy. Mathematically this is equivalent to the convexity property of the exponential

$$\langle e^A \rangle \geq e^{\langle A \rangle} \tag{1}$$

To simplify calculations we shall look at an Ising model in dimension d where the partition function in an external field is (N = number of sites)

II. MEAN FIELD APPROXIMATION

Take the classical Heisenberg model as an example. When the coordination number of the lattice gets very large (as is the case when the dimension d of the lattice goes to infinity) it is appealing to think of the numerous neighbouring variables as creating a mean field with relatively small fluctuations acting on a given spin. This leads to an independent spin model interacting with an external field determined consistently. There are means of increasing complexity to implement this simple idea.

$$Z = 2^{-N} \sum_{\sigma_i = \pm 1} e^{\beta \sum_{(ij)} \sigma_i \sigma_j + h \sum_i \sigma_i} \tag{2}$$

We can also write

$$Z = Z_H \left\{ Z_H^{-1}\, 2^{-N} \sum_{\sigma_i=\pm 1} e^{H\sum_i \sigma_i + \left[\beta \sum_{(ij)} \sigma_i\sigma_j + (h-H)\sum_i \sigma_i\right]} \right\} \tag{3}$$

$$Z_H = 2^{-N} \sum_{\sigma_i=\pm 1} e^{H\sum_i \sigma_i} = (\cosh H)^N$$

The term in brackets can be thought as an average with a weight proportionalto $e^{H\sum_i \sigma_j}$.

Therefore, applying (1)

$$Z \geq \operatorname*{Sup}_H \; (\cosh H)^N \, e^{<\beta \sum_{(ij)} \sigma_i\sigma_j + (h-H) \sum_i \sigma_i>_H} \tag{4}$$

If we define the free energy as

$$F = \lim_{N\to\infty} \frac{1}{N} \ln Z \tag{5}$$

and notice that

$$m = <\sigma_i>_H = \tanh H \tag{6}$$

then

$$F \geq \operatorname*{Sup}_H \left\{ \ln \cosh H + \beta d (\tanh H)^2 + (h-H)\tanh H \right\} \tag{7}$$

For a vanishing external field h the coefficient of the H^2 term in the r.h.s. is $(\beta d-1/2)$ and it is easy to check that when $\beta<\beta_c=1/2d$ the origin is an absolute maximum. This means that the mean field vanishes in this phase, as does the spontaneous magnetization. Furthermore for infinitesimal h we find

$$\beta < \beta_c = \frac{1}{2d} \qquad\qquad F(h) = \frac{1}{(1-2\beta d)}\,\frac{h^2}{2} + \ldots \tag{8}$$

so that the magnetization behaves linearly with h

$$m(h) = \frac{1}{1-2\beta d}\, h + \ldots \tag{9}$$

and the zero field susceptibility

$$\chi = \left.\frac{\partial^2 F}{\partial h^2}\right|_{h=0} = \frac{1}{1-2\beta d} \tag{10}$$

has a simple pole divergence at $\beta=\beta_c$. Of course this has to be corrected for the small fluctuations ($o(1/d)$) of H. For $\beta>\beta_c$ we find that (7) exhibits a doubly degenerate maximum in the absence of h, with the degeneracy lifted by an infinitesimal external field. The situation is typical of a second order transition. For β close to β_c the mean field and spontaneous magnetization have a square root behaviour

$$H \sim m \sim \sqrt{3}\ \sqrt{2\beta d-1} \tag{11}$$

Remarks. (a) By a Legendre transform we can replace the field H by the magnetization to maximize

$$\mathcal{F}(m) = -\ell n2 + \beta d m^2 - \sum \frac{1\pm m}{2}\ \ell n\ \frac{1\pm m}{2} = (-)\text{energy}+\text{entropy} \tag{12}$$

(b) The variational principle determines the best possible factorized density matrix. If boundary conditions are such that odd expectation values vanish (for instance periodic boundary conditions) this density matrix in the uniform case describes a mixture for $\beta>\beta_c$. However an infinitesimal external field selects one of the components and odd expectation values have non zero limits.

(c) We can also compute correlation functions using an external varying external field, obtaining therefore a varying mean field.

(d) The variational method can be turned into a variational perturbative method. We can correct the factorized trial density matrix by small correlations treated perturbatively. This can also be done using diagrammatic techniques described for instance in reference [2]. It will also reappear in the sequel. In this way we can obtain systematic corrections as a $1/d$ expansion. Let us quote here the first few terms obtained for the n-vector model [3](n=1 corresponds to the Ising model)

$$\frac{n}{2\beta_c d} = 1 - \frac{1}{2d} - \frac{1}{(2d)^2}\left[2-\frac{2}{n+2}\right] - \frac{1}{(2d)^3}\left[7-\frac{8}{n+2}\right] + \dots \tag{13}$$

B. Field Equations

Although the application of field equations may appear almost like a joke for Ising like systems we want to introduce it here because it turns out to be useful in a more intricate context. The idea is to express the invariances of the a priori measure. In the Ising case we can flip an arbitrary spin without affecting this measure. This leads at once to the identity

$$1 = \langle e^{-2\beta\, \sigma_i \sum_{j(i)} \sigma_j} \rangle \tag{14}$$

where j(i) denote the neighbours of a site i. Set

$$H_i = \beta \sum_{j(i)} \sigma_j \tag{15}$$

in such a way that

$$1 = \langle \cosh 2H_i \rangle - \langle \sigma_i \sinh 2H_i \rangle \tag{16}$$

Suppose that boundary conditions allow for non vanishing odd expectation values (or else add an infinitesimal external field h_i which amounts to displace $H_i \to H_i + h_i$). Since H_i is a sum of $q \equiv 2d$ random (but in principle correlated) variables we can assume as $q \to \infty$ that fluctuations are relatively unimportant as before. For a uniform configuration this transforms the equation (16) in the mean field relation

$$\begin{aligned} &(\cosh 2H-1)/\sinh 2H \equiv \tanh H = \langle \sigma \rangle \\ &H = 2d\beta \langle \sigma \rangle \end{aligned} \tag{17}$$

and the first correction to this picture involves correlations $\langle \sigma_i \sigma_j \rangle - \langle \sigma_i \rangle \langle \sigma_j \rangle$ among neighbours.

C. Random Field Transform

A last method which, although more formal, turns out to allow a systematic investigation of corrections, is based on the Laplace transform

$$e^{\frac{\beta}{2} \sum_{ij} \sigma_i J_{ij} \sigma_j} = \left[\det \beta J\right]^{-\frac{1}{2}} \int \mathcal{D}\left(\frac{h_i}{\sqrt{2\pi}}\right) e^{-\frac{1}{2\beta} hJ^{-1}h + h\sigma} \tag{18}$$

and we can arrange things so that J is invertible.

The sum over σ's becomes trivial in the partition function which becomes

$$Z = \left[\det\beta J\right]^{-\frac{1}{2}} \int \mathcal{D}\left(\frac{h_i}{\sqrt{2\pi}}\right) e^{-\frac{1}{2\beta} hJ^{-1}h + \sum \ell n \cosh h_i} \tag{19}$$

Formulas (18) and (19) explain the name given to this method and it is clear that the variable conjugate to h is the magnetization.

Then the mean field approximation is converted into the saddle point evaluation of the integral plus loop expansion to obtain correction. It is amusing to note that the large temperature (small β) expansion is nothing but the ordinary perturbative expansion of (19) as is readily seen by rescaling h into $\sqrt{\beta}h$.

Again the saddle point is given by the equation

$$(J^{-1}h)_i = \beta \tanh h_i \Leftrightarrow h_i = \beta J_{ij} \tanh h_j \tag{20}$$

which for nearest neighbour interactions and a uniform solution $h_i=H$ reduces to the form

$$H = 2d\beta \tanh H \tag{21}$$

Including one loop correction yields what would be called in field theory an effective potential given a prescribed uniform value of the field H

$$\begin{aligned} NF = \ell nZ &= -\frac{1}{2\beta} HJ^{-1}H + \sum \ell n \cosh H - \frac{1}{2} \operatorname{tr} \ell n(I+\beta J(1-\tanh H^2)) \\ &= -\frac{1}{2\beta} HJ^{-1}H + \sum \ell n \cosh H + \frac{1}{4} \beta^2 (1-\tanh H)^2 \sum J_{ij} J_{ji} + \dots \end{aligned} \tag{22}$$

The signal for spontaneous symmetry breaking will be the existence of a non trivial extremum given by

$$0 = -\frac{1}{\beta} J^{-1} H + \tanh H - 2d\beta^2 (1-\tanh H^2)^2 \tanh H + \dots \tag{23}$$

where the terms kept here are sufficient to compute the 1/d correction. There will indeed a non trivial solution when β exceeds a critical value given by

$$\frac{1}{2d\beta_c} = 1 - \frac{1}{2d} + \dots \tag{24}$$

in agreement with (13). Further corrections can be computed in similar fashion.

As we know careful examination of the Feynman diagrams appearing in the above perturbative expansion exhibit infrared divergences in dimension $d \leq 4$. It is then necessary to develop the whole machinery of renormalization group arguments, applying for instance to the $\varepsilon \equiv 4-d$ expansion, to obtain the deviation of critical indices from their "classical" mean field value. We refer for instance to references [4] and [5] for a thorough discussion of these matters.

D. Gauge Invariant Systems

The step from global to local symmetries viciates a number of intuitive arguments reviewed above. As we shall see however mean field theory will however survive in slightly modified form.

Again to avoid unnecessary complications we present the calculations for a Z_2-model with variables $\sigma_{ij}=\pm 1$ attached to links and interacting by quartets along elementary plaquettes. There are Nd such variables and the partition function reads (p stands for plaquette)

$$Z = 2^{-Nd} \sum_{\sigma_{ij}=\pm 1} e^{\beta \sum_p \sigma_p}$$
$$\sigma_p = \sigma_{12}\sigma_{23}\sigma_{34}\sigma_{41} \tag{25}$$

Elitzur [6] proved a theorem which looks at first disastrous for mean field ideas. Its content states in essence that there is no spontaneous breakdown of gauge invariance. Contrary to global symmetric systems the pattern of phase transitions cannot be discussed by isolating for low enough temperature pure phases for which non gauge invariant operators have nonvanishing expectation values (would be analogs of phases where magnetization has a definite non zero value in a given direction). A hint is provided by the fact that due to local invariance boundary conditions are ineffective. Elitzur's theorem is in fact stronger and for the Z_2 case goes as follows. Suppose one attempts to break the invariance in the bulk by adding a field h_{ij} coupled to σ_{ij} link-wise. Assume for definiteness that h is uniform and let us estimate the mean value of σ_{01}

$$\langle\sigma_{01}\rangle_h = \left(\sum e^{\beta\sum\sigma_p + h\sum\sigma_\ell}\sigma_{01}\right) / \sum e^{\beta\sum\sigma_p + h\sum\sigma_\ell} \tag{26}$$

The theorem asserts that this vanishes as h goes to zero. Let us perform in this expression a gauge transformation $\sigma_{ij} \to \varepsilon_i\sigma_{ij}\sigma_j$ where $\varepsilon_i=\pm 1$. Actually let $\varepsilon_i=1$, $i\neq 0$. Gauge averaging is allowed both in the numerator and denominator. Hence

$$|\text{Numerator}| = \left|\frac{1}{2}\sum_{\varepsilon_o=\pm 1}\sum e^{\beta\sum_p \sigma_p + h\sum'\sigma_\ell + \varepsilon_o h\sum_1^{2d}\sigma_{oj}}\varepsilon_o\sigma_{01}\right|$$
$$\leq \sum e^{\beta\sum_p\sigma_p + h\sum'\sigma_\ell}|\sinh 2dh|$$

$$\text{Denominator} = \frac{1}{2}\sum_{\varepsilon_o=\pm 1}\sum e^{\beta\sum_p \sigma_p + h\sum{}'\sigma_\ell + \varepsilon_o h\sum_1^{2d}\sigma_{oj}}$$

$$\geq \sum e^{\beta\sum_p \sigma_p + h\sum{}'\sigma_\ell}\ \frac{1}{2}e^{-2d|h|}$$

Hence

$$|\langle\sigma_{01}\rangle_h| \leq 2\, e^{2d|h|}\,|\sinh 2dh| \underset{h\to 0}{\to} 0 \tag{27}$$

The same argument works for other gauge groups. Hence the conclusion follows that a transition in such systems cannot be signaled by the non vanishing of a local non gauge invariant operator. We must rather look for the difference in behaviour of non local gauge invariant properties as suggested by Wilson, such as the loop operator

$$\langle W(C)\rangle = \langle\prod_C \sigma_\ell\rangle \tag{28}$$

The product (to be ordered and traced in the general case) of link observables along a closed curve C. We can draw a parallel between the global and local case as follows

	global	local
variable	σ_i	σ_{ij}
invariance	$\sigma_i \to \varepsilon\,\sigma_i$	$\sigma_{ij} \to \varepsilon_i\sigma_{ij}\varepsilon_j$
correlations	$\langle\sigma_i\sigma_j\rangle$	$\langle\prod_C \sigma_{ij}\rangle$
high temperature $\beta\to 0$	$\langle\sigma_o\sigma_L\rangle \underset{L\to\infty}{\sim} e^{-aL}$	$\langle\prod_C \sigma_{ij}\rangle \underset{C\to\infty}{\sim} e^{-KA}$
Low temperature $\beta\to\infty$	$\langle\sigma_o\sigma_L\rangle \underset{L\to\infty}{\sim} m^2$	$\langle\prod_C \sigma_{ij}\rangle \underset{C\to\infty}{\sim} e^{-K'\|C\|}$

With A and $|C|$ the minimal area enclosed by C and its perimeter. The The interpretation of the low temperature behaviour of spin systems is that by external means we can separate pure phases with $\langle\sigma\rangle=\pm m$. In the absence of any infinitesimal perturbation the system remains in a mixed state characterized by the large distance behaviour of correlations. Or stated slightly differently in large enough dimension the entropy per site varies from $\ell n2$ at large temperature to zero at low temperature. The analog statement for gauge systems is that at high temperature it varies from $d\ell n2$ (links are disordered) to $\ell n2$, the entropy corresponding the gauge group (to be factored out as β ranges from zero to infinity. But the Elitzur's theorem prevents us from interpreting the low temperature behaviour as meaning $\langle\sigma_\ell\rangle$ = cste which would naturally account for the perimeter-law. It does suggest however something of the form $\langle\sigma_\ell\rangle = \overline{\varepsilon_i\varepsilon_j m} = 0$ where the bar stands for an average over the gauge group. For the purpose of computing gauge invariant quantities this would be sufficient however since the gauge average would cancel out and m could be defined as $\lim_{|C|\to\infty} \langle W(C)\rangle^{1/|C|}$, i.e. in the previous notation m stands for $e^{-K'}$. Of course we have to worry how uniform the limit $|C|\to\infty$ is, since the space of curves is a very large one. But in principle the same problem exists for correlation functions in the spin case except it looks less severe in that there are "fewer" directions in which we can separate the points. We shall stick to this interpretation of "ordering" in gauge systems and progressively justify it using the mean field techniques. In the first place the convexity inequality remains solid mathematics. Hence

$$NF(\beta) \geq \sup_{H_{ij}} \left\{ \ln\cosh H_{ij} - H_{ij}\tanh H_{ij} + \beta\sum_{p}\left(\prod \tanh H_{ij}\right)\right\} \quad (29)$$

The right hand side is obviously gauge invariant. In this sense the approximate density matrix is only defined up to a gauge transformation which does not affect the computation of invariant averages.

We shall look for a uniform solution (up to a gauge transformation) in which case this simplifies into

$$d^{-1}F(\beta) > \sup_{H} \left\{ \ln\cosh H - H\tanh H + \beta\frac{d-1}{2}(\tanh H)^4 \right\} \quad (30)$$

Due to the four links surrounding a plaquette the energy term $(\tanh H)^4$ cannot complete with the H^2 entropy term in the neighbourhood of the origin. As a result we find a first order transition when the secondary maximum depicted schematically on figure 1 reaches the x axis

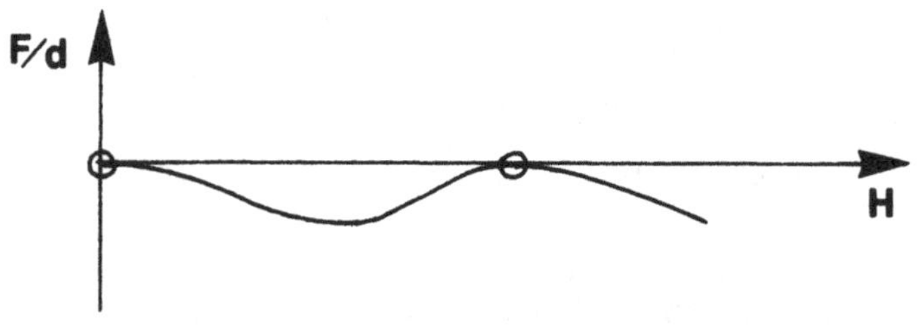

Figure 1

so that H jumps discontinuously from zero to a finite value for a critical value β_c given by the equations

$$\ln\cosh H_c - H_c\tanh H_c + \beta_c(d-1)/2\,(\tanh H_c)^4 = 0$$

$$H_c = 2\beta_c(d-1)\tanh H_c^3 \quad (31)$$

Numerically this yields

$$2\beta_c(d-1) = 2.75 \tag{32}$$

with a jump to $m_c = e^{-K'_c} = \tanh H_c = 0.66$

We find a strong area law in large dimension (i.e. in terms of a parameter $\tilde{\beta} = 2\beta(d-1)$) with the string tension, the coefficient K in the expression e^{-KA}, being infinite. Indeed for small β $e^{-K} = \tanh \beta \sim \beta$ so $K = \ell n \frac{1}{\beta} \simeq \ell n \frac{(d-1)2}{\tilde{\beta}} \to \infty$. Beyond β_c we have a perimeter law.

In order to check the correcteness of what has been obtained it is possible to compare it with the term by term large d limit of the exact low temperature expansion. Since the order of the limits $d \to \infty$, $\beta \to \infty$ has been interchanged it is somehow gratifying to see that both expressions do agree :

$$d^{-1}F = -\ell n2 + \frac{x}{4} + e^{-2x} + (3x-1/2)\, e^{-4x} + \ldots \tag{33}$$

$$x = 2\beta(d-1)$$

We may understand how this comes about clearly in a system with discrete symmetry. In the low temperature expansion we classify states according to the number of frustrated plaquettes. In the limit $d \to \infty$, these are obtained from an "ordered" or "cold" state with all link variables equal to the identity and flipping them one by one ; this cannot conflict with gauge equivalents which require acting on at least $2d \to \infty$ links at once. Therefore we procede *as if* there was indeed a non-vanishing ordering value for a link variable. This argument is more delicate in the case of a continuous group, which involves a low temperature series in powers of x^{-1} instead of exponentials. But again one can check the agreement.

The picture that emerges is that a pure gauge theory undergoes a first order transition from a confining to deconfined phase in large enough dimension. In the opposite direction it becomes a trivial model in two dimensions where no transition occurs. So one might suspect that there exists a critical dimension where the latent heat (the discontinuity in the derivative $\partial F/\partial \beta$) vanishes. For instance it turns out that Z_2 gauge models have a second order transition in 3 dimensions and a first order one in four and higher dimensions. Similarly a U(1) gauge model seems to undergo a continuous transition in four dimensions and a first order one in higher dimensions. Finally we would like to think that a pure SU(n) model has a limiting continuous behaviour when d=4 and $\beta \to \infty$ and a first order discontinuous behaviour for $d \geq 5$.

The analysis of the gauge invariant field equations

$$1 = \langle \cosh 2H_\ell \rangle - \langle \sigma_\ell \sinh 2H_\ell \rangle$$

$$H_\ell = \beta \sum_{p(\ell)} \sigma_{\ell'} \sigma_{\ell''} \sigma_{\ell'''} \tag{34}$$

with $p(\ell)$ all the plaquettes bounded by the link ℓ, yields the same equations as before except for an extra assumption require to relate $\langle H_\ell \rangle$ to $\langle \sigma_\ell \rangle$ due to the cubic character of the second relation in (34). For a uniform solution we find

$$m = \tanh H \quad , \qquad H = 2\beta(d-1)(\tanh H)^3 \tag{35}$$

and procede as we did using convexity. However in the small β phase when H and m go to zero we can use an ingenious idea due to Drouffe, Parisi and Sourlas [7] to devise a new approximation scheme which originates in some regularity observed in the high temperature expansion. They noticed that each time the power of β (or $\tanh \beta$) is increased by four units the polynomial in d which multiplies it in the series for the free energy (or a similar quantity) is increased by unity. The simple geometric reasons derives from the fact that the high temperature expansion involves surfaces made of plaquettes (each carrying a factor $\tanh \beta$). Therefore if you step four powers in $\tanh \beta$ you can replace a plaquette (factor $\tanh \beta$) by a deformation along a cube in 2(d-2) dimensions (factor $2(d-1)\tanh\beta^5$). Hence the stated property.

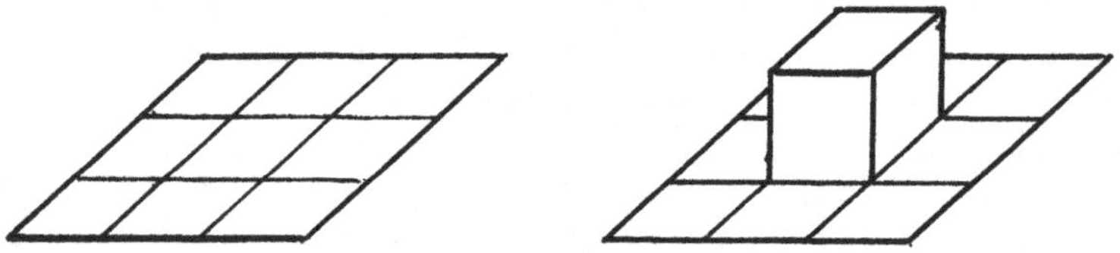

Figure 2

If you think of a given plaquette decorated in this way by a tree of cube deformations by repeating the process you come to the recursive equation

$$p = \tanh \beta + 2(d-2)\, p^5 \tag{36}$$

In the case of an arbitrary gauge group one would replace $\tanh \beta$ by

$$\tanh \beta \;\rightarrow\; \frac{\partial}{\partial \beta} \ln \int du\; e^{\beta \chi(u)} \tag{37}$$

a quantity proportional to β for small β where these approximations make sense. Equation (36) imply that p and β scale as $d^{-1/4}$. Thus if we keep $\beta d^{1/4}$ fixed while letting $d \rightarrow \infty$ we need not distinguish between β and $\tanh \beta$. The free energy obtained in this way is

$$\left[F \,/\, \frac{d(d-1)}{2} - \ln\cosh\beta\right] = \frac{(d-2)}{3}\, p^6 - 2(d-2)^2\, p^{10} \tag{38}$$

If we rescale all quantities according to

$$\tilde{\beta} = [2(d-2)]^{1/4}\,\beta \qquad \tilde{p} = [2(d-2)]^{1/4}\, p$$
$$\tilde{f} = [2(d-2)]^{1/2} \left\{F \,/\, \frac{d(d-1)}{2} - \ln\cosh\beta\right\} \tag{39}$$

we find the parametric representation

$$\tilde{\beta} = \tilde{p}(1-\tilde{p}^4)$$
$$\tilde{f} = \frac{1}{6}\,\tilde{p}^6\,(1-3\tilde{p}^4) \tag{40}$$

This non trivial behaviour in the confining phase leads to a singularity in the metastable region where $\tilde{\beta}$ reaches a maximum as a function of $\tilde{p}$ - and $\tilde{f}$ has a cusp for a value of β of order $d^{-1/4}$ much larger than β_c of order d^{-1}. One might speculate whether this singularity has a continuation in lower dimensions where it might hit the limit of the confining region.

To conclude on the subject of mean field we can also apply the random field method in this case. But here the Laplace transform cannot be computed in closed form, except to assert that we can find some $u(h,\beta)$ to write

$$Z = 2^{-Nd} \sum_{\sigma_\ell=\pm1} \exp \beta \sum \sigma_p =$$
$$= 2^{-Nd} \sum_{\sigma_\ell=\pm1} \int \mathcal{D}\, h_\ell \exp\left\{-u(h,\beta) + \sum h_\ell \sigma_\ell\right\} \tag{41}$$

This equation in fact defines $u(h,\beta)$ independently of the summation over the σ_ℓ's. Summing over σ_ℓ, the hunt for a saddle point leads to

$$\frac{\partial u}{\partial h_\ell} = \tanh h \tag{42}$$

We have to supplement it with some determination of u and the best we can do is again use a saddle point method to determine u from the equation

$$e^{\beta \sum mmmm} = \int \mathcal{D}\, h\; e^{-u(h,\beta) + \sum hm} \tag{43}$$

where now the m's are unconstrained variables. Note that this step would lead to the exact answer in the quadratic case we were dealing with in the spin case. It has in fact nothing to do with gauge invariance at this stage, and we would encounter similar problems had the spin interactions involved more than two spins at once. This was also reflected in the previous method when we had to figure out how to deal with the cubic second equation in (34). All this amounts to say that one is lead to a double saddle point method in the integral

$$Z = \int \mathcal{D}\, h\, \mathcal{D}\, m\; e^{-\sum hm + \beta \sum mmmm + \sum \ln\cosh h} \tag{44}$$

the variables to vary being h's and m's. Indeed the saddle point conditions are the same as before. And again the advantage is that we can study systematic corrections. As a bonus we find also that the extrema are gauge degenerate which accounts for Elitzur's theorem without affecting gauge invariant quantities. So in computing the latter we could factor out the gauge group by averaging over some gauge fixing condition, a la Faddeev and Popov, obtain their determinant and procede without encountering zero modes (in the case of a continuous group) in the perturbative corrections.

Let us now turn to another situation where semi-classical make their way i.e. the large n limits.

III. LARGE n-LIMITS

A. Vector Models

When the symmetry group of a model (or the size of the representation to which basic fields belong) becomes very large we expect interesting simplifications to occur [8]. Let us illustrate this in the case of the classical Heisenberg model with

$$Z = \int \prod_i d\vec{S}_i \; \exp n\,\beta \sum_{(ij)} \vec{S}_i . \vec{S}_j \tag{45}$$

with $\vec{S}_i$ a unit n-dimensional vector and dS the rotational invariant normalized measure on the unit sphere. The reason for the factor n in front of the action will soon be clear. Let us write the field equations by rotating by an infinitesimal amount a given spin $\vec{S}_i$ through

$$\vec{S}_i \to \vec{S}_i + \varepsilon\, A\vec{S}_i \tag{46}$$

where A is an arbitrary antisymmetric matrix. Performing this operation on the numerator of a correlation function $<\vec{S}^\gamma_o . \vec{S}^\delta_x>$ we get

$$0 = \delta_{i,o} <AS^\gamma_o S^\delta_x> + \delta_{ix} <S^\gamma_o AS^\delta_x> + h\beta \sum_{j(i)} <S^\gamma_o S^\delta_x (\vec{S}_j . A\vec{S}_i)> \tag{47}$$

Or stating that A is arbitrary and tracing the indices we find the rotationally invariant equation

$$(1-n)<\vec{S}_o \vec{S}_x>(\delta_{i,x}-\delta_{io})+n\beta \sum_{j(i)} <\left\{(\vec{S}_o\vec{S}_j)(\vec{S}_x\vec{S}_i)-(\vec{S}_o\vec{S}_i)(\vec{S}_x\vec{S}_j)\right\}> \tag{48}$$

This is exact for any n and any d but hardly useful in that it relates two and four point functions. Of course we could repeat the operation and get an infinite string of coupled equations. However we shall convince ourselves below that in the limit $n\to\infty$, β finite, expectation values of products of invariant quantities (scalar products) factorize

$$<(\vec{S}_1.\vec{S}_2)(\vec{S}_3.\vec{S}_4)\ldots> = <\vec{S}_1.\vec{S}_2><\vec{S}_3.\vec{S}_4>\ldots + O(1/n) \tag{49}$$

Thus taking this for granted we find a closed formula in the large n limit

$$\beta^{-1}(\delta_{i,o}-\delta_{i,x})<\vec{S}_o.\vec{S}_x>+\sum_{j(i)}(<\vec{S}_o.\vec{S}_j><\vec{S}_x.\vec{S}_i>-<\vec{S}_o.\vec{S}_i><\vec{S}_x.\vec{S}_j>) \tag{50}$$

This becomes algebraic for the Fourier transform G(q) defined through

$$<\vec{S}_{x_1}.\vec{S}_{x_2}> = \int_{-\pi}^{+\pi}\frac{d^dq}{(2\pi)^d}\, e^{iq.(x_1-x_2)}\, G(q) \tag{51}$$

$$\beta^{-1}(G(k)-G(q))+ 2\sum_{a=1}^{d}(\cos q_a-\cos k_a)G(q)G(k) = 0 \tag{52}$$

As long as G(q) has an inverse this says that

$$G^{-1}(q) = \beta(\xi^{-2} + 2\sum_{1}^{d}(1-\cos q_a)) \tag{53}$$

with ξ^{-2} a positive constant to be determined by the condition that $\vec{S}_i^2 = 1$ i.e.

$$\beta = \int_{-\pi}^{+\pi}\frac{d^dq}{(2\ )^d}\,\frac{1}{\xi^{-2}+2\sum_{a=1}^{d}(1-\cos q_a)} \tag{54}$$

As β increases starting from zero the correlation length ξ increases from $\xi \sim \beta^{1/2}$ to infinity for a critical value β_c given by

$$\beta_c = \frac{1}{2}\int_{-\pi}^{+\pi}\frac{d^dq}{(2\pi)^d}\,\frac{1}{\sum_{1}^{d}(1-\cos q_a)} = \frac{1}{2}\int_{0}^{\infty}d\alpha\, e^{-\alpha d}\,[J_o(i\alpha)]^d \tag{55}$$

Let us note that, given the change in normalisation, this is consistent with mean field theory in large dimension

$$2d\ \beta_c = 1 + \frac{1}{2d} + \frac{3}{(2d)^2} + \ldots \tag{56}$$

At the other extreme as d gets closer to d=2, $\beta_c\to\infty$ as $\beta_c\sim\frac{1}{d-2}$.

So in two dimensions there is only a symmetric phase and as we know this is correctly the lower critical dimension for such models.

When $d > 2$ and $\beta > \beta_c$ there appears a phase with spontaneous magnetization. This means that

$$G(q) = (2\pi)^d \, \delta^d(q) \, m^2 + G'(q) \tag{57}$$

in which case

$$1-m^2 = \int_{-\pi}^{+\pi} \frac{d^d q}{(2\pi)^d} \, G'(q) \tag{58}$$

and $G'(q)$ is given by equation (52) as

$$G'(q) = \frac{1}{\beta \, 2 \sum_1^d (1-\cos q_a)} \tag{59}$$

This is to be expected due to the presence of (n-1) massless Goldstone modes (the longitudinal mode is negligible in this limit). From the normalization condition we get

$$\beta > \beta_c \qquad 1-m^2 = \frac{\beta_c}{\beta} \tag{60}$$

consistent with a classical behaviour $m \sim (\beta-\beta_c)^{1/2}$ near the transition point and with the fact that the magnetization goes to unity as $\beta \to \infty$. From (54) and (55) we can obtain the behaviour of ξ as β approaches β_c from below. Indeed

$$\beta_c - \beta = \xi^{-2} \int_{-\pi}^{\pi} \frac{d^d q}{(2\pi)^d} \; \frac{1}{2 \sum_1^d (1-\cos q_a)} \cdot \frac{1}{\left[\xi^{-2} + 2 \sum_1^d (\cos q_a)\right]} \tag{61}$$

This has the interesting consequence that for $d > 4$, ξ grows like $(\beta_c-\beta)^{-1/2}$ which is the mean field prediction. However below four dimensions we encounter the usual infrared behaviour characteristic of the decisive role of fluctuations which modifies the relation between ξ and $(\beta_c-\beta)$

$$\begin{aligned} d > 4 \qquad & \xi \sim (\beta_c-\beta)^{-1/2} \\ 4 > d > 2 \qquad & \xi \sim (\beta_c-\beta)^{-1/d-2} \end{aligned} \tag{62}$$

All the above calculations are based on the factorization property (49) assumed to be valid for large n. This can be justified in either strong or weak coupling. Let us show directly why one expects it to work by justifying a saddle point approximation. To do so we explicit the constraint $\vec{S}^2 = 1$ using the equivalent of Lagrange multipliers as follows. Let Γ_n stand for the area of the unit (n-1)-dimensional sphere. Then

$$Z = \left(\frac{2n}{\Gamma_n}\right)^N \int \pi \frac{d\alpha_i}{2\pi} \, e^{in\sum\alpha_i} \int \pi dS_i \, e^{in\alpha_i \vec{S}_i^2 + n\beta\sum S_i S_j}$$

$$= \left(\frac{2n^{1/2}}{\Gamma_n}\right)^N \int \frac{d\alpha_i}{2\pi} \, e^{n\sum i\alpha_i - \frac{n}{2} \mathrm{Tr} \, \ell n \, [i\alpha_i - \frac{\beta}{2} \Delta]} \tag{63}$$

Where $\Delta = \sum_a e^{iP_a} + e^{-iP_a} = 2 \sum_a \cos P_a$.

Due to the large factor n in front of the exponential we do find that a semi-classical approximation is justified. Thus indeed eq.(49) holds. Moreover if we set

$$i\alpha_i = \frac{\beta}{2} [2d + \xi_i^{-2}] \tag{64}$$

the stationarity equation coincides with the gap equation (55). Furthermore we are now in principle in a position to compute corrections.

The same reasoning applies to the field theoretic version where one would trade a spin $\vec{S}_i$ on discrete sites for a continuous n-component field $\vec{\varphi}$ and study the equivalent vacuum functional

$$Z = \int \mathcal{D}\vec{\varphi} \, e^{-\int d^d x \, \frac{1}{2}(\partial\vec{\varphi})^2 + \frac{M_o^2}{2} \vec{\varphi}^2 + \lambda(\vec{\varphi}^2)^2} \tag{65}$$

When we"undo" the quartic term by introducing an auxiliary field we realize that the saddle point equations are equivalent to resumming geometric series of perturbation theory strings of bubble diagrams, which is an equivalent way of finding why the large n limit simplifies.

Similarly a number of other models involving an O(n) or SU(n) symmetry group with fields in the fundamental representation, like the two dimensional Gross-Neveu or CP(n) models can be treated by the same techniques [8].

B. Matrix Models

The vector models for n large simplified tremendously, yet their solution retains the flavor of the exact properties, reveals the critical role of dimensions four and two, and is the starting basis for a perturbative 1/n treatment. It was then very natural to hope that similar phenomena also take place for large n SU(n) gauge theories. Unfortunately this is not so, eventhough it has been argued that a number of simplifications do occur and yield some insight into applications to particle physics. Since there is no room here to discuss these matters in detail. I shall content myself to expose the bare roots of the problem.

We just recalled that in the limit $n \to \infty$ the only surviving small coupling diagrams of the vector models resum as geometric series of bubbles. G.'t Hooft [9] observed analogously that with a proper rescaling of the coupling constant the dominant diagrams in a gauge theory are the planar ones. This makes a huge difference since the structure of these diagrams is vastly more complicated and it looks at first that one needs to be very lucky to find a trick which resums them. It turns out that such tricks do exist in specific and non trivial cases as we shall see but no one has yet found a general answer.

To see what is involved it turns out that gauge invariance does not play a major role. What is relevant is that the fields belong to the adjoint representation of the symmetry group. This yields a wide variety of group invariants which is a source of difficulties.

A simple enough model, sufficient to uncover these various aspects, is just an analog of the spin model with $n \times n$ matrices (hermitian or symmetric say) rather than n-vectors as fields. Take the hermitian case to be specific, then

$$Z = \int dM_i \; e^{-\sum V(M_i) + \beta \sum_{(ij)} \mathrm{tr}\, M_i M_j} \tag{66}$$

The action is to be invariant under the transformations

$$M_i \to U M_i U^+ \tag{67}$$

with U belonging to U(n) (or SU(n) since any phase is cancelled). We take therefore the potential V of the form

$$V(M) = \mathrm{tr}\, v(M) \tag{68}$$

where v is a polynomial. We shall for definiteness pick

$$V(M) = \frac{1}{2}\,\mathrm{tr}\, M^2 + \frac{g}{n}\,\mathrm{tr}\, M^4 \tag{69}$$

where the scale is inserted for future convenience. Of course V(M) could contain products of traces and so on. The above choice (68) (69) is tailored to resemble somehow the gauge invariant case. Note that due to hermiticity the "kinetic" term $\mathrm{tr}\, M_i M_j$ is real.

The measure dM has to respect invariance according to (67). Up to a factor we choose

$$dM = \prod_{\alpha} dM_{\alpha\beta} \prod_{\alpha<\beta} d\,\mathrm{Re}\, M_{\alpha\beta}\, d\,\mathrm{Im}\, M_{\alpha\beta} \tag{70}$$

To see the planarity condition emerge as $n \to \infty$ we observe that it is sufficient to study the simple integral

$$Z_n = \int dM\; e^{-1/2\mathrm{tr}M^2 - g/n\mathrm{tr}M^4} \tag{71}$$

in a perturbation expansion in powers of g using Wick's theorem. Then

$$Z_n/Z_o = 1 - \frac{g}{n} <\mathrm{tr}\ M^4>_o + \frac{g^2}{2n^2} <(\mathrm{tr}\ M^4)^2>_o + \ldots \qquad (72)$$

with $Z_o = \int dM\ e^{-1/2\mathrm{tr}M^2}$ and the propagator simply given by

$$<M_{\alpha\beta}\ M^*_{\gamma\delta}> = \delta_{\alpha\gamma}\ \delta_{\beta\delta} \qquad (73)$$

The trick to keep track of the flow of indices is to represent this propagator by opositely oriented double lines (Fig.3a) as was suggested by 't Hooft. Then the quartic vertex tr M^4 is shown on Fig. 3b, involving 4 distinct indices.

When we compute for instance g/n $<\mathrm{tr}\ M^4>_0$, Wick's theorem yields 3 contributions, 2 in the form of the planar

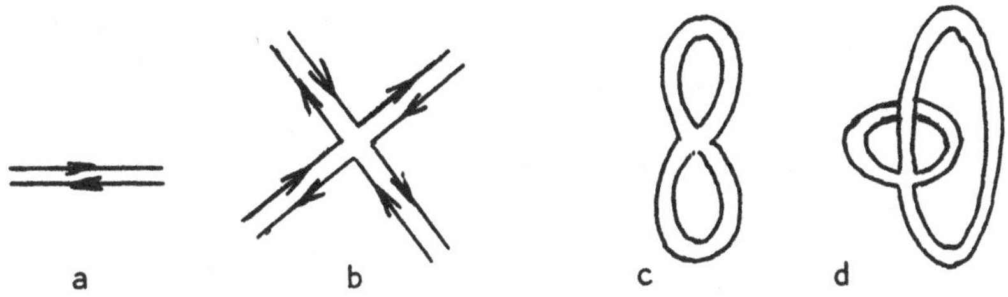

Figure 3

graph shown on Fig.3c contributing a factor g/n n^3 due to the 3 closed index circuits and 1 from graph 3d, with a factor g/n.n, a non-planar graph with only one index loop. Thus

$$\frac{g}{n} <\mathrm{tr}\ M^4> = n^2\ (2g + \frac{g}{n^2}) \qquad (74)$$

The distinction in the behaviour for n large according to the topology of the graph is already apparent here and will extend to higher orders. One notices of course that nothing would change had we discussed a more ambitious model in higher dimensions. This would simply amount to decorate these expressions by Feynman integral without affecting the counting of powers in n.

For higher orders, we follow 't Hooft's method assigning to each connected vacuum graph a closed orientable surface. To each closed index loop corresponds an oriented face. To each collapsed double line an edge joining two vertices. The opposite orientation of double lines ensures the consistency of the orientation of the surface. Note that for a graph as shown on Fig.3c the external closed line is associated to a face which means that it is to be thought as drawn on a sphere rather than a plane. We stick to standard nomenclature by calling it planar.

Let v be the number of vertices (contribution $(-g/n)^v$), e the number of propagators or edges, and f the number of faces or closed index loops (contribution nf). Each propagator joins two vertices, therefore

$$e = 2v \tag{75}$$

The graph will therefore carry a power of n given by

$$n^{f-v} = n^{f-e+v} \tag{76}$$

and we recognize that this power is nothing but the Euler characteristic of the surface

$$\chi = f-e+v = 2-2H \tag{77}$$

i.e. a topological invariant also related to the classification of closed orientable surfaces in terms of handles H to be added to a sphere with $\chi = 2$ obtained by gluing together two polygons along their edges. If these polygons have s edges then $f = 2$, $e = v = s$, hence $\chi = 2$. For an ordinary torus, take a square and identify opposite edges. This yields $f = 1$, $e = 2$, $v = 1$, hence $\chi = 0$. And so on.

Hence the conclusion ; the dominant connected vacuum diagrams are all planar and of order n^2. The next correction of order n^0 is obtained from the diagrams drawn on a torus etc... Let us call the dominant contribution $n^2 E_o(g)$

$$E_o(g) = \lim_{n\to\infty} n^{-2} \ell n \frac{Z_n(g)}{Z_o} \tag{78}$$

It remains to be seen how one can effectively compute $E_o(g)$. In the simple integral case (zero dimensional field theory !) of equation (71) there exists a simple answer derivable by various means. The underlying reason is that again our favorite saddle point method is available due to a simple counting argument. Indeed the integration in (71) can be performed in two steps. Any hermitian matrix can be written

$$M = U \Lambda U^+ \tag{79}$$

with U unitary and Λ the diagonal matrix of real eigenvalues. Similarly the measure factorizes into an angular term dU the Haar measure on the unitary group and a term only involving the eigenvalues λ_α

$$dM = \text{cst } dU \prod_{\alpha<\beta} (\lambda_\alpha - \lambda_\beta)^2 \prod_\alpha d\lambda_\alpha \tag{80}$$

The appearance of a Jacobian involving a product of factors $(\lambda_\alpha - \lambda_\beta)^2$ for each pair of eigenvalues can be traced to the increase in ambiguity in the factorization (79) when two eigenvalues coincide.

The corresponding singularity is the exact analog of what happens when one goes from cartesian to polar coordinates. In fact since only pairs of eigenvalues occur one can think of 2×2 hermitian matrices and simple dimensional arguments yield the required power 2 in $(\lambda_\alpha-\lambda_\beta)^2$. Once angles are explicitely integrated out, Z/Z_0 can be written in terms of the n eigenvalues as

$$Z_n = Z_0 \int \pi\, d\lambda_\alpha\, e^{-\sum\left(\lambda_\alpha^2/2 + g/n\, \lambda_\alpha^4\right) + \sum\limits_{\alpha \neq \beta} \ln|\lambda_\alpha-\lambda_\beta|} \tag{81}$$

This amazingly interpreted as a Coulomb repulsive one-dimensional gas in a container, with the number of particles n much smaller than the expected behaviour of $\ln Z/Z_0$ proportional to n^2. The latter factor can be made explicit by rescaling the λ's by a factor $n^{1/2}$ to rewrite

$$Z = \text{cste} \int \pi\, d\lambda_\alpha\, e^{-n^2\left\{\frac{1}{n}\sum\limits_1^n\left(\frac{\lambda_\alpha^2}{2} + g\lambda_\alpha^4\right) - \frac{1}{n^2}\sum\limits_{\alpha\neq\beta} \ln|\lambda_\alpha-\lambda_\beta|\right\}} \tag{82}$$

We are now clearly invited to apply semi-classical methods to determine the normalized density of eigenvalues. Had the number of variables be comparable in size (n^2) to the total free energy this would have been foolish since the fluctuations would contribute a comparable amount as the leading term.

In this simple case the details of the solution to the classical equations are of no great interest. Let us simply quote the results [10]. If $u(\lambda)d\lambda$ denotes the density of eigenvalues between λ and $\lambda+d\lambda$ the result is

$$\begin{cases} u(\lambda) = \frac{1}{2\pi}(1+8ga^2 + 4g\lambda^2)\sqrt{4a^2-\lambda^2} \\ 12g\, a^4 + a^2 - 1 = 0 \end{cases} \tag{83}$$

and

$$\begin{aligned} E_0(g) &= -\frac{1}{2}\ln a^2 + \frac{1}{24}(a^2-1)(9-a^2) \\ &= -\sum_{v=1}^{\infty} (-12g)^v \frac{(2v-1)!}{v!(v+2)!} \end{aligned} \tag{84}$$

A side observation is that the integral just computed defines a generalization of Gaussian ensembles of random (hermitian) matrices and (83) is the corresponding generalization of Wigner's semicircle law. Moreover it is somehow comforting that the leading large n evaluation of $\ln Z_n(g)/Z_0$ is already in quite remarkable agreement with the result of an exact evaluation for small enough n throughout the all range of g. This is examplified in the following table for n=3.

	g=1	g=10
$1/9\ell n\, Z_3(g)/Z_o$	0.4249	0.8868
$E_o(g)$	0.4197	0.8807

(85)

and this agreement could be made more spectacular by including $1/n^2$ corrections.

Now its seems somehow miraculous that one can similarly solve this large n limit problem in a one dimensional field theory - i.e. quantum mechanics for the previous arguments seem to rule out the use of semiclassical methods in this case as they would already rule them had we considered a case of only two coupled matrices.

The solution in the quantum mechanical case emerges as the result of the following trick. We want to compute the ground state energy of the Hamiltonian

$$H = -\Delta + \mathrm{tr}\left(\frac{1}{2} M^2 + \frac{g}{n} M^4\right) \tag{86}$$

acting on wave functions $\psi(M)$. We will make the assumption that the ground state wave function is pure S-wave i.e. depends only (in symmetric way) on the eigenvalues λ_i of M. Of course

$$\Delta \equiv \sum_\alpha \frac{\partial^2}{\partial M_{\alpha\alpha}} + \frac{1}{2}\sum_{\alpha<\beta} \frac{\partial^2}{\partial \mathrm{Re} M^2_{\alpha\beta}} + \frac{\partial^2}{\partial \mathrm{Im} R^2_{\alpha\beta}} \tag{87}$$

Now this looks like a typical n-boson problem and we expect the ground state energy to be of order n^2. Using polar coordinates this ground state energy is obtained as

$$E(g) = n^2 E_o(g) + \ldots =$$

$$\operatorname*{Inf}_{\psi} \frac{\int \pi\, d\lambda_\alpha \prod_{\alpha<\beta} (\lambda_\alpha-\lambda_\beta)^2 \sum \frac{1}{2}\left(\frac{\partial\psi}{\partial\lambda_\alpha}\right)^2 + \sum_\alpha \left(\frac{\lambda_\alpha^2}{2} + \frac{g}{n}\lambda_\alpha^4\right)\psi^2}{\int \pi\, d\lambda_\alpha \prod_{\alpha<\beta} (\lambda_\alpha-\lambda_\beta)^2 \psi^2} \tag{88}$$

Now we are naturally lead to recast the problem as an equivalent n-fermion problem by defining the antisymmetric wave function

$$\phi(\lambda) = \prod_{\alpha<\beta} (\lambda_\alpha-\lambda_\beta)\, \psi(\lambda) \tag{89}$$

and as a result it is readily seen that the individual fermions decouple. We find an independent n-fermion state built out of the successive energy levels of the one particle hamiltonian

$$h = -\frac{1}{2}\frac{\partial^2}{\partial\lambda^2} + \frac{\lambda^2}{2} + \frac{g}{n}\lambda^4 \tag{90}$$

So that the exact solution, no matter what the value of n is, can be expressed in terms of the ordered eigenvalues e_k of h as

$$E(g) = \sum e_k \, \theta(e_F - e_k)$$
$$n = \sum \theta(e_F - e_k) \qquad (91)$$

where e_F is the Fermi level. When $n \to \infty$ we use the semi-classical approximation to evaluate this sums which are dominated by large quantum numbers so that rescaling $\lambda \to n^{1/2}\lambda, e_F = n\varepsilon$

$$E_0(g) = \varepsilon - \int \frac{d\lambda}{3\pi} (2\varepsilon - \lambda^2 - 2g\lambda^4)^{3/2} \, \theta(2\varepsilon - \lambda^2 - 2g\lambda^4)$$
$$1 = \int \frac{d\lambda}{\pi} (2\varepsilon - \lambda^2 - 2g\lambda^4)^{1/2} \, \theta(2\varepsilon - \lambda^2 - 2g\lambda^4) \qquad (92)$$

This can be easily evaluated numerically and we can compare (for n=1 !) the planar approximation E_0 to the exact result [11] . See how good this looks (and 1/n corrections are easily computed)

g	$E_0(g)$	$E(g)$
0.01	0.505	0.507
0.1	0.542	0.559
0.5	0.651	0.696
1.0	0.740	0.804
50	2.217	2.500
1000	5.915	6.694

(93)

Even for large g this is not so bad

$$E_0 \sim 0.58993 \, g^{1/3} \qquad E(g) \sim 0.66799 \, g^{1/3} \qquad (94)$$

Unfortunately this is as far as one can go with exact results. Of course one can study variants like transforming the one matrix problem into the one plaquette problem of lattice gauge theory as done by Gross and Witten and similar things. Also there has been a fairly extensive discussion initiated by Witten and Polyakov of the factorization of expectation values of invariant operators, a property proved using for instance perturbation theory [12]. This suggested the existence of a "master field" (or class of master fields using group transformations) to compute these expectation values. But none has succeded in finding these master fields. A different class of attempts uses factorization in the field equations which remain largely untractable. But perhaps the last word has not been said and some one will come with a brilliant new idea.

We now turn to a very different cycle of concepts using the duality between order and disorder variables.

IV. DUALITY

The original idea of duality is due Kramers and Wannier in the context of the 2-d Ising model has been generalized in various ways. It yields powerful results for systems with an abelian symmetry group whether global or local and relates high and low temperature regimes.

We start with the 2d-Ising case. We expand the partition function in the strong coupling (high temperature) series, $t = \tanh\beta \ll 1$

$$(\cosh\beta)^{-2N} Z(\beta) = \sum_{\text{configuration}} t^{L} \tag{95}$$

where a configuration is the choice of a (finite) set of L links in such a way that an even number (0,2,4) meet at each vertex. Each conguration is in one to one correspondence with a pair of configurations on the dual lattice of centers of plaquettes. In one of them all far away plaquette carry a value $\tau_i = +1$ and we change the sign τ each time we cross a link of the original set. The other configuration is obtained by reversing all the τ's. The condition that an even number of links meet at each vertex ensures the consistency of the procedure. As you circle around such a vertex the variables τ change sign an even number of times, so return to their initial value. L is the number of changes of signs,

$$L = \sum_{(ij)} \frac{1-\tau_i\tau_j}{2} \tag{96}$$

The sum runs over nearest neighbours on the dual isomorphic lattice. One ends up with a one to one map between a high temperature and a low temperature expansion. To be precise let

$$e^{-2\tilde{\beta}} = \tanh\beta \tag{97}$$

then

$$\begin{aligned}(\cosh\beta)^{-2N} Z(\beta) &= 1/2 \sum_{\tau_i=\pm 1} e^{-\tilde{\beta} \sum_{(ij)} (1-\tau_i\tau_j)} = \\ &= 2^N e^{-2N\tilde{\beta}} Z(\tilde{\beta})\end{aligned} \tag{98}$$

The corresponding relation for the free energy is

$$F(\beta) = \ln \sinh 2\beta + F(\tilde{\beta}) \tag{99}$$

Since from (3) $\sinh 2\beta \sinh 2\tilde{\beta} = 1$, relation (99) is reflexive as duality should be : the dual of the dual is the original. A candidate for a unique critical point, later confirmed by Onsager's solution, is the self dual one $\beta = \tilde{\beta}$ hence

$$\tanh \beta_c = \sqrt{2}-1 \tag{100}$$

One can go much deeper using this duality idea in the analysis of the 2d Ising model. The ultimate stage has been reached by Mc Coy, Perk and Wu who succeded in writing analogous relations to (99) for the correlation functions [13]. Let me quote here the results for the two point function. Let

$$C(m,n,\beta) = \langle\sigma_{oo}\ \sigma_{mn}\rangle \tag{101}$$

we shall suppress the variable β and define $\tilde{C}(m,n,\beta) \equiv C(m,n,\tilde{\beta})$ as the correlation function at the dual temperature. Then one shows that

$$\sinh 2\beta[C^2(m,n)-C(m,n+1)C(m,n-1)] + \sinh 2\tilde{\beta}\ [\tilde{C}^2(m,n)-\tilde{C}(m+1,n)\tilde{C}(m-1,n)] = 0 \tag{102}$$

and an other relation with the roles of m and n interchanged as required by the equality of interactions in the two directions. Similarly one has

$$\sinh 2\beta\ [C(m+1,n+1)C(m,n)-C(m+1,n)C(m,n+1)] = \sinh 2\tilde{\beta}\ [\tilde{C}(m+1,n+1)\tilde{C}(m,n)-\tilde{C}(m+1,n)\tilde{C}(m,n+1)] \tag{103}$$

Observe that (102) and (103) express the invariance under duality (a change in sign is involved in (102)) of certain quadratic combinations of two point correlations. Combining (102) and (103) and using the asymptotic behaviour of correlations one proves that the ratio

$$G(m,n) = (\sinh 2\beta)^{1/2}\ C(m,n)/(\sinh 2\tilde{\beta})^{1/2}\ \tilde{C}(m,n) \tag{104}$$

satisfies a discrete analog of the Painleve (or sinh-Gordon) equation !!

$$2\cosh 2\beta\ \cosh 2\tilde{\beta}\ G(m,n)\ [1-G(m,n+1)G(m,n-1)G(m+1,n)G(m-1,n)] =$$
$$[1+G^2(m,n)][G(m,n+1)+G(m,n-1)-G(m,n+1)G(m,n-1)(G(m+1,n)+G(m-1,n)) +$$
$$G(m+1,n)+G(m-1,n)-G(m+1,n)G(m-1,n)(G(m,n+1)+G(m,n-1))] \tag{105}$$

and this together with (102) and (103) determines the correlation function using known boundary conditions.

Although the present derivations of equations (102) and (103) using the transfer matrix formalism are not too complicated it remains a challenge to find an elementary proof along the lines used for the free energy.

Let us pause to discuss the elements entering the self duality. First we need a similar dual lattice. For instance had we started with a triangular or hexagonal lattice dual to each other an extra transformation (the so called star triangle transformation) would be needed to obtain self duality. In the second place (and this is not too apparent in the Z_2 case) an isomorphism between the symmetry group and its dual, i.e. the set of its irreducible representations. Fortunately the dual of an abelian group is itself an abelian group though in general not isomorphic. The case of Z_n groups (n-th roots of unity) is a favourable case of isomorphism. When this is not the case we simply find a mapping between two different models.

The situation is far more complicated and to my knowledge unsolved in the non abelian case. If we return to the argument given at the beginning of this section,we see that we had to solve what amounted to a cohomology condition. A configuration of links could be thought as a form and the condition that an even number meet at each vertex as stating that the differential of this form vanishes (actually the whole formulation should be transposed to the dual lattice). As a result the form was the differential of a function-in the previous case the configuration of dual variables $\{\tau_i\}$ up to an overall ambiguity in sign. In a non-abelian case we have to replace the choice of a link by an assignment of a group representation. The closure condition requires that the product of representations occurring at each vertex combines to the identity. Who can "solve" for this condition ?

Before leaving two dimensions it is instructive to look at an other non self-dual case, where duality relates two models with apparently very different physics.

Start with a so-call XY or O(2) spin model. The variable on each site is a unit two dimensional vector (or equivalently an angle φ) so that

$$Z = \int \prod_i \frac{d\varphi_i}{2\pi} \, e^{\beta \sum_{(ij)} v(\varphi_i - \varphi_j)} \tag{106}$$

Here v is some periodic function (period 2π) with a maximum at the origin. For large β this is a simple spin wave model while we know from the work of Kosterlitz and Thouless that for some finite β a transition occurs. Correlation become of finite range (spontaneous mass generation) with correlation length growing like $\exp \frac{\text{cste}}{(\beta-\beta_c)^{1/2}}$ near the transition.

For the purpose of illustrating the duality idea it is best to use the Villain form where we replace $e^{\beta v(\varphi)}$ by a sum of Gaussians

$$e^{\beta v(\varphi)} \to \sum_{k=-\infty}^{+\infty} e^{-\beta/2(\varphi-2\pi k)^2} \tag{107}$$

Clearly this is as close to a Gaussian as is allowed by the periodicity condition. Indeed it is the solution of the heat equation on a circle rather than a line and is one of Jacobi's θ-functions.

This is of no specific importance here. What is important is that while for large β equation (107) is a good representation and peaks very strongly around $\phi=0$ (mod 2π) so that the leading term corresponds to $k=0$ and the other terms are down by factors $e^{-2\pi|k|\beta}$ this is obviously not so when $\beta\to 0$ and the function becomes almost constant on the circle. An other representation is then useful, obtained using the Poisson summation formula which is nothing but the expansion in terms of irreducible characters. Using

$$e^{-\beta\varphi^2/2} = \int \frac{dp}{\sqrt{2\pi\beta}}\, e^{-p^2/2\beta + ip\varphi} \tag{108}$$

we find

$$\sum_k e^{-\beta/2(\varphi-2\pi k)^2} = \frac{1}{\sqrt{2\pi\beta}} \sum_{n=-\infty}^{+\infty} e^{in\varphi}\, e^{-n^2/2\beta} \tag{109}$$

This ties in nicely with the fact that Fourier transforms of Gaussians are Gaussians. It is well adapted to a high temperature expansion which clearly exhibits that this phase is symmetric

$$Z = (2\pi\beta)^{-N} \sum_{n_{ij}} e^{-\sum_{(ij)} n_{ij}^2/2\beta} \prod_i \delta\left(\sum_{j(i)} n_{ij}\right) \tag{110}$$

Having selected on each oriented link a relative integer n_{ij} we have the constraint that the flow entering each vertex be zero. This is analogous to $\mathrm{div}B=0 \leftrightarrow B=\mathrm{curl}A$. That is $h_{ij}=h_\alpha-h_{\alpha'}$ where α,α' are sites of the dual lattice with the link $(\alpha\alpha')$ dual to the link (ij) (a convention on orientation is included). We have then to sum over all h_α integers but one. The result is a discrete Gaussian model on the dual lattice

$$Z = (2\pi\beta)^{N} \sum_{h_\alpha} e^{-\sum_{(\alpha\alpha')} (h_\alpha-h_{\alpha'})^2/2\beta} \tag{111}$$

We find a typical interchange of high and low temperature between the XY model and the discrete Gaussian (a variant of the Solid on Solid model which would have a linear instead of quadratic interaction $(h-h')^2 \to |h-h'|$). Thus the XY transition can be identified with the roughening transition for a large enough value of β for which the difference in heights between far away points becomes infinite

$$\lim_{x\to\infty} \langle (h_o-h_x)^2 \rangle = \infty \tag{112}$$

We could go further by expressing spin correlations in this langage relating the difference in heights to vortices, using an other transformation to the two-dimensional Coulomb gas, but this would lead us to far away [14].

Of course duality can be extended to higher dimensions. For instance it is easy to see that in 3d, where links are dual to plaquettes the Ising model is dual to a Z_2 gauge model, and that a coupled Z_2 gauge invariant model is self dual with a phase diagram schematically drawn on Fig. 4.

coupled model in 3 dimensions

$$Z = \sum e^{\beta_g \sum \sigma_p + \beta_m \sum \sigma_i \sigma_{ij} \sigma_j}$$

O 2nd order transition

duality

$x \to (1-y)$

$y \to (1-x)$

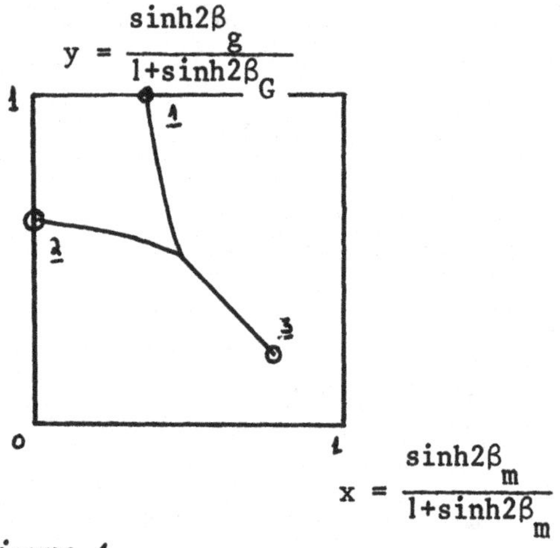

Figure 4

The circles indidate 2nd order transition. The nature of the transition lines is as yet unknown, (1) is the pure Ising transition, (2) its dual in the pure gauge field. The line ending in (3) is analogous to a liquid gas transition, presumably of first order.

In four dimensions the Z_2 gauge model is self dual and Monte Carlo simulations [15] have confirmed that the fixed point $\tanh\beta_c=\sqrt{2}-1$ looks like a unique first order transition point.

In general self duality does not imply that transitions occur at fixed points. An example is afforded by Z_n clock models in d=2, or their gauge analog in 4d. For n large enough they exhibit three phases interchanged by duality with the middle one having infinite correlation length and the two transitions occurring at values β_1 of order 1 and β_2 of order n^2 so that the low temperature phase with discrete excitations disappears in the limit $n\to\infty$ (U(1) limit). This seems to indicate that compact QED in four dimensions has a continuous transition. It is somehow reminiscent of the two dimensional case where non linear effects in the XY model have a hard time to beat infinite correlations (without long range order) while non-linear σ-models only exhibit a symmetric phase. We hope that there is an analog to this situation in four dimensional gauge models

Duality as was already mentioned can be used to explore disorder variables. For instance in the three dimensional Z_2 gauge case let us look for the meaning of the Wilson loop average corresponding to a curve C. From the Ising point of view the curve C lies on its dual lattice and the action of W amounts to frustrate the model along an arbitrary surface bounded by C, i.e. to change the couplings from ferro- to antiferromagnetic for all links crossing this surface. The particle physicists string tension can then be identified with the ordinary surface tension of coexisting phases (+ and - domains). This is how surface roughening made its way in the computation of loop averages.

In four dimensions we have even a more interesting situation. A loop variable and its dual can be considered to be part of the same model, very much as in the 2d-Ising case. This is how 't Hooft loops the dual of Wilson loops come into play. Interpreted in a transfer matrix formalism the two corresponding operators do not commute rather their commutation involves $(-1)^s$ where s is the intertwinning number of the corresponding curves.

So from duality we can start guessing (and sometimes proving) a lot about the behaviour of abelian systems though it is not of course exclusive of other approaches.

V. EPILOGUE - RANDOM CURVES VERSUS RANDOM SURFACES

Throughout these lectures we have emphasized similarities between global and local symmetries. There are also some fundamental differences which at the lowest level manifest themselves in the difference between the behaviour of random curves and random surfaces the building blocks of these theories. This is apparent for instance when we deal with high temperature expansions. Take for instance an Ising model in large dimension. Then the correlation function between far away points can be approximated by a sum over path ($t \equiv \tanh\beta$)

$$\langle\sigma_o\sigma_x\rangle \sim \sum_{L_j \partial L=\{o,x\}} t^{|L|} \sim \int_{-\infty}^{+\infty} \frac{d^d p}{(2\pi)^d} \frac{e^{ip.x}}{1-2t\sum_1^d \cos p_a} \qquad (113)$$

The approximation involved imply the neglect of incomplete cancellation of closed paths between numerator and denominator when computing correlations as average values and hence improper weighting of self intersections. But in d large enough the paths become brownian with a simple recurrence relation in the form of a different equation yielding the second form quoted in (113). The susceptibility is then given by a geometric series, each term being derived from the previous one by remarking that there are $\sim$ 2d directions to proceed at each step, hence a factor 2dt

$$\chi = \sum_x \langle\sigma_o\sigma_x\rangle \sim \frac{1}{1-2dt} \qquad (114)$$

This of course entails the usual critical value $\beta_c \sim t_c \sim 1/2d$. It expresses however slightly more if we relate it to the fact that the transition is of second order. For large enough paths and β close to β_c we have a universal (renormalized) theory of Brownian motion independent of lattice details that can be expressed using continuous variables. Approximating in(113) $2\sum_1^d(1-\cos p_a)$ by p^2 and t by β it reads

$$\begin{aligned}\langle\sigma_o\sigma_x\rangle &\simeq \int \frac{d^dp}{(2\pi)^d}\, e^{ip.x}\int_o^\infty dL\; e^{-L[(1-2\beta d)+\beta p^2]}\\ &\sim \int_o^\infty \frac{dL}{2\beta}\,\langle x|\; e^{-L(1-2\beta d)/2\beta-L\,P^2/2}\,|0\rangle\\ &\sim \int_o^\infty \frac{dL}{2\beta}\int_{\substack{x(o)=0\\ x(L)=x}} \mathcal{D}x\; e^{-\int_o^L d\tau\,[\dot{x}^2/2+(1-2\beta d/2\beta)]}\end{aligned} \tag{115}$$

L plays the role of a proper time with $L\sim\vec{x}^2$. The Brownian approximation is an other disguise of mean field theory.

The question now is whether we can extend this to surfaces instead of curves. This has proved to be quite non trivial. A first disastrous signal comes from the fact that the general behaviour of gauge theories in large dimensions (where we could hope that self intersecting problems are washed out) involves a discontinuous transition from area to perimeter behaviour. Consequently we have no candidate there for a universal (cut off independent) theory of random surfaces. The "many fingers" picture of the Drouffe, Parisi, Sourlas approximation looks highly cut off dependent and develops in a kind of "foamy" object.

Recently Polyakov has argued that a reasonable picture only emerges at lower dimensions but then we are back to self interactions complexities. One would perhaps also put some hope in large n-approximations.

The "natural" highly non renormalizable formulation would involve a parameter domain $\mathcal{D}$(the inside of a unit circle say) and a mapping $\vec{\varphi}$ from $\mathcal{D}$ to d-dimension space constrained by the condition that φ maps the boundary $\partial\mathcal{D}$ on a given curve C. Then

$$\langle W(C)\rangle \overset{?}{=} \int \mathcal{D}\vec{\varphi}\;\; e^{-K_o\iint_{\mathcal{D}} dx_1dx_2\,\sqrt{\det h_{ab}}} \tag{116}$$

the action being the area of the surface $\vec{\varphi}(\mathcal{D})$ times a constant K_o. Indeed the two tangent vectors $\partial\vec{\varphi}/\partial x_1\, dx_1$ and $\partial\vec{\varphi}/dx_2\, dx_2$ span an area given by

$$\begin{aligned}|\partial_1\vec{\varphi}\wedge\partial_2\vec{\varphi}|dx_1dx_2 &= \sqrt{\det h}\; dx_1dx_2\\ h_{ab} &= \partial_a\vec{\varphi}.\partial_b\vec{\varphi}\end{aligned} \tag{117}$$

note that h_{ab} appears as the induced metric on the parameter space. Formally (116) is reparametrization independent, a necessary condition for the theory to make sense. Proposals have been made to replace the action in (116) by different ones selected to insure the equivalence of saddle points i.e. given by the area of minimal surfaces. This is in the spirit of the relation between (113) and (115).

A naîve attempt at evaluating (116) would involve a saddle point method where the action is expanded around a minimal surface with (d-2) transverse fluctuations taken into account. For a curve C in the form of a rectangle $T \times R$ with $T \to \infty$ first, this leads including "leading" (i.e. one loop) corrections to

$$\lim_{T\to\infty} -\frac{1}{T} \ln W = K_o R + \lim_{T\to\infty} \frac{d-2}{2T} \sum_{m,n=1}^{\infty} \ln\left(\frac{m^2\pi^2}{T^2} + \frac{n^2\pi^2}{R^2}\right) + \ldots$$

$$= (K_o + \delta K)R + \text{cste} - \frac{(d-2)\pi}{24}\frac{1}{R} + \ldots \tag{118}$$

i.e. the string tension is renormalized and there appears a universal dimensionless 1/R correction [16]. Further terms would seem to involve an increasing number of undetermined (divergent) parameters.

The way in which (118) is derived assumes specific (transverse) small deformations of the surface. The corresponding factor (d-2) implies the absence of corrections in 2 dimensions. This neglects stretching and overlapping which occur even in two dimensions. This could conceivably lead to a contribution analogous to (118) but with a d-independent coefficient, therefore modifying the coefficient of the 1/R term. We might parenthetically remark that there is a close relationshing between the 1/R term in (118) and the conformal anomaly in two dimensions. This comes about because both are statements about the Green function of a two dimensional "free field" propagator when one discusses departures from infinite flat space, here we include finite boundary conditions. For the conformal anomaly the two dimensonal space has curvature and the effect switches from infrared to ultraviolet contributions.

Of course random surface theory has a strong analogy with the dual resonance model interpreted in terms of time evolving strings where the same Nambu integral (113) appears [17][18].

The challenge of making sense or modifying in a meaningful way the surface theory is still with us !

REFERENCES

[1] K.Wilson, Phys.Rev. D10, 2445 (1974)

[2] R.Balian, J.M.Drouffe, C.Itzykson, Phys.Rev. D11, 2104 (1975)

[3] M.E.Fisher, D.S.Gaunt, Phys.Rev. 133A, 224 (1964)
R.Abe, Prog.in Theor. Phys. 47, 62 (1972)

[4] K.G.Wilson, J.B.Kogut, Physics Report, 12C, 77 (1974)

[5] E.Brézin, J.C.Le Guillou, J.Zinn-Justin, in "Phase transitions and critical phenomena", vol.VI, Domb and Green eds, Academic Press 1977

[6] S.Elitzur, Phys.Rev. D12, 3978 (1975)

[7] J.M.Drouffe, G.Parisi, N.Sourlas, Nucl.Phys. B161, 397 (1979)

[8] H.E.Stanley, Phys.Rev. 176, 718 (1968)
For a review see the 1979 Erice Lectures of S.Coleman "1/n"

[9] G.'t Hooft, Nucl.Phys. B72, 461 (1974)

[10] E.Brézin, C.Itzykson, G.Parisi, J.B.Zuber, Comm.Math.Phys. 59, 35 (1978)

[11] F.T.Hioe, E.W.Montroll, J.Math.Phys. 16, 1945 (1975)

[12] E.Witten, Nucl.Phys. B160, 57 (1979)

[13] B.Mc Coy, J.H.Perk, T.T.Wu, Phys.Rev.Lett. 46, 757 (1981)

[14] J.V.Jose, L.P.Kadanoff, S.Kirkpatrick, D.R.Nelson, Phys.Rev. B16, 1217 (1977)

[15] M.Creutz, L.Jacobs, C.Rebbi, Phys.Rev.Lett. 42, 1390 (1979)

[16] M.Luscher, K.Symanzik, P.Weisz, Nucl.Phys. B173, 365 (1980)

[17] Y.Nambu, Copenhagen Symmer Symposium 1970

[18] A.M.Polyakov, Phys.Lett. B103, 207 (1981)

QUANTUM SCATTERING TRANSFORMATION

L. D. Faddeev

Steklov Mathematical Institute
Academy of Sciences
SU-191011 Leningrad

In one-dimensional mathematical physics we know three successful lines of development: the Bethe - Hulthen theory of quantum magnetics, the Onsager - Baxter theory of planar models in statistical mechanics and the Gardner - Greene - Kruskal - Miura - - Zakharov - Shabat inverse scattering method in classical field theory. Recently it has been realized that after quantization of the latter method, a unified approach to all three of these lines emerges. This new method - the quantum inverse method or quantum scattering transformation (QST) - is now three years old and continues to develop rapidly in several centers (Leningrad Steklov Institute, Fermilab, Batavia, Freiburg University, etc.). There exist several surveys of the basic ideas of QST [1), 2), 3), 4)]. In this text I shall mention some very recent developments made in the Leningrad group. Needless to say, their authors helped me in compiling this review.

1. Basic Formulae of QST

To fix ideas and notations I shall present a brief list of the main formulae of the method in a half-abstract way. We shall use the language of a quantum field-theoretical model on the line. The parallel classical statistical mechanics treatment is given in lectures of M. Gaudin [5)] ; see also [6)].

From the beginning we shall consider the model given on a lattice of finite length N . Let $\mathfrak{h}_n$ be the Hilbert space for the representation of the field operators X_n^α on lattice site number n . Operators X_n^α for different n are supposed to commute. (This assumption of ultralocality is fulfilled in the most interesting examples. For possible generalizations see [7)].)

Then the full Hilbert space of the model has the form

(1) $$\mathcal{H}_N = \bigotimes_n h_n .$$

The limit $N \to \infty$ (and $\Delta \to 0$, where Δ is a lattice distance for the continuous models) is to be taken later.

The Hamiltonian H of the model, acting in $\mathcal{H}$, is embedded into a family of commuting operators (conservation laws), and QST gives a simultaneous diagonalization for all of them.

The generating object for QST is a matrix operator $L_n(\lambda)$ acting as a matrix in an auxiliary space V , the matrix elements being operators in the ring generated by the X_n^α . They depend on a parameter λ which in what follows will play the role of momentum for quasi-particles. In the most simple but interesting enough case V is two-dimensional, $V = \mathbb{C}^2$, and below we shall often illustrate QST on this example.

The main property of $L_n(\lambda)$ is a set of commutation relations which we shall write in a condensed form as follows:

(2) $$R(\lambda,\mu)\left(L_n(\lambda) \otimes L_n(\mu)\right) = \left(L_n(\mu) \otimes L_n(\lambda)\right) R(\lambda,\mu)$$

Here $R(\lambda,\mu)$ is a c-number matrix acting in $V \otimes V$. In the case where $V = \mathbb{C}^2$, R is a (4×4)-matrix. A typical example is given by a matrix of the form (abc-form)

(3) $$R = \begin{pmatrix} a & 0 & 0 & 0 \\ 0 & b & c & 0 \\ 0 & c & b & 0 \\ 0 & 0 & 0 & a \end{pmatrix} .$$

The property discussed is of course very specific and imposes lots of conditions both on $L_n(\lambda)$ and $R(\lambda,\mu)$. For instance (modulo some degenerate cases) $R(\lambda,\mu)$ is to satisfy a Yang-Baxter condition, which following Korepin [8] we shall write in the form

(4) $$\begin{aligned}&\left(R(\lambda,\mu) \otimes I\right)\left(I \otimes R(\lambda,\sigma)\right)\left(R(\mu,\sigma) \otimes I\right)\\ &= \left(I \otimes R(\mu,\sigma)\right)\left(R(\lambda,\sigma) \otimes I\right)\left(I \otimes R(\lambda,\mu)\right) ,\end{aligned}$$

all entries here being matrices in $V \otimes V \otimes V$. Classification of the possible solutions of this equation is an interesting mathematical problem which has quite a few applications not only in QST (see [9] - [11]).

The matrix $L_n(\lambda)$ plays the role of the Lax operator $L(\lambda)$ in the classical inverse scattering method. The role of the scattering data is played by the monodromy matrix $T_N(\lambda)$, which is an ordered product of all $L_n(\lambda)$:

$$T_N(\lambda) = \overleftarrow{\prod_n} L_n(\lambda) \tag{5}$$

It is a matrix in V whose matrix elements are operators in $\mathfrak{h}_N$. In the case where $V = \mathbb{C}^2$ let us denote them as follows:

$$T_N(\lambda) = \begin{pmatrix} A_N(\lambda) & B_N(\lambda) \\ C_N(\lambda) & D_N(\lambda) \end{pmatrix} \tag{6}$$

From the main property (2) of $L_n(\lambda)$ it follows that $T_N(\lambda)$ satisfies the same commutation relations:

$$R(\lambda,\mu)\left(T_N(\lambda) \otimes T_N(\mu)\right) = \left(T_N(\mu) \otimes T_N(\lambda)\right) R(\lambda,\mu) \tag{7}$$

In particular, this set contains the following relations:

$$[F_N(\lambda), F_N(\mu)] = 0 \quad \text{where} \quad F_N(\lambda) = \mathrm{tr}\, T_N(\lambda) = A_N(\lambda) + D_N(\lambda) \tag{8}$$

$$A_N(\lambda) B_N(\mu) = B_N(\mu) A_N(\lambda) \frac{1}{c(\mu,\lambda)} - B_N(\lambda) A_N(\mu) \frac{b(\mu,\lambda)}{c(\mu,\lambda)} \tag{9}$$

$$D_N(\lambda) B_N(\mu) = B_N(\mu) D_N(\lambda) \frac{1}{c(\lambda,\mu)} - B_N(\mu) D_N(\lambda) \frac{b(\lambda,\mu)}{c(\lambda,\mu)} \tag{10}$$

The next simplifying property of $L_n(\lambda)$ is the existence of a local vacuum, namely a state $\omega_n \in \mathfrak{h}_n$ such that $L_n(\lambda)$ is triangular when applied to it,

$$L_n(\lambda)\,\omega_n = \begin{pmatrix} \alpha(\lambda) & * \\ 0 & \beta(\lambda) \end{pmatrix} \omega_n \, , \tag{11}$$

$\alpha(\lambda)$ and $\beta(\lambda)$ being c-number functions. From this property we see that the reference state

$$\Omega_N = \bigotimes_n \omega_n \tag{12}$$

is an eigenstate for both $A_N(\lambda)$ and $D_N(\lambda)$,

$$A_N(\lambda)\Omega_N = (\alpha(\lambda))^N \Omega_N , \quad D_N(\lambda)\Omega_N = (\beta(\lambda))^N \Omega_N , \tag{13}$$

and is annihilated by $C_N(\lambda)$. From the commutation relations it follows that the vector

$$\Psi(\{\lambda\}) = B(\lambda_1) \dots B(\lambda_n)\, \Omega_N \tag{14}$$

is an eigenvector for the commuting set of operators $F_N(\lambda)$,

$$F_N(\lambda)\, \Psi(\{\lambda\}) = \Lambda(\lambda,\{\lambda\})\, \Psi(\{\lambda\}), \tag{15}$$

if the numbers λ_j are all different and satisfy equations of the form

$$\left(\frac{\alpha(\lambda_j)}{\beta(\lambda_j)}\right)^N = \prod_{\substack{l=1 \\ l\neq j}}^{n} \frac{c(\lambda_l,\lambda_j)}{c(\lambda_j,\lambda_l)} , \quad 1 \leq j \leq n , \tag{16}$$

the eigenvalue $\Lambda(\lambda,\{\lambda\})$ being given by

$$\Lambda(\lambda,\{\lambda\}) = \alpha(\lambda)^N \prod_{l=1}^{n} \frac{1}{c(\lambda_l,\lambda)} + \beta(\lambda)^N \prod_{l=1}^{n} \frac{1}{c(\lambda,\lambda_l)} . \tag{17}$$

The last two formulae are quite general and can appear even in cases where we have no local vacuum. Together with the definition of the state $\Psi(\{\lambda\})$ they present an algebraic generalization of the Bethe Ansatz known from the theory of magnetics. It is clear that if there exists ξ such that $\beta(\xi) = 0$ then $L_n(\xi)$ will become degenerate and $\ln F_N(\lambda)$ will be additive for λ near ξ. In more detail, the logarithmic derivatives of $F_N(\lambda)$ at $\lambda = \xi$ have the following eigenvalues:

$$\left(\frac{d}{d\lambda}\right)^k \ln F_N(\lambda)\Big|_{\lambda=\xi} \quad \Psi(\{\lambda\})$$

$$= \left[N\left(\frac{d}{d\lambda}\right)^k \ln \alpha(\lambda)\Big|_{\lambda=\xi} + \sum_{l=1}^{n} \left(\frac{d}{d\lambda}\right)^k \ln \frac{1}{c(\lambda_l,\lambda)}\Big|_{\lambda=\xi} \right] \tag{18}$$

$$\cdot \Psi(\{\lambda\})$$

The first term on the rhs of (18) is a contribution from the reference state, while the second is additive and can be interpreted as the contribution from the quasiparticles with the following dispersion law for the one-(quasi)particle states:

$$h_k(\mu) = \left(\frac{d}{d\lambda}\right)^k \ln \frac{1}{c(\mu,\lambda)}\Big|_{\lambda=\xi} \tag{19}$$

The Hamiltonian H of the model is contained in the family of logarithmic derivatives considered above. It is a quasilocal function of the field operators X_n^α .

In the more general case where dim $V > 2$ one can still have an R-matrix of abc-form, but the coefficients a,b and c will be matrices by themselves. The same is true for the "eigenvalue" $\Lambda(\lambda,\{\lambda\})$, so that in order to get the ordinary eigenvalue of $F_N(\lambda)$ one must diagonalize Λ. Now Λ itself looks like a trace of a monodromy matrix $\prod_l c(\lambda_l, \lambda)^{-1}$ for a new dynamical system, and to diagonalize it, we can once again use the procedure described above. In such a way a hierarchy of Bethe Ansätze appears, which can be used for the complete diagonalization of the original problem. Such a situation usually appears when the one-site system of operators X_n^α corresponds to a system with several degrees of freedom (colour). I refer to the papers by Kulish [12), 13)] and Takhtajan [14)].

At this point we can discuss the infinite-volume limit. There are two possibilities:

1. Ω_N is a ground state of the Hamiltonian.
2. The ground state is a state $\Psi(\{\lambda\})$ with some special distribution of λ_j (vacuum polarization).

In the first case there exist the limits

$$A(\lambda) = \lim_{N\to\infty} A_N(\lambda)\,\alpha(\lambda)^{-N}, \quad B(\lambda) = \lim_{N\to\infty} B_N(\lambda) \tag{20}$$

which satisfy the simple commutation relations

$$A(\lambda)\,B(\mu) = B(\mu)\,A(\lambda)\,\frac{1}{c(\mu,\lambda)}\,. \tag{21}$$

The operators

$$R(\lambda) = B(\lambda)\,A(\lambda)^{-1} \tag{22}$$

play the role of the Zamolodchikov operators and satisfy the commutation relations

$$R(\lambda)\,R(\mu) = R(\mu)\,R(\lambda)\,S(\lambda,\mu)\,, \tag{23}$$

where

$$S(\lambda,\mu) = \frac{c(\mu,\lambda)}{c(\lambda,\mu)}\,. \tag{24}$$

The states

$$\begin{aligned} \Psi_{in}(\{\lambda\}) &= R(\lambda_n)\ldots R(\lambda_1)\,\Omega_\infty\,, \\ \Psi_{out}(\{\lambda\}) &= R(\lambda_1)\ldots R(\lambda_n)\,\Omega_\infty\,, \end{aligned} \tag{25}$$

for λ_j arranged in increasing order ($\lambda_1 < \lambda_2 < \ldots < \lambda_n$), are normalized in- and out-states, and $S(\lambda,\mu)$ is a two-particle S-matrix in terms of which one can wirte the full S-matrix in factorized form.

In the second case, and in typical,examples, the λ_j for the ground state are distributed quasicontinuously over some interval J. Their distribution in the limit $N \to \infty$ is given by a density function $\rho(\lambda)$ which satisfies a linear integral equation written in terms of $\alpha(\lambda)$, $\beta(\lambda)$ and $c(\lambda,\mu)$. An example where the ground state is a sea of quasiparticles is given by the equation

$$\rho(\lambda) = \psi(\lambda) + \int_J \phi(\lambda,\mu)\,\rho(\mu)\,d\mu\,, \tag{26}$$

$$\psi(\lambda) = \frac{d}{d\lambda}\ln\frac{\alpha(\lambda)}{\beta(\lambda)}\,, \quad \phi(\lambda,\mu) = \frac{d}{d\lambda}\ln\frac{c(\mu,\lambda)}{c(\lambda,\mu)}\,.$$

For the characterization of the low-lying states one introduces the function $\sigma(\lambda,\mu)$ which plays the role of the distribution function of quasiparticles in the one-particle state, μ being the corresponding one-particle momentum. In fact, $\sigma(\lambda,\mu)$ is a resolvent of the kernel $\phi(\lambda,\mu)$. A formal derivation of this result was first given in [15] for the example of the XXX-model. However, as was shown recently by Takhtajan and myself [16], one must be careful when defining the true physical eigenstates of $F_N(\lambda)$ when $N \to \infty$.

In terms of $\sigma(\lambda,\mu)$ one can now express additive characteristics of the spectrum and scattering. For instance, the one-particle eigenvalues and the two-particle S-matrix are given by

$$\tilde{h}_k(\lambda) = \int_J h_k(\mu)\,\sigma(\lambda,\mu)\,d\mu\,, \tag{27}$$
$$\ln \tilde{S}(\lambda) = \int_J \ln S(\mu)\,\sigma(\lambda,\mu)\,d\mu\,.$$

Moreover it is possible to write down the expression for the vacuum and the analogue of the operator $R(\lambda)$. Indeed, the limits

$$\tilde{A}(\lambda) = \lim_{N\to\infty} A_N(\lambda)\,a(\lambda)^{-N} \exp\left\{-N\int_J \ln\frac{1}{c(\mu,\lambda)}\,\rho(\mu)\,d\mu\right\} \tag{28}$$
$$\tilde{B}(\lambda) = \lim_{N\to\infty} B_N(\lambda)$$

exist, and the vacuum is formally given by the expression

$$\Omega_{\text{ground}} = \lim_{N\to\infty} \exp\left\{N\int_J \ln B(\lambda)\,\rho(\lambda)\,d\lambda\right\}\Omega_N\,. \tag{29}$$

The Zamolodchikov one-particle creation operator $\tilde{R}(\lambda)$ is given by

$$\tilde{R}(\lambda) = \exp\left\{\int_J \ln B(\mu)\,\sigma(\lambda,\mu)\,d\mu\right\}\tilde{A}(\lambda)^{-1}, \tag{30}$$

and its commutation relation

$$\tilde{R}(\lambda)\,\tilde{R}(\mu) = \tilde{R}(\mu)\,\tilde{R}(\lambda)\,\tilde{S}(\lambda,\mu) \tag{31}$$

is consistent with the interpretation of $\tilde{S}(\lambda,\mu)$ as a two-particle S-matrix.

The discussion of the last formulae is given in [16] for the example of the XXX-model; however, their general validity is clear from the derivation given. The formulae for $\tilde{S}(\lambda,\mu)$ appeared first in Korepin's paper [17]. Thus in both cases the limit $N \to \infty$ can be performed, and the spectrum of all commuting conservation laws allows for a particle interpretation.

2. Examples

In the original papers [18], [19], QST was developed for the example of the nonlinear Schroedinger equation (N.S.)

$$i\frac{\partial\psi}{\partial t} = -\frac{\partial^2\psi}{\partial x^2} + 2\varkappa|\psi|^2\psi \tag{32}$$

and the Sine-Gordon equation (S.G.)

$$\frac{\partial^2 u}{\partial t^2} - \frac{\partial^2 u}{\partial x^2} + \frac{m^2}{\beta}\sin\beta u = 0\,. \tag{33}$$

The lattice regularization was introduced only approximately. In fact the simplicity of the N.S. model in the case without vacuum polarization allowed for its investigation directly in the continuum limit, i.e. without any recourse to the lattice [20)-22)]. For more complicated models, however, this regularization seems to be indispensible, especially in the case when vacuum polarization is present. Moreover, as we shall see below, the lattice formulation is quite universal and allows for a simple derivation of commuting conservation laws (trace identities) and for a quantum analogue of the Gelfand - Levitan - Marchenko equation. Last but not least it opens the possibility to introduce genuine mathematical rigour (this is not discussed below).

Here we shall illustrate the formulae of § 1 for the simplest example of the spin 1/2 XXX-model mentioned before. The field operators X_n^α are the spin operators S_n^α given in terms of Pauli matrices

$$S_n^\alpha = \frac{1}{2}\sigma^\alpha\,, \quad \alpha = 1,2,3\,, \tag{34}$$

and the quantum space $\mathfrak{h}_n$ for site number n is $\mathbb{C}^2$. The commutation relations

(35) $$[S_n^\alpha, S_m^\beta] = i\varepsilon^{\alpha\beta\gamma} S_n^\gamma \delta_{nm}$$

are ultralocal. We shall use the periodic boundary conditions

(36) $$S_n^\alpha = S_{n+N}^\alpha .$$

The Hamiltonian has the form

(37) $$H = \pm \sum_n (S_n^\alpha S_{n+1}^\alpha - \frac{1}{4}),$$

the different signs corresponding to the ferromagnetic (-) and antiferromagnetic (+) cases.

The matrix $L_n(\lambda)$ is given by

(38) $$L_n(\lambda) = \lambda I \otimes I + i \sum_\alpha S_n^\alpha \otimes \sigma^\alpha ,$$

where σ^α acts in the auxiliary space $V = \mathbb{C}^2$. In matrix form $L_n(\lambda)$ can be written as

(39) $$L_n(\lambda) = \begin{pmatrix} \lambda I + iS_n^3 & i(S_n^1 - iS_n^2) \\ i(S_n^1 + iS_n^2) & \lambda I - iS_n^3 \end{pmatrix} .$$

It is clear that the vector

(40) $$\omega_n = \begin{pmatrix} 1 \\ 0 \end{pmatrix}$$

plays the role of a local vacuum and

(41) $$\alpha(\lambda) = \lambda + \frac{i}{2} , \quad \beta(\lambda) = \lambda - \frac{i}{2} = \overline{\alpha(\bar\lambda)}$$

The R-matrix has abc-form and without loss of generality we can put a = 1; then b and c are given by

(42) $$b(\lambda,\mu) = \frac{i}{\lambda-\mu+i} , \quad c(\lambda,\mu) = \frac{\lambda-\mu}{\lambda-\mu+i} .$$

The Bethe Ansatz equations look as follows:

$$\left(\frac{\lambda_j + \frac{i}{2}}{\lambda_j - \frac{i}{2}}\right)^N = \prod_{\ell \neq j} \frac{\lambda_j - \lambda_\ell + i}{\lambda_j - \lambda_\ell - i} \quad . \tag{43}$$

The first two conservation laws

$$P_N = \frac{1}{i} \ln F_N(\lambda)\Big|_{\lambda = \frac{i}{2}} \;,\quad H_N = \pm\left(\frac{i}{2}\frac{d}{d\lambda} \ln F_N(\lambda)\Big|_{\lambda=\frac{i}{2}} - \frac{N}{2}\right) \tag{44}$$

have the meaning of quasimomentum and energy. From the expression for the general eigenvalue of $F_N(\lambda)$ we find that the quasi-particle state

$$\Psi(\lambda) = B(\lambda)\,\Omega_N \tag{45}$$

has the following eigenvalues for P_N and H_N:

$$p(\lambda) = \frac{1}{i} \ln \frac{\lambda + \frac{i}{2}}{\lambda - \frac{i}{2}} \;(\mathrm{mod}\; 2\pi) \;,\quad h(\lambda) = \mp \frac{1}{2} \frac{1}{\lambda^2 + \frac{1}{4}} \tag{46}$$

Here λ is allowed to assume only quantized values

$$\lambda_j = \frac{2\pi n_j}{N} \;,\quad n_j \in \mathbb{Z} \;. \tag{47}$$

The detailed investigation of the limit $N \to \infty$ in the ferromagnetic case, where Ω_N is the physical ground state, and in the antiferromagnetic case, where vacuum polarization appears, is given in [16]. Some generally accepted results for the antiferromagnetic case were to be corrected.

Korepin and Izergin [30),23)] have recently given matrices $L_n(\lambda)$ for certain lattice models which become the N.S. and S.G. models in the limit $\Delta \to 0$, where Δ is the lattice distance. This allows to apply all the machinery developed for lattice models to these models. In particular, the ultraviolet cutoff problem - which in the previous literature [19),24),25)] had not been treated in a satisfactory way - allows for a natural solution.

Let us present some formulae for these models, which we shall denote as LNS and LSG models, correspondingly.

For the LNS model, the field operators are ordinary Schrödinger operators ψ_n^*, ψ_n with ultralocal commutation relations

$$[\psi_n, \psi_m^*] = \delta_{nm}\Delta , \tag{48}$$

where we have explicitely introduced Δ. The matrix $L_n(\lambda)$ has the form

$$L_n(\lambda) = \begin{pmatrix} 1 - \frac{i\lambda\Delta}{2} + \frac{\kappa}{2}\psi_n^*\psi_n & -i\sqrt{\kappa}\,\psi_n^*\varrho_n \\ i\sqrt{\kappa}\,\varrho_n\psi_n & 1 + \frac{i\lambda\Delta}{2} + \frac{\kappa}{2}\psi_n^*\psi_n \end{pmatrix},$$

$$\varrho_n = \left(1 + \frac{\kappa}{2}\psi_n^*\psi_n\right)^{1/2} . \tag{49}$$

κ plays the role of the coupling constant.

At this point it is worthwhile to comment on the connection of this matrix with the one for the XXX model. The operators

$$t_1 = \frac{i}{\sqrt{\kappa}\Delta}\left(\varrho_n\psi_n + \psi_n^*\varrho_n\right)$$

$$t_2 = \frac{1}{\sqrt{\kappa}\Delta}\left(\varrho_n\psi_n - \psi_n^*\varrho_n\right) \tag{50}$$

$$t_3 = -\frac{2}{\kappa\Delta}\left(1 + \frac{\kappa}{2}\psi_n^*\psi_n\right)$$

are nothing but the Holstein - Primakoff representation of the spin operators for full spin t^2 equal to

$$t^2 = l(l+1) \quad , \quad l = -\frac{2}{\kappa\Delta} . \tag{51}$$

With the formal identification $S_n \leftrightarrow t_n$ the two L-matrices differ only by a normalization:

$$L_n^{XXX} = \frac{2i}{\Delta}\sigma_3 L_n^{LNS} \tag{52}$$

This is not unexpected because in the classical continuous version the N.S. and XXX models are equivalent [26].

The R-matrix has abc-form with the entries

(53) $$a = 1 \ , \quad b = \frac{-i\varkappa}{\lambda - \mu - i\varkappa} \ , \quad c = \frac{\lambda - \mu}{\lambda - \mu - i\varkappa} \ ,$$

and the local vacuum is given by the oscillator vacuum ω_n,

(54) $$\psi_n \omega_n = 0 \ ,$$

lying in $\mathfrak{h}_n = L^2(\mathbb{R})$ with the local eigenvalues

(55) $$\alpha(\lambda) = 1 - \frac{i\lambda\Delta}{2} \ , \quad \beta(\lambda) = 1 + \frac{i\lambda\Delta}{2} \ .$$

All these formulae turn to those of the XXX model after the substitution

(56) $$\lambda\Delta \to 4\lambda \ , \quad \ell = -\frac{2}{\varkappa\Delta} \to \frac{1}{2} \ .$$

The conservation laws are given in terms of the logarithmic derivatives of the trace of the monodromy matrix in the vicinity of $\lambda = 2i/\Delta$. Unfortunately, these conservation laws are only quasilocal and contain exponentially decreasing interactions between arbitrary sites in the lattice. In the continuum limit $\Delta \to 0$, however, one acquires the usual N.S. model conservation laws.

For the LSG model, the field operators are also given by canonical operators p_n, q_n with ultralocal commutation relations

(57) $$[p_n, q_m] = -i\delta_{nm} \ .$$

Introduce Weyl type operators

(58) $$\pi_n = \exp\left\{\frac{i\beta}{2} p_n\right\} \ , \quad \varphi_n = \exp\left\{\frac{i\beta}{2} q_n\right\} \ ,$$

and write

(59) $$u_n = (1 + 2\tau \cos \beta q_n)^{1/2} \ , \quad \tau = \left(\frac{m\Delta}{4}\right)^2 \ ,$$

where m and ß are the mass and the coupling constant. Then the corresponding matrix $L_n(\lambda)$ looks as follows:

(60)
$$L_n(\lambda) = \begin{pmatrix} \pi_n^{-1/2} u_n \pi_n^{-1/2} & \tau^{1/2}(\lambda \varphi_n^{-1} - \lambda^{-1}\varphi_n) \\ \tau^{1/2}(\lambda^{-1}\varphi_n^{-1} - \lambda\varphi_n) & \pi_n^{1/2} u_n \pi_n^{1/2} \end{pmatrix}$$

As in [19] the local vacuum exists only for the product of two adjacent matrices $L_{n+1}(\lambda)L_n(\lambda)$, and the corresponding eigenvalues are given by

(61)
$$\alpha(\lambda) = \frac{\tau}{b}(\lambda^2 e^{-i\gamma} + b)(\lambda^{-2} e^{i\gamma} + b) \;, \quad \beta(\lambda) = \alpha(\tfrac{1}{\lambda}) \;,$$
$$b = 2\tau \Big/ 1 + \sqrt{1-4\tau^2} \;, \qquad \gamma = 8\beta^2 .$$

The R-matrix coincides with that for the continuous model and has abc-form with a = 1 and

(62)
$$b = \frac{i \sin\gamma}{\sinh(\alpha + i\gamma)} \;, \quad c = \frac{\sinh\alpha}{\sinh(\alpha + i\gamma)} \;; \quad \alpha = \ln\frac{\lambda}{\mu} .$$

The quasilocal conservation laws are obtained by expansion of the $\ln F_N(\lambda)$ in the vicinity of $\lambda^2 = b^{\pm 1} e^{i\gamma}$. In particular, the quasiparticle has the following dispersion law for the energy

(63)
$$h(\mu) = \frac{m^2\Delta}{4} \sin\gamma \left[e(\mu) + e(\tfrac{1}{\mu}) \right]$$
$$e(\mu) = \frac{(1 - 2\mu^2 b\cos\gamma + \mu^4 b^2(1 + 2\cos\gamma))\,\mu^2}{(1+\mu^2 b e^{3i\gamma})(1+\mu^2 b e^{-3i\gamma})(1+\mu^2 b e^{i\gamma})(1+\mu^2 b e^{-i\gamma})}$$

In the continuum limit $\Delta \to 0$, we get the relativistic formula

(64)
$$h_c(\mu) = \frac{m^2\Delta}{2} \sin\gamma \cosh\alpha \;, \quad \alpha = \ln\mu^{-2}$$

which was used in [19]. In contrast to $h_c(\mu)$, the function $h(\mu)$ is positive only for finite α of order $\ln \Delta^{-1}$, so that we get a natural Fermi momentum $\Lambda \sim \Delta^{-1}$ which is defined through the equation $h(\Lambda) = 0$. This cutoff is to be used for the definition of the true ground state before going to the limit $N \to \infty$, $\Delta \to 0$. Realisation of this program is now in progress.

Let us mention also that for special values of the coupling constants κ or γ the one-site Hilbert space $\mathfrak{h}_n$ has a finite-dimensional subspace $\tilde{\mathfrak{h}}_n$ invariant with respect to the action of the field operators X_n^α. For LNS these values are given by

(65)
$$\kappa_\ell = -\frac{2}{\ell\Delta} \;, \quad 2\ell \in \mathbb{Z}_+ \;,$$

and for LSG they are as follows:

(66) $$\gamma_{p,q} = 2\pi \frac{p}{q} \quad , \quad p,q \in \mathbb{Z}_+ , \; (p,q) = 1$$

In the first instance the subspace $\tilde{\mathfrak{h}}_n$ is the space of the spin l representation of the group SU(2); in the second example π_n and φ_n realize the twisted representation of the group $\mathbb{Z}_q \otimes \mathbb{Z}_q$. This fact can be used to prove the completeness of Bethe vectors in a way analogous to that used in [16] for the XXX model.

We also see that the L - matrices for the XXX model can be generalized to any value l of the spin. In particular, there appears a generalization of the Heisenberg Hamiltonian, considered above, for higher spins. The example of spin l = 1 was introduced by Zamolodchikov and Fateev [27]:

(67) $$H_1 = \sum_n (q_n - q_n^2) \quad \text{where } q_n = S_n^\alpha S_{n+1}^\alpha$$

The formula for arbitrary spin was given by Kulish and Sklyanin:

(68) $$H_l = \sum_n \sum_{j=1}^{2l} \Big[\Big(\prod_{k=0}^{2l} \frac{q_n - x_k}{x_j - x_k} \Big) \Big(\sum_{k=1}^{j} \frac{1}{k} \Big) \Big] \quad \text{where}$$
$$q_n = S_n^\alpha S_{n+1}^\alpha$$
$$x_j = \frac{1}{2} j(j+1) - l(l+1) \quad , \quad 0 \leq j \leq 2l$$

The ground state and the low-lying excitations were recently investigated by Takhtajan [28] along the lines of [16]. It is interesting to mention that for all spins l the only low-lying excitation is a kink with spin 1/2.

The group-theoretical origin of these generalizations has been analyzed by Kulish and Sklyanin in [3],[29].

Let us conclude this section with the following statement made by Korepin [30]: for all L - matrices introduced here one can define the analogue of the determinant. In fact, for some special η the matrix

(69) $$\Delta(\lambda) = \sigma_2 L_n(\lambda) \sigma_2 L_n(\lambda + \eta)^T$$

is proportional to the unit matrix in V, the factor $d(\lambda)$ being a c-number. It is natural to call $d(\lambda)$ the quantum determinant of the matrix $L_n(\lambda)$ (note the shift in the λ-variable in the last factor in (69)). One can rewrite this result in the form

$$L_n(\lambda)^{-1} = \frac{1}{d(\lambda)} \sigma_2 L_n(\lambda+\eta)^T \sigma_2 \tag{70}$$

which will be useful in the next section.

3. The Gelfand-Levitan-Marchenko equation

In classical field theory the inverse problem of reconstructing the field variables from given scattering data is solved by means of the Gelfand-Levitan-Marchenko (GLM) equation. This equation is equivalent to the factorization (Riemann-Hilbert) problem for the monodromy matrix. In QST the corresponding factorization can be based on the last formula of the preceding section. For definiteness we consider the N.S. model, beginning with its lattice formulation and subsequently going to the limit $\Delta \to 0$. The presentation follows the work of Smirnov [31].

Let $L_n(\lambda)$ be the L-matrix for the LNS model. Korepin's formula looks as follows:

$$L_n(\lambda)^{-1} = \frac{1}{d(\lambda)} \sigma_2 L_n(\lambda+i\kappa)^T \sigma_2 \ ,$$

(71)

$$d(\lambda) = \left(1-\frac{i\lambda\Delta}{2}\right)\left(1+\frac{i\lambda\Delta}{2}-\kappa\Lambda\right)$$

We introduce the Jost matrices

$$T^{(+)}_{N,k}(\lambda) = \prod_{n=k}^{N} L_n(\lambda) \ , \quad T^{(-)}_{N,k}(\lambda) = \prod_{n=-N}^{k-1} L_n(\lambda) \ , \tag{72}$$

where we have changed notations slightly so that the running variable n takes values from -N to +N and the length of the lattice is 2N + 1. It is evident that

$$T_N(\lambda) = T^{(+)}_{N,k}(\lambda)\, T^{(-)}_{N,k}(\lambda) \tag{73}$$

for any k. Using Korepin's formula we obtain the factorization relation

$$T^{(+)}_{N,k}(\lambda) = \frac{1}{d(\lambda)^{N+k-1}} T_N(\lambda)\,\sigma_2\, T^{(-)}_{N,k}(\lambda+i\varkappa)^T\,\sigma_2 \tag{74}$$

The first row of this equation (with λ changed to $\lambda-i\varkappa$) reads as follows:

$$\begin{aligned} &A_N(\lambda-i\varkappa)^{-1}\begin{pmatrix} A^{(+)}(\lambda-i\varkappa) \\ B^{(+)}(\lambda-i\varkappa)\end{pmatrix} \\ &= \frac{1}{d(\lambda+i\varkappa)^{N+k-1}}\left[\begin{pmatrix} D^{(-)}_{N,k}(\lambda) \\ -B^{(-)}_{N,k}(\lambda)\end{pmatrix} + B_N(\lambda)A_N(\lambda)^{-1}\begin{pmatrix} -C^{(-)}_{N,k}(\lambda) \\ A^{(-)}_{N,k}(\lambda)\end{pmatrix}\right]\end{aligned} \tag{75}$$

Here we have also used the relation

$$A_N(\lambda-i\varkappa)^{-1} B_N(\lambda-i\varkappa) = B_N(\lambda)\,A_N(\lambda)^{-1} \tag{76}$$

which follows from Korepin's formula. This relation is the quantum analogue for the relation for the Jost solutions (see for example [32])

$$\frac{1}{a(\lambda)}\, g(x,\lambda) = \tilde{f}(x,\lambda) + \frac{b(\lambda)}{a(\lambda)}\, f(x,\lambda)$$

which plays the basic role for the derivation of the GLM equation in the classical case. Note the shift in the spectral parameter on the lhs of the quantum relation.

One can perform the limit $N \to \infty$ using the fact that the limits

$$\begin{aligned} T(\lambda) &= \lim_{N\to\infty} V^{-N}(\lambda)\,T_N(\lambda)\,V^{-N}(\lambda) \\ T^{(+)}_k(\lambda) &= \lim_{N\to\infty} V^{-N+k}(\lambda)\,T^{(+)}_{N,k}(\lambda) \\ T^{(-)}_k(\lambda) &= \lim_{N\to\infty} T^{(-)}_{N,k}(\lambda)\,V^{-N-k+1}(\lambda) \end{aligned} \tag{77}$$

exist, where

$$V(\lambda) = \mathrm{diag}\left(1 - \frac{i\lambda\Delta}{2}, 1 + \frac{i\lambda\Delta}{2}\right) \tag{78}$$

The existence of these limits can be established in the same way as is done in [1),20)] for the continuum case. Defining the normalized Jost solutions

$$\phi_k(\lambda) = \begin{pmatrix} A_k^{(+)}(\lambda) \\ B_k^{(+)}(\lambda) \end{pmatrix}, \quad \tilde{\phi}_k(\lambda) = \begin{pmatrix} C_k^{(+)}(\lambda) \\ D_k^{(+)}(\lambda) \end{pmatrix},$$

$$\chi_k(\lambda) = \begin{pmatrix} -C_k^{(-)}(\lambda) \\ A_k^{(-)}(\lambda) \end{pmatrix}, \quad \tilde{\chi}_k(\lambda) = \begin{pmatrix} D_k^{(-)}(\lambda) \\ -B_k^{(-)}(\lambda) \end{pmatrix}, \tag{79}$$

we can rewrite the factorization relation as follows:

$$\tilde{\chi}_k(\lambda) + \left(\frac{1 - \frac{i\lambda\Delta}{2}}{1 + \frac{i\lambda\Delta}{2}}\right)^k R(\lambda)\,\chi_k(\lambda) = A(\lambda - i\kappa)^{-1}\,\phi_k(\lambda - i\kappa) \tag{80}$$

Introducing $x = \Delta$ and performing the limit $\Delta \to 0$, we get the final relation

$$\tilde{\chi}(x,\lambda) + e^{-i\lambda x} R(\lambda)\,\chi(x,\lambda) = A(\lambda - i\kappa)^{-1}\,\phi(x,\lambda - i\kappa). \tag{81}$$

One also has the involution relations

$$\phi(x,\lambda) = -i\sigma_2\,\tilde{\phi}^*(x,\bar{\lambda}), \quad \chi(x,\lambda) = -i\sigma_2\,\tilde{\chi}^*(x,\bar{\lambda}), \tag{82}$$

where * means componentwise conjugation. The operators $A(\lambda)$, $\phi(x,\lambda)$ and $\chi(x,\lambda)$ are analytic for $\mathrm{Im}\,\lambda > 0$, and in the limit $|\lambda| \to \infty$ they behave as follows:

$$A(\lambda) \to I, \quad \phi(x,\lambda) \to \begin{pmatrix} I \\ 0 \end{pmatrix}, \quad \chi(x,\lambda) \to \begin{pmatrix} 0 \\ I \end{pmatrix} \tag{83}$$

All these properties are derived by the same methods as in the classical theory [32)].

For the repulsive case ($\kappa > 0$), the operator $A(\lambda)^{-1}$ is also analytic for $\mathrm{Im}\,\lambda > 0$. Thus the factorization relation leads to the equation

$$\tilde{\chi}(x,\lambda) = \begin{pmatrix} I \\ 0 \end{pmatrix} + \frac{1}{2\pi i}\int_{-\infty}^{\infty} \frac{R(\mu)\,\chi(x,\mu)\,e^{-i\mu x}}{\lambda - \mu}\,d\mu\,. \tag{84}$$

Together with the involution relation, this gives a system of linear equations which determines $\chi(x,\lambda)$ from $R(\mu)$. The field operators $\psi^*(x), \psi(x)$ are easily written when $\chi(x,\lambda)$ is known. This equation is exactly the GLM equation in QST. It was first derived in [33] without any recourse to the lattice; but it was not clear how to modify the derivation for the attractive case. Smirnov's method makes such a modification straightforward. (During the session of the Institute I have learned from Dr. Göckeler that he has generalized the method of [33] to the attractive case; see [35].)

In the attractive case, the operator $A(\lambda)^{-1}$ has singularities in the upper half-plane, namely on the lines $\mathrm{Im}\,\lambda = -n\varkappa/2$, $n = 1,2,..$; these correspond to bound states of n particles. The GLM equation is modified by an additional term

$$\frac{1}{2\pi i} \sum_{n=1}^{\infty} \int_{\Gamma_n} \frac{\mathrm{disc}\, A(\mu - i\varkappa)^{-1} \phi(x, \mu - i\varkappa)}{\lambda - \mu} d\mu \tag{85}$$

on the rhs, where

$$\mathrm{disc}\, A(\lambda)^{-1} = A(\lambda + i0)^{-1} - A(\lambda - i0)^{-1} \tag{86}$$

and Γ_n is the line $\mathrm{Im}\,\lambda = -n\varkappa/2$.

The next step is to express $\mathrm{disc}\, A(\lambda - i\varkappa)^{-1} \phi(x, \lambda - i\varkappa)$ in terms of $\chi(x,\mu)$ in order to close the system. Using the obvious relation

$$A^{(+)}(x,\lambda)\, A^{(-)}(x,\lambda) = A(\lambda) = B^{(+)}(x,\lambda)\, C^{(-)}(x,\lambda) \tag{87}$$

and the equality

$$A(\lambda)\ \mathrm{disc}\, A(\lambda)^{-1} = 0 , \tag{88}$$

one can obtain

$$\mathrm{disc}\, A(\lambda - i\varkappa)^{-1} \phi(x, \lambda - i\varkappa) = G(x,\lambda)\, \chi(x,\lambda) , \tag{89}$$

where

$$G(x,\lambda) = \mathrm{disc}\, A(\lambda - i\varkappa)^{-1} B^{(+)}(x, \lambda - i\varkappa)\, A^{(-)}(x,\lambda)^{-1} . \tag{90}$$

We shall show that the x-dependence of $G(x,\mu)$ is trivial. Indeed, we have the identity

$$A(\lambda)\, B^{(+)}(x,\mu-i\kappa)\, A^{(-)}(x,\mu)^{-1}$$

(91)
$$= \frac{\lambda-\mu+i\kappa}{\lambda-\mu} B^{(+)}(x,\mu-i\kappa)\, A^{(-)}(x,\mu)^{-1} A(\lambda)$$
$$+ \frac{i\kappa}{\lambda-\mu} A(\mu-i\kappa)\, A^{(-)}(x,\mu-i\kappa)^{-1} B^{(+)}(x,\lambda)\, A^{(-)}(x,\lambda)\, A^{(-)}(x,\mu)^{-1}$$

which can be derived using the known commutation relations [20] between the matrix elements of $T^{(\pm)}(x,\lambda)$. Combining the last formula with the definition of G, one gets

(92)
$$A(\lambda)\, G(x,\mu) = \frac{\lambda-\mu+i\kappa}{\lambda-\mu} G(x,\mu)\, A(\lambda).$$

Since $\ln A(\lambda)$ is the generating function for the conservation laws, it follows that

(93)
$$e^{iPt}\, G(x,\lambda)\, e^{-iPt} = e^{i\lambda t}\, G(x,\lambda) \quad ,$$

(94)
$$e^{iHt}\, G(x,\lambda)\, e^{-iHt} = e^{i\lambda^2 t}\, G(x,\lambda) \quad ,$$

where P and H are the momentum and energy operators. These relations show that $G(x,\lambda)$ is an angle-type variable. From the definition of G and the obvious property

(95)
$$e^{-iPt}\,\phi(x,\lambda)\, e^{iPt} = \phi(x+t,\lambda) \; , \; e^{-iPt}\,\chi(x,\lambda)\, e^{iPt} = \chi(x+t,\lambda)$$

of ϕ and χ, it follows that

(96)
$$G(x+t,\lambda) = e^{-i\lambda t}\, G(x,\lambda).$$

Introducing notations

$$G_0(\lambda) = B(\lambda)\, A(\lambda)^{-1} = R(\lambda)$$

(97)
$$G_n(\lambda) = G(0,\lambda-\tfrac{i\kappa}{2}n) \quad , \quad n=1,2,\ldots$$

$$\chi^{(n)}(x,\lambda) = \chi(x,\lambda-\tfrac{i\kappa}{2}n) \quad , \quad \tilde{\chi}^{(n)}(x,\lambda) = \tilde{\chi}(x,\lambda+\tfrac{i\kappa}{2}n)$$

let us rewrite the GLM equation in terms of the G_n:

$$\tilde{\chi}^{(n)}(x,\lambda) = \begin{pmatrix} I \\ 0 \end{pmatrix} + \frac{1}{2\pi i} \sum_m e^{-\frac{\varkappa}{2} m x_0} \int_{-\infty}^{\infty} \frac{G_m(\mu) \chi^{(m)}(x,\mu) e^{i\mu(x-x_0)}}{\lambda - \mu - \frac{i\varkappa}{2}(n+m)} d\mu \tag{98}$$

Here, x_0 is an arbitrary point of the real axis.

It is possible to write down the full system of commutation relations between all $G_n(\mu)$ and their conjugates.

In his recent paper [35], Göckeler has constructed the creation operator $R_n(\lambda)$ for the n-particle bound state carrying momentum λ. Comparison of the results of Göckeler and Smirnov leads to the conclusion that $G_n(\mu)$ is factorized in terms of the operators $R_n(\mu)$:

$$G_m(\mu) = R_{m+1}(\mu) R_m^*(\mu + i\varkappa/2) \tag{99}$$

The operator $R_m^*(\mu + i\varkappa/2)$ is not correctly defined. However, it enters the GLM equation only through the product $R_m^*(\mu + i\varkappa/2)\, \chi^{(m)}(x,\mu)$ which can be constructed by analytic continuation of the correctly defined quantity $R_m^*(\mu)\, \chi^{(m)}(x,\mu - i\varkappa/2)$. Iteration of the GLM equation then gives the expression for $\chi, \tilde{\chi}$ (and subsequently for ψ^*, ψ) in terms of $R_m(\mu)$.

The main interest in the GLM equation is connected with the problem of constructing the Green functions for the models under investigation. The program of solving this problem using the GLM equation for the NS model was formulated by Thacker, Wilkinson and Creamer in [37]; for a survey, see [36]. Interesting results are also given in [38] and [39]. However, the full problem is not yet solved, and it constitutes a serious challenge to the development of QST.

It is clear from the presentation that the Smirnov method can be used to derive the GLM equation for any model with L-matrices satisfying the Korepin relation. In particular, Smirnov himself is now investigating the XXZ model by this method.

Conclusion

As one can see from this text and from numerous references, QST is a rapidly developing method in quantum one-dimensional mathematical physics. It leads to the exact solution of interesting field-theoretical models, and it produces new exactly soluble models. It teaches us that the particle spectrum above the physical ground state can be considerably different from the predictions of perturbation theory. Moreover, the ground state itself has quite a complicated structure in terms of quasiparticles. All this means that QST presents us with a profilic playground in quantum field theory and mathematical physics.

References

1. L.D. Faddeev, Sov. Scient. Rev. C1, 107 (1980)
2. H.B. Thacker, Rev. Mod. Phys. 53, 253 (1981)
3. P.P. Kulish and E.K. Sklyanin, Quantum Spectral Transform Method, in: "Proceedings of the International Symposium on Integrable Quantum Fields", C. Montonen, ed., to be published
4. A.G. Izergin and V.E. Korepin, Fizika Elem. Chastits 13 (to be published)
5. M. Gaudin, Algebraic Aspects of Exact Models, in: these proceedings
6. L.D. Faddeev and L.A. Takhtajan, Uspekhi Mat. Nauk 34, 13 (1979)
7. S.A. Tsyplyaev, Teor. Mat. Fiz. 48, 24 (1981)
8. A.G. Izergin and V.E. Korepin, Commun. Math. Phys. 79, 303 (1981)
9. B. Berg, M. Karowski, V. Kurak and P. Weisz, Nucl. Phys. B 134, 125 (1978)
10. P.P. Kulish and E.K. Sklyanin, Zapiski Nauch. Semin. LOMI 95, 129 (1980)
11. A.A. Belavin, Nucl. Phys. B 180 (FS 2), 189 (1981)
12. P.P. Kulish, Doklady Akad. Nauk SSSR 255, 323 (1980)
13. P.P. Kulish and N. Yu. Reshetichin, Zh. Eksper. Teor. Fiz. 80. 214 (1981)
14. L.A. Takhtajan, Zapiski Nauch. Semin. LOMI 101, 158 (1981)
15. J. Des Cloiseaux and J. Pearson, Phys. Rev. 128, 2131 (1962)
16. L.D. Faddeev and L.A. Takhtajan, Zapiski Nauch. Semin. LOMI 109, 134 (1981)
17. V.E. Korepin, Teor. Mat. Fiz. 41, 169 (1979)
18. E.K. Sklyanin, Doklady Akad. Nauk SSSR 244, 1337 (1979)
19. L.D. Faddeev, E.K. Sklyanin and L.A. Takhtajan, Teor. Mat. Fiz. 40, 194 (1979)
20. E.K. Sklyanin, Zapiski Nauch. Semin. LOMI 95, 55 (1980)
21. H.B. Thacker and D. Wilkinson, Phys. Rev. D 19, 3660 (1979)
22. J. Honerkamp, P. Weber and A. Wiesler, Nucl. Phys. B 152, 266 (1979)
23. A.G. Izergin and V.E. Korepin, Lett. Math. Phys. 5, 199 (1981)
24. H. Bergknoff and H.B. Thacker, Phys. Rev. D 19, 3666 (1979)
25. V.E. Korepin, Commun. Math. Phys. 76, 165 (1980)
26. V.E. Zakharov and L.A. Takhtajan, Teor. Mat. Fiz. 38, 17 (1979)
27. A.B. Zamolodchikov and V.A. Fateev, Yadernaya Fiz. 32, 581 (1980)
28. L.A. Takhtajan, Phys. Lett. A (to be published)
29. P.P. Kulish, N. Yu. Reshetichin and E.K. Sklyanin, Lett. Math. Phys. 5 , 393 (1981)
30. A.G. Izergin and V.E. Korepin, Doklady Akad. Nauk SSSR 259, 76 (1981)
31. F. Smirnov, Doklady Akad. Nauk SSSR (to be published)

32. L.D. Faddeev, J. Sov. Math. 5 , 334 (1976)
33. D.B. Creamer, H.B. Thacker and D. Wilkinson, Phys. Rev. D 21, 1523 (1980)
34. M. Göckeler, Z. Physik C7, 263 (1981)
35. M. Göckeler, Z. Physik C11, 125 (1981)
36. H.B. Thacker, The Quantum Inverse Method and Green's Functions for Completely Integrable Field Theories, in: "Proceedings of the International Symposium on Integrable Quantum Fields", C. Montonen, ed., to be published
37. D.B. Creamer, H.B. Thacker and D. Wilkinson, Phys. Rev. D 23, 3081 (1981)
38. M. Jimbo and T. Miwa, Preprint RIMS-370, Kyoto University (1981)
39. J. Honerkamp, Nucl. Phys. B 190, 301 (1981)

PHASES IN GAUGE THEORIES

S. Yankielowicz

Tel-Aviv University, Israel*
and
CERN, Geneva

INTRODUCTION

Gauge theories are believed to belong to the class of theories which describe nature. At each energy (distance) scale nature can be described in terms of an effective gauge theory. For example at present available energies physics seem to be correctly described by the standard $SU_C(3) \times SU(2) \times U(1)$ theory. It is important to note that degrees of freedom which are considered elementary on a given scale may turn out to be composite at a higher energy scale. Given the (effective) gauge theory, its large distance properties (phase structure, symmetry realization etc.) are determined by the structure of the vacuum. In these lectures I would like to concentrate on two aspects of the vacuum structure in gauge theories (without and with fermions).

1. Phases in pure gauge theories
   In particular we'll be interested in the characterization of the different phases in terms of topological excitations (the 't Hooft picture) [1].
2. Dynamical Symmetry Breaking - chiral symmetry realization

This subject is rather wide and I'll concentrate on one specific aspect, i.e., the anomaly constraint equations in composite equations in composite theories.

The lectures are organized as follows. In the first part we discuss the 't Hooft picture of phases in pure guage theories [1]. This will serve us as a motivation to look at a simple toy model i.e. Z(N) gauge theories in which one can show that the different types of phases do exist [2]. In particular I'll discuss the

*Supported by the U.S.-Israel Binational Science Foundation

phases of the pure Z(N) gauge theory and the phases in the presence of a $\theta F\tilde{F}$ term [3]. (In this part I'll follow the recent papers by Carly and Rabinovici [4]. The emphasis will be on the description of the vacuum in terms of certain condensates. Given the condensate we can determine which degrees of freedom are confined and which are screened.

The second part will be devoted to the anomaly constraint equation in composite theories [5]. The main issue here is the question of chiral symmetry realization. I'll discuss both the vector case (QCD like theories) and the chiral case.

I PHASES IN PURE GUAGE THEORIES - THE T'HOOFT DESCRIPTION

Recently [1,3] 't Hooft has argued that essentially three types of phases can exist in a gauge theory:

(i) Confinement - electric superconductors:
This phase is believed to be realized for example in QCD or the technicolor sector of technicolor theories. In this phase the vacuum is a condensate of magnetic monopoles and as a result there exists a linear potential between external electric charges.

(ii) Higgs - superconductor:
This phase is realized in theories where a Higgs phenomenon takes place such as the Weinberg Salam theory and Grand Unification theories. The Higgs phase is dual to the confining phase in the sense that the vacuum is a condensate of electric charges, hence magnetic monopoles are confined.

(iii) Coulomb phase:
This is the phase one encounters in QED. It is characterized by the fact that neither electric nor magnetic charges condense.

It is clear from this description that in principle also other kinds of phases may occur where a bound state which carries both electric and magnetic charges condenses. Such phases occur in the presence of a $\theta F\tilde{F}$ term and are referred to by 't Hooft as oblique confinement.[3]) .

At this point it will be of interest to remember the picture advocated by 't Hooft. We are interested in the large distance properties of a given gauge theory. As a first step toward understanding the structure of the vacuum we should disentangle the relevant degrees of freedom.

According to 't Hooft the degrees of freedom in a SU(N) gauge theory are determined by a partial fixing of the gauge, i.e., by going to a unitary like gauge. The gauge is fixed up to the maximal abelian subgroup $[U(1)]^{N-1}$. To be more precise, one chooses a composite field X which serves as a composite Higgs. The demands on X are:

1. X transforms covariantly under gauge transformation, $X' = \Omega X \Omega^{-1}$.

2. X belongs to the adjoint representation of SU(N).

X can be represented as a matrix. The eigenvalues of this matrix are of course gauge invariant. The unitary like gauge is defined by diagonalizing X

$$X = \begin{pmatrix} \lambda_1 & 0 \\ 0 & \lambda_n \end{pmatrix} \quad (1.1)$$

The gauge is not determined completely since it is clear that X is still invariant under

$$\Omega = \begin{pmatrix} e^{i\phi_1} & & \\ & \ddots & \\ & & e^{i\phi_N} \end{pmatrix} , \quad \sum \phi_i = 0 \quad (1.2)$$

The subgroup of Ω is the maximal abelian (cartan subalgebra) $[U(1)]^{N-1}$. The gauge condition (eq.(1.1)) becomes singular at points when two of the eigenvalues coincide. It has been shown by 't Hooft that those singularities correspond to magnetic monopoles with respect to the $[U(1)]^{N-1}$ subgroup. We conclude, therefore, that within this gauge, the effective theory which describes the large distance behaviour of our SU(N) gauge theory, is a $[U(1)]^{N-1}$ gauge theory enriched with monopoles. The gluons and quarks carry electric charge with respect to the U(1)'s, while the monopoles carry magnetic charge with respect to the U(1)'s. The values of these charges must satisfy the Dirac quantization condition [6] in order for the theory to be consistent. Now it is clear that three natural possibilities may occur and they will characterize the different phases that can exist a priori in the gauge theory.

(i) Confinement - the field of a magnetic monopole develops non-vanishing vacuum expectation value
(ii) Higgs Mode - one of the purely electrically charged objects which is Lorentz scalar develops non-vanishing vacuum expectation value.
(iii) Coulomb mode - all particles get positive mass square and no condensation occurs.

It is also clear that mixed modes can appear when some bound state which carries both electric and magnetic charges condenses. 't Hooft refers to such a phase, which may occur in the presence of a $\theta F\tilde{F}$ term, as an oblique confinement. The important point is that all phases are characterized by the points in the electric-magnetic charge lattice, which develop vacuum expectation value. Given the electric and magnetic charges of the condensate we can determine which particles are confined and which particles are screened. The Wilson loop operator [7] for the confined quantum numbers will have an area law behaviour while for the other quantum numbers there will be a perimeter law behaviour. As we'll discuss later on, particles whose charge have zero Dirac unit ($n = 0$) with respect to the condensate are screened by the Higgs mechanism much in the same way that it happens in superconductivity [8]. Particles with $n \neq 0$ must have flux lines attached to them; hence they will be confined.

II Z(N) GAUGE THEORIES AS LABORATORY MODELS

It has been stressed before that Z(N) gauge theories serve as a relevant and interesting laboratory model [1,2]. If one adopts the picture advocated by 't Hooft [3] then the relevance of the Z(N) gauge theories can be pushed even further. The Z(N) gauge theory, as will be explained in a moment, is equivalent to a $U(1)$ gauge theory with a conserved charge N particle. Since the pure compact $U(1)$ gauge theory is equivalent to a non compact $U(1)$ gauge theory plus monopoles [9], we see that the Z(N) theory is equivalent to a non compact $U(1)$ gauge theory plus charges plus monopoles. All the degrees of freedom which are important, according to the 't Hooft picture, are found in the Z(N) gauge theory. Therefore, we expect to find all the different phases which may occur in a gauge theory already in this system.

We start with the partition function of the compact $U(1)$ theory (compact QED)

$$(2.1) \qquad Z_{U(1)} = \int_{-\pi}^{\pi} \prod_{r,\mu} d\,\theta_\mu(r)\; e^{\frac{1}{g^2}\sum_{r,\mu,\nu} \cos\theta_{\mu\nu}(r)}$$

where $\theta_\mu(r)$ is an angle variable which resides on the link $\ell = (r,\mu)$ which starts at the point r and goes in the μ direction. We denote by $\theta_{\mu\nu}(r)$ the lattice curl

$$(2.2) \qquad \theta_{\mu\nu}(r) = \theta_\mu(r) + \theta_\nu(r+e_\mu) - \theta_\mu(r+e_\nu) - \theta_\nu(r) \equiv \Delta_\nu\theta_\mu(r) - \Delta_\mu\theta_\nu(r)$$

The notation is summarized in Fig. 1. The local $U(1)$ symmetry involves shifting all variables attached to the point r by the amount $\theta(r)$

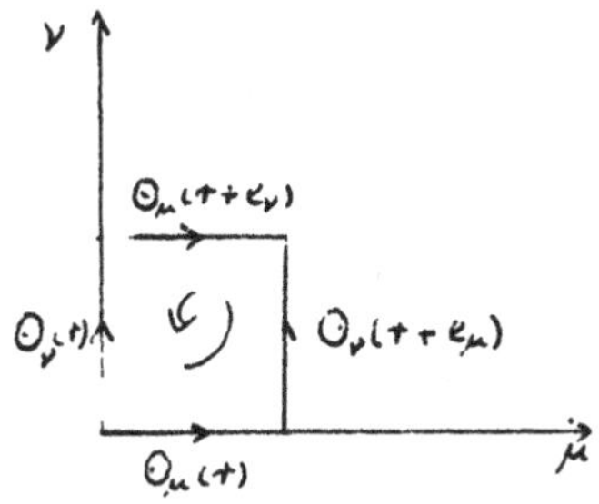

Fig. 1 , The lattice curl

(2.3) $$\theta_\mu(r) \to \theta_\mu(r) - \theta(r)$$
$$\theta_\nu(r) \to \theta_\nu(r) - \theta(r)$$

Transforming simultaneously at the point r and $r+e_\mu$ gives

(2.4) $$\theta_\mu(r) \to \theta_\mu(r) + \theta(r+e_\mu) - \theta(r) \equiv \theta_\mu(r) - \Delta_\mu \theta(r)$$

The compact $U(1)$ theory is known to undergo a phase transition at a finite g^2. The large g^2 phase is a confining phase and the Wilson loop order parameter shows an area law behaviour. The small g phase is a massless Coulomb phase.

The Z(N) gauge theory is defined by taking the $U(1)$ action and restricting the angle variables to take just N discrete values. To make the computations easier we shall work in the Villain approximation in which $Z_{U(1)}$ of eq(2.1) is replaced by

(2.5) $$Z^V_{U(1)} = \int_{-\pi}^{\pi} \prod_{r,\mu} d\theta_\mu(r) \prod_{r,\mu\nu} \sum_{n_{\mu\nu}(r)=-\infty}^{\infty} e^{-\frac{1}{2g^2}\left(\theta_{\mu\nu}(r) - 2\pi n_{\mu\nu}(r)\right)^2}$$

The $n_{\mu\nu}(r)$ are integer variables associated with each plaquette. The approximation amounts to expansion of the $\cos\theta_{\mu\nu}$ around all of its minima. Note that both the symmetry and the periodicity of the original theory are preserved in the Villain approximation. Hence, we believe that both are in the same universality class. For the Z(N) theory we can write [2)].

(2.6) $$Z^V_N = \Big(\prod_{r,\mu} \sum_{\ell_\mu(r)=-\infty}^{\infty}\Big)\Big(\prod_{r,\mu\nu} \sum_{n_{\mu\nu}(r)=-\infty}^{\infty}\Big) \int_{-\pi}^{\pi} [d\theta_\mu(r)]\, \delta(\Delta_\mu \ell_\mu) \times$$
$$e^{iN\sum_{r,\mu} \ell_\mu(r)\theta_\mu(r)}\; e^{-\frac{1}{2g^2}\sum_{r\mu\nu}\left(\theta_{\mu\nu}(r) - 2\pi n_{\mu\nu}(r)\right)^2}$$

The effect of summation over the integer link variables $\ell_\mu(r)$ is to discretize Θ_μ in units of $2\pi/N$. Note that the partition function in eq.(2.6) is just the partition function for the compact $U(1)$ theory (eq.2.5) in the presence of an external conserved electric current (ℓ_μ) which carries a charge which is multiple of N. The fact that ℓ_μ has to be conserved results from the demands that the partition function should be gauge invariant. The summation over ℓ_μ is therefore a summation over closed electric loops. We conclude that a Z(N) theory is equivalent to a compact $U(1)$ theory plus electric charge N conserved current. The next step is to identify the monopoles in our system. The most general antisymmetric integer field $n_{\mu\nu}$ can be written as

$$n_{\mu\nu}(r) = \Delta_\mu n_\nu - \Delta_\nu n_\mu + \frac{1}{2}\varepsilon_{\mu\nu\lambda\kappa} M_{\lambda\kappa} \tag{2.7}$$

where n_μ and $M_{\lambda\kappa}$ take integer values. If it were not for the second term we would have that $\Delta_\mu \tilde{n}_{\mu\nu} = 0$ ($\tilde{n}_{\mu\nu} = \frac{1}{2}\varepsilon_{\mu\nu\alpha\beta} n_{\alpha\beta}$ is the dual of $n_{\mu\nu}$). In order to have monopoles we must have $\Delta_\mu \tilde{n}_{\mu\nu} \neq 0$. Now we are in a position to define the monopole current which resides on links of the dual lattice

$$m_\mu \equiv \frac{1}{2}\varepsilon_{\mu\nu\lambda\kappa}\Delta_\nu n_{\lambda\kappa} = \Delta_\nu \tilde{n}_{\mu\nu} \tag{2.8}$$

Note that the monopoles are associated with the periodicity (compactness) of the theory. It is easy to check using eqs.(2.7) and (2.8) that

$$m_\mu = \Delta_\nu M_{\mu\nu} \tag{2.9}$$

Since $M_{\mu\nu}$ is antisymmetric

$$\Delta_\mu m_\mu = 0 \tag{2.10}$$

i.e. the monopole current is conserved. Equation (2.9) can be inverted thus revealing the Dirac string attached to the monopole

$$M_{\mu\nu}(R) = \hat{n}^\mu (\hat{n}\,\Delta)^{-1} m^\nu(R) - (\mu \leftrightarrow \nu) \tag{2.11}$$

where $\hat{n}^\mu$ is a constant unit vector on the lattice; and $(\hat{n}\cdot\Delta)^{-1}$ is a line integral in the $\hat{n}$ direction. We can now introduce non compact variable A_μ

$$A_\mu(r) = \Theta_\mu(r) - 2\pi n_\mu(r) \tag{2.12}$$

and rewrite the partition function in eq.(2.6)

$$Z = \sum_{\{\ell_\mu(r)\}} \sum_{\{m_\mu(R)\}} \int_{-\infty}^{+\infty} [dA_\mu] \, \delta(\Delta_\mu \ell_\mu) \, \delta(\Delta_\mu m_\mu) \times$$
$$e^{-\frac{1}{2g^2} \sum_{r,\mu\nu} \left(F_{\mu\nu} - \varepsilon_{\mu\nu\lambda\kappa} \hat{n}_\lambda (\hat{n}\Delta)^{-1} m_\kappa \right)^2} \, e^{iN \sum_{r,\mu} \ell_\mu A_\mu} \tag{2.13}$$

Now it is clear that the Z(N) gauge theory can be written as a non compact U(1) plus (conserved) electric charge current and (conserved) magnetic charge current.

Next we choose a gauge (Feynman gauge) to perform the quadratic integration over the A_μ's

$$Z = \sum_{\{\ell_\mu(r)\}} \sum_{\{m_\mu(R)\}} e^{-\frac{1}{2}(Ng)^2 \sum_{r,r'} \ell_\mu(r) \, G(r-r') \, \ell_\mu(r')} \times$$
$$e^{-\frac{1}{2}\left(\frac{2\pi}{g}\right)^2 \sum_{R,R'} m_\mu(R) \, G(R-R') \, m_\mu(R')} \times \tag{2.14}$$
$$e^{iN \sum_{R,\mu} m_\mu(R) \, \Theta_{\mu\nu}(R-r) \, \ell_\nu(r)}$$

$G(r)$ is the lattice propagator (Coulomb Green's function)

$$-\Delta^2 G(r) := -\sum_\mu \left(G(r+e_\mu) + G(r-e_\mu) - 2G(r) \right) = \delta(r) \tag{2.15}$$

and

$$\Theta_{\mu\nu}(R-r) = 2\pi \, \varepsilon_{\mu\nu\lambda\kappa} \, \hat{n}_\lambda (\hat{n}\Delta)^{-1} \Delta_\kappa^{(r)} \, G(R-r) \tag{2.16}$$

represents the interaction between electric and magnetic loops. We have recasted the theory into a Coulomb gas of conserved electric and magnetic currents. The first term represents the Coulomb interaction between the electric loops, while the second term represents the Coulomb interaction between the magnetic loops. Each closed loop can be viewed as a particle-antiparticle pair being created at a given point propagating (along the loop) and reannihilating at another point. Clearly the electric charge unit is $q = Ng$ while the magnetic charge unit is $2\pi/g$. The Dirac quantization condition [5] is therefore satisfied.

(2.17) $$qh = N_g \frac{2\pi}{g} = 2\pi N$$

We can perform a duality transformation, i.e. the theory is symmetric under the interchange

(2.18) $$\ell_\mu \to - m_\mu \qquad m_\mu \to - \ell_\mu$$
$$N_g \longleftrightarrow 2\pi/g$$

which is analogous to the well known duality in electrodynamics ($E \to B$, $B \to -E$). The selfdual point is $g^2 = 2\pi/N$

For large enough N the Z(N) theory has three phases. The heuristic argument goes as follows: if there were only two phases then the selfduality would give the critical point at $(g_c^2)_N = 2\pi/N$ For $g^2 > 2\pi/N$ the Wilson loop will have an area law behaviour $\exp(-b_N A)$. Let us denote by $(g_c^2)_{U(1)}$ the critical point of PQED. Then for large enough N , $2\pi/N < (g_c^2)_{U(1)}$. Therefore, in the region $2\pi/N < g^2 < (g_c^2)_{U(1)}$ the Z(N) theory will have an area law behaviour while the U(1) theory exhibits a perimeter law behaviour. For $N \to \infty$ the Z(N) theories go smoothly into the U(1) theory; hence as $N \to \infty$ the area law behaviour of the Z(N) theory in this region should tend towards a perimeter law. This means that as we increase N the system becomes mere ordered (for fixed g^2 which plays the role of temperature). This is, however, unreasonable since increasing N means increasing the number of allowed configurations and thus increasing the entropy. Having more entropy cannot improve the range of correlation. Our original assumption that there are only two phases must therefore be wrong. There must be more than two phases. A rigorous proof to this effect is given by Elitzur , Pearson and Shigemitsu [2)].

We would like next to understand the different phases in terms of condensates. Let us draw a charge diagram in which we denote by points the charge combinations our system can have. (Fig.2) The electric charge e is measured in units of Ng and the magnetic charge h in units of $2\pi/g$. The partition function (eq.2.14) indicates that the fundamental excitations are electric loops with any integer charge and magnetic loops with any integer charge.

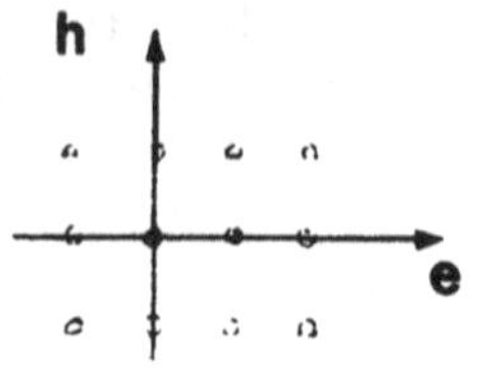

Fig. 2. The charge diagram of the Z(N) theory.

Note that we do not a priori claim that the system will necessarily have bound states carrying e.g $e=2$ $h=1$. It is up to the dynamics to decide it. The charge diagram only gives all the values of the charges which are allowed from quantum number consideration.

Consider a loop of length L which carries electric charge ℓ and magnetic charge m. In the naive energy-entropy consideration which we'll carry, we'll neglect the interaction term in eq(2.14). Moreover, to estimate the energy we pay we'll take from eq(2.14) just the self energy mass terms ($r = r'$, $R = R'$)

(2.19) $$\text{Energy} \approx \frac{1}{2}\left[(Ng)^2 \ell^2 + \left(\frac{2\pi}{g}\right)^2 m^2\right] G(0)\, L$$

The entropy can be estimated by taking the log of the number of loops of length L which pass through a given point.
A random walk estimate gives

$$\text{Entropy} \approx \log \mu^L = L \log \mu$$

where μ is a geometrical constant ($\mu = 7$ since at each step on the lattice the loop can choose seven different directions). It will be favourable to excite such loops from the vacuum provided the loss in energy is small compared to the gain in entropy

(2.20) $$(Ng)^2 \ell^2 + \left(\frac{2\pi}{g}\right)^2 m^2 \leq \frac{2 \log \mu}{G(0)} \equiv C$$

The equality in eq.(2.20) is an equation for an ellipse which can be drawn on the charge diagram (Fig.2). All points which are inside the ellipse can (according to this very simple considerations) condense. It is clear that for large g in order to satisfy the inequality (eq.2.20) we must have $\ell = 0$ but m can be different from zero. We conclude that the large g phase corresponds to a vacuum which is a condensate of magnetic charge. Hence, this phase is an electric confining phase. It is easy to check that the Wilson's loop for any external charge shows an area law behaviour. The physics leading to the confinement is dual to the physics which leads to (magnetic) confinement of magnetic monopoles in superconductors. If we insert external charges to the confining phase in which magnetic charge is condensed, the electric flux emanating from them will be confined to tubes giving rise to a linear potential. For small g necessarily $m = 0$ and $\ell \neq 0$. In this phase electric charge condenses. It is the Higgs phase which is nothing but a superconductor phase. Here magnetic monopoles are confined. For a large enough N, it is clear that for intermediate g values in order to satisfy the inequality (2.20) we must take $\ell = m = 0$. In this case we do not have any condensate. This phase is the Coulomb phase with massless photon. The estimate of the critical N gives $N_c \approx 5$. For $N > N_c$ the Z(N) theory has three phases: Higgs, Coulomb and Confinement.

Z(N) Gauge Theories in the Presence of a Θ-Angle

As we discussed in the introduction, 't Hooft has recently [3] advocated that in the presence of an (instanton) angle Θ more and new types of phases can occur. The physics underlying these phases is intimately connected to the observation of Witten [10] that in the presence of Θ monopoles can carry non integer electric charge proportional to Θ. Witten's argument starts from the generalized Dirac condition for two dyons [11]

(2.21) $$e_1 g_2 - e_2 g_1 = 2\pi n$$

where we denote by e and g the electric and magnetic charges. If we take the electron (e,o) and any dyon (q,g) we get

(2.22) $$eg - q\,0 = eg = 2\pi n$$

There is no restriction on the electric charge of the dyon. What we get, however, is the known relation $g = \frac{2\pi}{e} n$. If we take now two dyons with charges $(q, \frac{2\pi}{e})$ and $(q', \frac{2\pi}{e})$ we get

(2.23) $$e_1 g_2 - e_2 g_1 = (q - q')\frac{2\pi}{e} = 2\pi n$$

which immediately yields that

(2.24) $$q - q' = ne$$

i.e. only electric charge difference of dyons must be an integer. The electric charge itself need not be integer.

(2.25) $$q = ne + \delta e$$

However if the theory is invariant under CP then the existence of a dyon $(q, 2\pi/e)$ implies the existence of dyon $(-q, 2\pi/e)$ (recall that E and B have different transformation properties). Now the generalized Dirac condition gives

(2.26) $$e_1 g_2 - e_2 g_1 = q\frac{4\pi}{e} = 2\pi n$$

and q is determined. We conclude that δ in eq.(2.25) measures the amount of CP violation. If the only CP violation comes via a $\Theta F\tilde{F}$ term then Witten [10] shows that

(2.27) $$q = ne - \frac{\Theta}{2\pi} e$$

Both Witten's and 't Hooft's results can be demonstrated within the Z(N) gauge theories once we introduce a $\theta F\tilde{F}$ term. I shall follow in this point the paper of Cardy and Rabinovici [4] and refer interested readers to this paper for details. The first problem is how to introduce the $\theta F\tilde{F}$ term. The idea of Cardy and Rabinovici is to note that in the continuum for an abelian $U(1)$ theory

$$\frac{1}{2}\int F\tilde{F}\,d^4x = \int \partial_\mu A_\nu \tilde{F}_{\mu\nu} = \int A_\nu \partial_\mu \tilde{F}_{\mu\nu} = \int A_\nu m_\nu \tag{2.28}$$

where m_ν denotes the magnetic current. I have been here very careless about boundary terms and introduction of magnetic charge, but one can argue that the result obtained is correct. The effect of the $\theta F\tilde{F}$ is essentially to couple A_μ to the monopole current. Since this term is of the form $A_\mu j_\mu$ we should not be surprised that as a result the monopole becomes a dyon and carries an electric charge proportional to θ.

To put this new term on the lattice poses a problem since θ_μ and m_μ are defined on the lattice and the dual lattice correspondingly. The term we add to the action is therefore

$$A_\theta = i\,\frac{N\theta}{2\pi}\sum_{r,R} f(r-R)\,(\theta_\mu(r) - 2\pi n_\mu(r))\,m_\mu(R) \tag{2.29}$$

where $f(r-R)$ is some short range function whose precise form does not affect the infra-red physics. To understand the coefficient in A_θ let us think of $f(r-R)$ as a δ-function. The coefficient in A_θ is chosen to ensure the periodicity of the action in θ. It is easy to check that under the transformation $\theta \to \theta+2\pi$ the full action (eqs.(2.13),(2.29)) goes to itself (change the summation variables (ℓ,m) to $(\ell-m,m)$). Note again that A_θ is of the form $A_\mu m_\mu$ i.e. the monopole carries electric charge $\theta/2\pi$ which is precisely Witten's result (eq.(2.27)).

Next, the steps leading to eq.(2-19) can be repeated with the new term eq.(2.29) added to the previous action.

$$\begin{aligned} Z = \sum_{\{\ell_\mu(r)\}}\sum_{\{m_\mu(R)\}} & \exp\Big[-\frac{2\pi}{g^2}\sum_{R,R'} m_\mu(R)\,G(R-R')\,m_\mu(R')\Big] \times \\ & \exp\Big[-\frac{N^2g^2}{2}\sum_{r,r'}\sum_{R,R'}\big(\ell_\mu(r) + \tfrac{\theta}{2\pi}f(r-R)\,m_\mu(R)\big)\,G(r-r')\big(\ell_\mu(r') + \tfrac{\theta}{2\pi}f(r'-R')\,m_\mu(R')\big)\Big] \times \\ & \exp\, iN\sum_{\mu,R} m_\mu(R)\,\Theta_{\mu\nu}(R-r)\,\ell_\nu(r) \end{aligned} \tag{2.30}$$

Several remarks are in order

1. If we ignore the short range function f, (which we'll do from this point on) the action is manifestly periodic in θ.
2. The parity transformations correspond to $\ell_\mu \to -\ell_\mu$, $m_\mu \to m_\mu$. The action is invariant under parity at $\theta = 0$ and $\theta = \pi$ only.
3. The Schwinger interaction term in eq.(2.30) does not include θ. This is related to the fact that the generalized Dirac condition is associated with this term. (Recall the Witten's argument that leads to eq.(2.27) and which is certainly consistent with Dirac's condition). For details on this important point see [4] and [10].

At this point we can repeat our naive energy entropy consideration for the action in eq.(2.30). Consider now a loop of length L carrying an electric and magnetic charge $(\ell+\frac{\theta}{2\pi}m, m)$. The estimate of the energy will be done again by neglecting all terms beside the self energy mass term.

$$\text{energy} \approx \left(\frac{2\pi^2}{g^2}m^2 + \frac{N^2g^2}{2}\left(\ell+\frac{\theta m}{2\pi}\right)^2\right) G(0)L$$

The estimate of entropy is just as before

$$\text{entropy} \approx L \log \mu \qquad (\mu \approx 7)$$

The energy entropy consideration gives

$$\frac{m^2}{T} + \left(\ell+\frac{\theta}{2\pi}m\right)^2 T \leq \frac{\log 7}{\pi G(0)} \equiv C \quad , \; T = \frac{Ng^2}{2\pi} \tag{2.31}$$

As before the condensation criterion eq.(2.31) defines the interior of an ellipse in the plan of the charge lattice. However now the points on the charge diagram, which correspond to possible excitations in our system, are tilted compared to the $\theta = 0$ case. The situation is depicted in Fig.3. The ellipse has e and h as principle axes. The ratio of the axis is T while the area of the ellipse is $\pi C/N$.

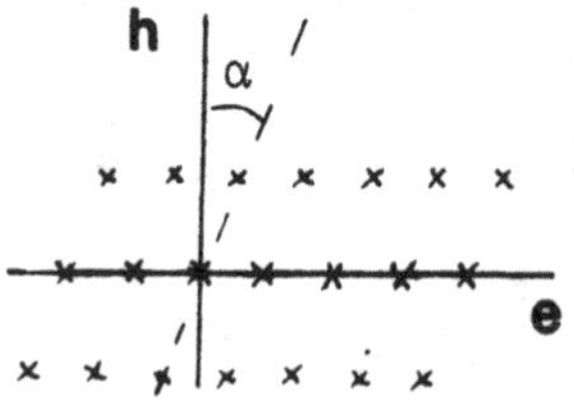

Fig.3 The lattice charge diagram for $\theta \neq 0$ $\left(\tan\alpha = \frac{\theta}{2\pi}\right)$.

For large N it is clear that no point will be inside the ellipse besides the origin unless T is very large or very small. We conclude that for intermediate coupling we are going to find a Coulomb phase. For small N there will always be points inside the ellipse i.e. no Coulomb phase. It is also clear that now a situation may arise where the condensate has both ℓ and m different from zero. This kind of phase is now referred to by 't Hooft as an oblique confinement phase. To demonstrate the situation choose $\theta = \pi$. The lattice charge diagram is shown in Fig.4 taken from ref. [4]. For $\theta=\pi$, $\ell+\frac{\theta}{2\pi}m = \ell+\frac{m}{2}$. The (ℓ,m) values corresponding to the three possible condensates drawn in Fig.4 are:

		(ℓ,m)
(2.32)	C	(1,0)
	M	(0,1)
	O	(-1,2)

The phase in which C condenses is a phase where electric charge ℓ condenses. This is a Higgs phase. The phase where M condenses corresponds to a condensation of magnetic charge i.e. it is a confining phase. The new type of phase is the phase where O condenses since O carries both ℓ and m different from zero.

Once we have established which is the condensate, we would like to understand the properties of this phase. In particular we would like to know which quantum numbers are confined and which quantum numbers are screened. Suppose we are in a phase in which the condensate carries both electric and magnetic quantum numbers (ℓ_c,m_c). This corresponds to a given point on the charge diagram. Let us draw the straight line which connects the origin to the condensate point. Following 't Hooft we'll refer to this straight

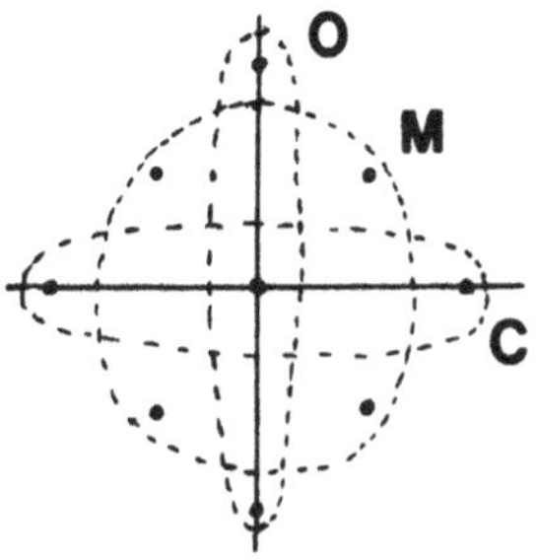

Fig 4. The lattice charge diagram for $\theta=\pi$.

line as the subspace of the condensate. All points on the charge diagram which lie on this subspace are characterized by

$$\ell'/m' = \ell_c/m_c \quad . \tag{2.33}$$ 14

It is straightforward to calculate the behaviour of the Wilson's loop corresponding to a particle carrying the charges (ℓ, m). It turns out that for particles which are outside the condensate subspace we observe an area law behaviour i.e. these quantum numbers are confined. Particles which reside in the condensate subspace show a perimeter law behaviour, i.e. they are screened. The string tension is proportional to $(m_c\ell - m\ell_c)^2$. We refer the readers to[4] for details. The physics behind this result is not difficult to understand. The subspace of the condensate is characterized by the fact that all of its points satisfy the generalized Dirac condition eq.(2.26) with the condensate for n = o. They have zero Dirac unit with respect to the condensate. Therefore if we introduce into the vacuum a particle-antiparticle pair with charges in this subspace there will be no string between them. All other points on the charge diagram satisfy the generalized Dirac condition with the condensate with non-vanishing unit η. This implies that particles with these quantum numbers will be confined.

<u>Summary</u>

1. All phases in a gauge theory can be characterized by designating the points on the charge diagram that develop vacuum expectation value, i.e. by designating the quantum numbers which condense.
2. The relative Dirac unit for all pairs of those points which condense must vanish.
3. All particles whose charge lie on the subspace of the condensate show short range interactions. Their gauge fields are screened by the Higgs mechanism (much in the same way that the photons "acquire a mass" in superconductivity).
4. All particles that have non vanishing Dirac unit with respect to the points in the condensate's subspace, behave like a monopole in a superconductor i.e. they are confined.
5. For the SU(2) case the maximal abelian subgroup is $U(1)$ and the subspace is a straight line through the origin. For the SU(N) case the maximal subgroup is $[U(1)]^{N-1}$ and the subspace is N - 1 dimensional linear space.
6. The confining phase corresponds to a condensate with $\ell = 0$, $m \neq 0$. The Higgs phase corresponds to a condensate with $\ell \neq 0$, $m = 0$. The Coulomb phase has no condensate i.e. $\ell = m = 0$.
7. In the presence of θ angle new phases of the oblique confinement type ($\ell \neq 0$ $m \neq 0$) appear. The oblique confining phases have some interesting and unexpected features which may be relevant to particle physics. We refer the reader to 't Hooft's recent paper [3] for details, possible scenarios and speculations.

III CHIRAL SYMMETRY REALIZATION - COMPOSITE QUARKS AND LEPTONS

In this part of the lectures we'll be interested in systems which involve both gauge and fermionic matter degrees of freedom. As indicated in the introduction there is now a vast number of questions which one would like to investigate. Here I'll concentrate on a particular one i.e. chiral symmetry realization. Once we introduce into the gauge theory (massless) fermions we automatically have a global flavour type of symmetry. The question of how this symmetry is realized in the vacuum is of central importance for any realistic or semi-realistic model. For example, within the framework of QCD this global symmetry is $SU_L(N_f) \times SU_R(N_f) \times U(1)$, where N_f refers to the number of flavours. If, as we believe, QCD describes correctly the hadronic world in which the pion is an (almost) Goldstone boson, then this symmetry must be broken to $SU_D(N_f)$ (the diagonal vector $SU(N_f)$). Also in theories with dynamical symmetry breaking (e.g. technicolor type of theories) the realization of chiral symmetries plays an essential role.

In this lecture I'll talk about models of composite quarks and leptons. The techniques and results are, however, applicable also for other cases of interest (QCD, models with dynamical symmetry breaking etc).

Are Quarks and Leptons Composite?

There are several reasons which make the idea of composite quarks and leptons attractive [12])

(i) Too many quarks and leptons - generation structure.

If we count the number of known quarks and leptons taking into account the color quantum number we have almost forty of them. It would certainly be more attractive if all of them were made of some other few more elementary degrees of freedom. Moreover, we all know that the spectrum of quarks and leptons reveals the structure of generations. Up to now we have seen (almost) three generations. Such a pattern of families usually hints toward the existence of some substructure.

(ii) Similarity between quarks and leptons

The known spectrum of quarks and leptons is organized into families. Within each family there is a remarkable similarity and connection between quarks and leptons. Both are doublet of the weak interactions and the sum of the charges within one generation must add up to zero in order to cancel anomalies. This hints again toward the possibility that quarks and leptons are made out of the same more fundamental degrees of freedom. If this is indeed the case then we may also understand in a simple way the charge quantization which is observed in nature.

(iii) Naturalness [13)]

This reason has to do with our theoretical prejudice. We are working within the framework of quantum field theory. Given the theory which may include some bare parameters, we should in principle be able to calculate all physical quantities of interest. What we believe is that physical results should be stable against small variation of the parameters in the theory. In particular if some physical masses turn out to be small, we would like it to be a natural result of the theory due to some symmetry and not a result of a fine turning (due to renormalization) of the bare parameters of the theory. By making particles composite one can improve on the range of naturalness. For example theories with elementary scalars (Higgs) are known to be unnatural (we disregard here the possibility of supersymmetry). In technicolor models one improves on the naturalness by making the Higgs composite.

Scale of Binding

If quarks and leptons are composite we would like to get the exponential limitations on the binding scale [12)]. There are two kind of experiments.

1. High momentum probes
   (i) High energy tests of QED
   (ii) Scaling in deep inelastic scattering
   (iii) e^+e^- annihilation

All these experiments result in bounds on the form factors of quarks and leptons.

2. Low momentum experiments
   (i) Anomalous magnetic moments $(g-2)$
   (ii) $\mu \to e\gamma$
   (iii) proton life time

The best unbiased (model independent) restriction on the binding scale Λ_H comes from g-2 experiments

$$\Lambda_H > (1-10)\ \mathrm{TeV}$$

Other experiments give $\Lambda_H \gtrsim 1000\ \mathrm{TeV}$
Down to distances $\sim 1/\Lambda_H$ quarks and leptons appear to thave no substructure.

Massless Composite Fermions

Relative to the binding scale Λ_H all known quarks and leptons appear to be massless. In view of our previous discussion, we should start with the underlying theory of the constituents of quark and lepton (here after to be called preons) and account in a natural way for the appearance of composite massless bound states. Normally we expect all masses to be of the order of the scale Λ_H. To get massless bound states we need a symmetry which will protect the bound states from being massive. The only symmetry which we know to protect the masslessness of fermionic states is chiral symmetry. We would therefore start with an underlying theory which has chiral symmetry built in. The problem that we face is therefore, under which conditions enough chiral symmetry survives spontaneous breakdown to ensure the appearance of zero mass fermionic states in the spectrum. The 't Hooft anomaly constraint equations give a necessary condition for this to happen [14].

Framework - Restrictions on Model Building

The underlying theory of the preons is going to be a gauge theory based on a group G_{HC} which we'll refer to as the hypercolor group. We shall assume that this theory has the property of confinement. The degrees of freedom which we have are gauge bosons and zero mass preons to ensure chiral symmetry. We shall take all the preons to be spin $\frac{1}{2}$ left handed. The preons are assigned to representations $R_1, R_2, \ldots, R_N$ of G_{HC} with multiplicities $N_1, N_2, \ldots, N_n$. The two requirements that we impose on the choice of representations are:

(i) There should be no hypercolor anomalies.
(ii) The theory should be assymptotically free.

Once we have assigned the representations the theory has automatically a global hyperflavor symmetry

$$G^*_{HF} = U(N_1) \otimes \quad \otimes U(N_n)$$

which is broken by hypercolor instantons' effects to

$$G_{HF} = SU(N_1) \otimes \quad \otimes SU(N_N) \times [U(1)]^{N-1} \times Z_N \tag{3.1}$$

This symmetry is realized inevitably in the fundamental representation of each SU(N). The generators of the flavor symmetry are integrals over gauge invariant currents J^a_μ which are exactly conserved.

The tree point function of the hyperflavor currents will have a c-number anomaly proportional to $d_{abc} = Tr(\lambda_a \{\lambda_b, \lambda_c\})$ where λ_a is the generator of G_F in a given representation.

$$\Gamma^{abc}_{\nu\lambda\mu} = \int d^4x\, d^4y\, e^{i(k_1 x + k_2 y)} \langle 0 | T\, J^a_\nu(x)\, J^b_\lambda(y)\, J^c_\mu(0) | 0 \rangle \tag{3.2}$$

The c-number anomaly relation is [15]

$$q^\mu \Gamma^{abc}_{\nu\lambda\mu} = 4\pi^2 d^{abc} \varepsilon_{\alpha\beta\nu\lambda} k_1^\alpha k_2^\beta \tag{3.3}$$

It is important to recall that this result which is obtained by computing the lowest order triangle diagram is exact to all orders in perturbation theory. This is known as the Adler Bardeen no renormalization theorem [16]. We shall assume that this relation survives confinement. An argument to support this assumption is that in the usual way one derives the anomaly relation by considering the triangle diagram, it is the large momenta and hence the short distances which are important. Confinement has to do, on the other hand, with the large distance behaviour of the theory. We have, therefore, an exact result (eq.(3.3)) which must be saturated by the physical states in our system. In a confining theory the anomaly equation must be saturated by the massless bound state spectrum [17] (singlets under G_{HC}). Three different situations may occur.

(i) Spontaneous symmetry breaking of G_{HF}. In this case we have Goldstone bosons which saturate the anomaly relation.

(ii) Chiral symmetry is not broken. Then necessarily massless [18] composite spin 1/2 [19] fermions should exist and they saturate the anomaly relation.

(iii) Combination of the two, i.e. the chiral symmetry is partially broken.

We are of course interested in case (ii). The hypercolor singlet, spin 1/2 fermions would be identified with quarks and leptons. The fermionic bound states fall into irreducible representations of G_{HF} and one can show [20] that they couple to flavor current via their charges. The idea is to identify $SU_C(3)\times SU(2)\times U(1)$ (color and electroweak interactions) within G_{HF} and gauge it. In this way the standard theory becomes the effective low energy theory of the massless fermionic bound states (quarks and leptons). Note that at scale Λ_H where hypercolor is strong we assume that all other interactions are negligible due to asymptotic freedom. We, therefore, investigate the formation of composite fermions at distances Λ_H^{-1} in the fully symmetric theory. We turn on color and weak interactions at larger distances and discuss then their effects. A particular important question is whether they can account for a small (compared to Λ_H) mass of quarks and leptons [21].

The 't Hooft constraint equations give us a necessary condition for chiral symmetry not to be broken. They arise from the demand that if chiral symmetry is not broken the anomaly on the preon level must be equal to the anomaly of the massless composite fermions [5].

$$(3.4) \qquad \sum_i \dim(R_i)\, tr_F\, \lambda_a \{\lambda_b \lambda_c\} = \sum_B tr_B\, \lambda_a \{\lambda_b \lambda_c\}\, \ell_B$$

where ℓ_B is the number of left handed minus the number of right handed composite fermions in the representation B of G_{HF}. Note that from its definition ℓ_B must be an integer. There are three types of anomaly equations that one may have according to which group the flavor current belongs $SU(N_i) \otimes SU(N_j) \otimes SU(N_k)$, $SU(N_i) \times SU(N_i) \times U(1)_j$ or $U(1)_i \otimes U(1)_j \otimes U(1)_k$.

Decoupling Requirements

Besides the 't Hooft anomaly constraint equations which provide a necessary condition for chiral symmetry not to be broken, there are other constraints known as the decoupling requirements. We shall distinguish between two possible scenarios.

1. Persistence mass condition [5]

When some chiral symmetry is broken in the underlying Lagrangian say by giving intrinsic mass to some preon $m_H \bar{\psi}_L \psi_R$, the flavor symmetry is broken $G_{HF} \to G'_{HF}$. The remaining chiral symmetry G'_{HF} should not prevent the composite fermions containing any number of massive preons from acquiring a mass. That means, that with respect to the remaining chiral symmetries G'_{HF}, the appropriate composite fermions containing the massive preon must appear in representations that are chiral pairs. This will enable us to form chiral invariant quadratic mass terms in the effective low energy Lagrangian. Equivalently we can describe the situation in the following way. Imagine giving to a certain preon a mass, thus breaking the symmetry G_{HF} down to G'_{HF}. The fermionic bound states which were classified according to the representations of G_{HF} can be still classified under G'_{HF}. We expect all bound states containing the massive preon, to become massive. (It must be eventually the case as m_H is increased due to the Appelquiste Carazonne Symanzik decoupling theorem [22]). This is possible only if every left handed bound state has a right handed partner in the same representation of G'_{HF} (parity doublet), so that together they can form a Dirac mass term in the effective low energy theory. Therefore for any representation $r' \in G'_{HF}$

$$\sum_{r \ni r'} \ell(r) = 0 \quad , \quad r \in G_{HF} \, , \, r' \in G'_{HF} \tag{3.5}$$

where $\ell(r)$ are the indices defined after eq. (3.4). We have to find a simultaneous solution to eq. (3.4) and (3.5) with the constraint that the $\ell(r)$'s should come out integers. Recently it has been shown that in vector like gauge theories with f massless preon flavours, the decoupling condition eq.(3.5) requires the massless fermionic bound states to form irreducible representations of the SU(f/f) superalgebra [23].

2. Phase Transitions [24]

In the persistence mass condition scenario we assumed that composites containing any massive preons should become automatically massive for any value of m_H (preon mass).
However due to the remaining symmetry G'_{HF}, the composites may remain massless even when some preons contained in them are given small mass. Of course the decoupling theorem [22] ensures us that as $m_H \to \infty$ the appropriate composites should also become very massive. As the mass of the preon is sent to infinity the theory may undergo a phase transition. The residual symmetry G'_{HF} should be broken spontaneously to allow the composites containing the massive preon to become massive. Below the transition point there can exist massless composites containing preons of small enough mass.

Solutions to the Anomaly constraint Equations

1. Vector like theories [5,23]

This class of theories is known also as left-right symmetric theories. For each left handed preon in the representation R of G_{HC}, there exists a left handed preon in the conjugate representation R* of G_{HC} (which is equivalent to a right handed preon in the same representation R). As an important example take $G_{HC} = SU(N)$ with preons belonging to the fundamental representation of G_{HC}.
There will be f left handed preons and f right handed preons.
The global hyperflavor symmetry is therefore $G_{HF} = SU_L(f) \otimes SU_R(f) \otimes U(1)$.
Note that those are QCD like theories. For N = 3 if you think about the preons as the quarks, and the composites as mesons and baryons; then you are dealing exactly QCD, addressing the problem of chiral symmetry breaking. Recall that if t'Hooft anomaly equations are not satisfied then we necessarily have spontaneous breakdown of chiral symmetry. It turns out that the anomaly equations themselves (eq. (3.4)) do admit solutions except

for the case in which f is a multiple of 3. However when we take into account the decoupling conditions as well, no solution exists unless $f = 2$. Both the persistence mass condition and the phase transition scenario support the idea that chiral symmetry in QCD like theories is spontaneously broken.

In the large N limit it was proved that chiral symmetry is broken presumably to the diagonal SU(f) subgroup [25]. The physics behind this result is quite easy to understand. No massless composite fermions (baryons) exist in the large N limit, hence the anomaly equations must be saturated by zero mass Goldstone bosons i.e. chiral symmetry must be spontaneously broken. A nice proof of the non existence of zero mass composite fermions (baryons in the case of QCD) was given by S. Coleman. Consider the composite fermion number (baryon number) two point function depicted in Fig. 5. In the large N limit the leading contribution is of order N. If massless composite fermions exist they would couple at $q^2 = 0$ via their charge which is equal to N (the composite fermion is made out of N preons). Hence the contribution from the zero mass composite fermions intermediate state would be proportional to N^2 . Since the imaginary part is positive definite no cancellation of this contribution can occur. We have arrived to a contradiction whose resolution is that no zero mass composite fermions exist in the large N limit. Note that the above arguments cannot exclude solutions without chiral symmetry breaking based on the group $G_{HC} = SU(N) \otimes SU(N)$. Such solutions which satisfy all the necessary constraints have been proposed recently [23].

2. Chiral Theories [26), 27)]

For the left right symmetric theories the strongest constraints came from the decoupling requirement. For chiral gauge theories in which gauge invariant mass terms cannot be usually written there are less constraints and therefore more solutions. A general method for constructing solutions to the 't Hooft anomaly constraint equations has been developed by us [26]. Let us recall the requirements that the underlying preon theory must satisfy:

1. Since it is a gauge theory based on the group G_{HC} , there should be no hypercolor anomalies.

2. All currents belonging to G_{HF} are exactly conserved even in the presence of the gauge fields. In particular no U(1)

Fig. 5. The baryon number two point function.

current which belongs to G_{HF} is allowed to have any hypercolor anomalies.

3. For currents belonging to G_{HF} we demand that the 't Hooft constraint equations would be satisfied. This ensures that the anomaly on the preon's level is equal to the anomaly on the composite level.

The method itself consists of the following steps:

1. Consider any group or superalgebra G and look at all subgroups of G which have the form A x B (A and B need not be simple). We'll try to check whether a solution exists in which we identify A with G_{HC} and B with G_{HF}.

2. Choose a complex representation r of G which is anomaly free (has a vanishing d symbol). If no such representation of G exists then no solution based on subgroups of G exists.

3. Decompose r under $A \times B \equiv G_{HC} \times G_{HF}$

$$r = \sum (c, f) \qquad c \in G_{HC}, \; f \in G_{HF} \tag{3.6}$$

The representations which appear in the decomposition can be separated into two sets, i.e. those of the form $(1, f)$ (singlets under G_{HC}) and those of the form (c, f) with $c \neq 1$. If any of these sets is empty no solution based on this subgroup of G exists. If both sets are not empty then a solution may exist in which the preons are assigned to the representations (c, f), and the composites (which must be singlets under G_{HC}) to $(1, \bar{f})$ ($\bar{f}$ is the representation conjugate to f).

$$\text{preons} = \sum_{\substack{(c,f) \\ c \neq 1}} (c, f) \qquad \text{composites} = \sum_{f} (1, \bar{f}) \tag{3.7}$$

Namely, if one starts with an underlying theory based on a gauge group G_{HC} and a global hyperflavor group G_{HF} with preons residing in the representations (c, f) and composite fermions in $(1, \bar{f})$, then this set of preons and bound states satisfies the anomaly constraint equations.

Not all solutions are of course physically acceptable. Starting with the underlying gauge theory the hyperflavor group by definition (eq.(3.1) contains only SU(n) and U(1) groups, with preons in the fundamental representation of SU(n). Solutions which do not satisfy it should be rejected (although they may be of interest for cases in which the chiral symmetry is only partly broken).

The proof that the procedure described above indeed gives a solution is very simple. Since we have started with an anomaly free representation

$$(3.8) \qquad \sum_{\substack{(c,f)\\ c\neq 1}} d_{abe}(c,f) + \sum_{f} d_{abe}(1,f) = 0$$

where d is the d-symbol of the appropriate representation. The indices a,b,e refer to generators of G. Three particular cases of eq. (3.8) can be considered.

(i) For a,b,e which correspond to the hypercolor currents the second term is zero (singlet under hypercolor). Equation (3.8) then guarantees that the gauge theory based on G_{HC} with preons in representations (c,f) is anomaly free.

(ii) For an index a in the hypercolor group and b,e indices of hyperflavor currents, again the second term in eq.(3.8) vanishes. The first term then guarantees that all currents in G_{HF} are exactly conserved (no hypercolor anomaly).

(iii) For a,b,e hypercolor indices the second term in eq.(3.8) can be taken to the right hand side and the sign swallowed by going from the representation $(1,f)$ to its conjugate $(1,\bar{f})$. The equation thus obtained guarantees that the anomaly constraint equations are satisfied (recall eq.(3.7).

Using this method a general survey had been carried out. The interested reader is referred to ref. 26 for details. Many solutions to the 't Hooft anomaly constraint equations have been found. Unfortunately most of them were uninteresting from the physical phenomenological point of view. A prerequist from a realistic solution is that the number of composite massless fermions (to be identified with quarks and leptons) should be larger then the number of preons. Moreover the hyperflavor group G_{HF} should be large enough to include $SU_c(3)\times SU(2)\times U(1)$ (and perhaps also some family group). The color SU(3) and electroweak SU(2) x U(1) should be gauged in the second stage as discussed before. An interesting semirealistic model which was studied in detail can be found in ref. 28. Finally let us recall that even if realistic solution is found, one would have then to address the difficult problem of giving small masses to quarks and leptons and account for their mass hierarchy [21). Extra constraints on composite models come from proton decay [30)]. Also the question of whether gauge bosons can be composite is very intriguing [29)]. Certainly much more work has to be carried out to see whether the whole idea of composite quarks and leptons proves itself viable.

Acknowledgement

I would like to thank the organizers for their warm hospitality and the nice atmosphere which prevailed during the Summer Institute.

I would like to thank Dr. E. Rabinovici for giving us the use of the results of ref. 4.

References

1. G. 't Hooft, Nucl. Phys. B 153, 141 (1979)
   G. 't Hooft, lectures given at the Schladming Winter-school 1980
   G. 't Hooft, lectures given at the 21 Scottish Summer-school 1980
   G. 't Hooft, Calt-68-819 (1981)
2. D. Horn, M. Weinstein, S. Yankielowicz, Phys. Rev. D 19,3715 (1979)
   S. Elitzur, R. Pearson, J. Shigemitsu, Phys. Rev. D19,3698 (1979)
   A. Guth, A. Ukawa, P. Windey, Phys. Rev. D 21, 1013 (1980)
3. G. 't Hooft, CALT-68-819 (1981)
4. J.L. Cardy, E. Rabinovici, Hebrew University preprint RI-521
5. G. 't Hooft, Proceedings of the Cargese Summer Institute (1979)
   Y. Frishman, A. Schwimmer, T. Banks, S. Yankielowicz, Nucl. Phys. B 177, 157, (1980)
6. P.A.M. Dirac, Proc. R. Soc. A 133, 60 (1931)
7. K.G. Wilson, Phys. Rev. D10, 2445 (1974)
8. For a review see F. Englert, Cargese Summer Institute (1977)
9. T. Banks, R. Meyerson, J. Kogut, Nucl. Phys. B 129, 493 (1977)
   R. Savit, Phys. Rev. Lett. 39, 55 (1977)
   M. Peskin, Ann. Phys. (N.Y.) 113, 121 (1978)
10. E. Witten, Phys. Lett. 86 B, 283 (1979)
11. J. Schwinger, Phys. Rev. 144, 1087 (1966); 173, 1536 (1968)
    D. Zwanziger, Phys. Rev. 176, 1480 (1968)
12. For a review see e.g. H. Harari, lectures given at SLAC Summer Institute 1980
13. K. Wilson, unpublished
    L. Susskind, Phys. Rev. D 20, 2619 (1979)
    G. 't Hooft, ref. 5
    T. Vettman, University of Michigan preprint 1980
14. G. 't Hooft, ref. 5
15. S.L. Adler, Phys. Rev. 177, 2426 (1969)
    J.S. Bell, R. Jackiw, Nuovo Cimento 60 A, 47 (1969)
16. S.L. Adler, W.A. Bardeen, Phys. Rev. 182, 1517 (1969)
17. G. 't Hooft, ref. 5
    Y. Frishman, A. Schwimmer, T. Banks, S. Yankielowicz, Nucl. Phys. B 177, 157, (1980)
    A.D. Dolgor, V.I. Zakharov, Nucl. Phys. B 27, 525 (1971)
18. G. 't Hooft, ref. 5
    Y. Frishman, A. Schwimmer, T. Banks, S. Yankielowicz, ref. 17

19. G. 't Hooft, ref. 5
S. Weinberg, E. Witten, Phys. Lett. 96 B, 59 (1980)
20. Y. Frishman, A. Schwimmer, T. Banks, S. Yankielowicz, ref. 17
21. S. Weinberg, Texas preprint 1981, Phys. Lett 102 B, 401 (1981)
S. Dimopolous, L. Susskind, ITP-681, Stanford preprint (1980)
22. T. Appelquist, I. Carrazone, Phys. Rev. D 11, 2856 (1975)
K. Symanzik, Comm. Math. Phys. 34, 7 (1973)
23. A. Schwimmer, RU-81-49, Rutgers University preprint 1981
24. S. Weinberg, ref. 21
I. Bars, S. Yankielowicz, Phys. Lett. 101 B, 159 (1981)
I. Bars, YTP81-19, Yale preprint (1981)
J. Preskill, S. Weinberg, Phys. Rev. D 24, 1059 (1981)
25. S. Coleman, E. Witten, Phys. Rev. Lett. 45, 120 (1980)
E. Witten, Nucl. Phys. B 156, 269 (1979)
G. Veneziano, Nucl. Phys. B 159, 213 (1979)
26. T. Banks, S. Yankielowicz, A. Schwimmer, Phys. Lett. 96 B, 67 (1980)
27. S. Dimopolous, S. Raby, L. Susskind, Nucl. Phys. B 173, 208, (1980)
28. I. Bars, S. Yankielowicz, ref. 24
I. Bars, YTP 81-19, Yale preprint (1981)
29. S.D. Bjorken, Ann. Phys. 24, 174 (1963)
T. Eguchi, Phys. Rev. D 14, 2755 (1976)
H. Harari, N. Seiberg, Phys. Lett. 98 B, 269 (1981); Phys. Lett. 100 B, 41 (1981)
T. Banks, A. Zaks, Nucl. Phys. B 184, 303 (1981)
D. Amati, R. Barbieri, A.C. Davis, G. Veneziano, Phys. Lett. 102 B, 408 (1981)
S. Weinberg, E. Witten, Phys. Lett. 96 B, 59 (1980)
30. R. Casalbuoni, R. Gato, Univ. de Genève preprint (1981)

UNIFICATION OF GAUGE AND GRAVITY INTERACTIONS FROM COMPOSITENESS

G. Veneziano

CERN, Geneva, Switzerland

Within our present understanding, it is conceivable that all elementary interactions can be described by gauge theories. This common structure opened the way towards unification, first with the successful electroweak theory of Glashow-Weinberg-Salam and then by its more speculative merging with the SU(3) theory of strong interactions in the so-called grand unified theories[1] (GUTs). The energy scale at which this GUT unification takes place, M_{GUT}, is very large, ranging between 10^{14} and 10^{17} GeV depending on the details of the actual theoretical scheme.

In spite of their doubtless appeal and partial success, GUTs are still facing hard, unresolved problems (origin of families, hierarchy problem, etc.). Even more frustrating has been the impossibility so far to include gravity in a unified scheme. Not only is gravity a gauge theory (of a space-time rather than of an internal symmetry), it is also expected to become strong and hence non-negligible at a scale (of order 10^{19} GeV) which is embarrassingly close to M_{GUT} itself.

On the other hand, whereas in usual GUTs electroweak and strong interactions are renormalizable interactions becoming equally weak at M_{GUT}, the gravitational interaction shows such a steady increase of strength with energy that its meaning as a quantum theory is itself doubtful. This difference is of course an obvious obstacle for unifying gravity with the other interactions. Another more formal difference is that, in the conventional approach, the gauge connection is considered as a fundamental field whereas in gravity it is given in terms of the fundamental metric or vierbein fields.

In this talk I shall propose and discuss a theoretical framework in which all gauge symmetries are introduced in the same way, i.e., in terms of fundamental fermionic fields only. Gauge bosons, as well as gravitons, will be composite and their interactions will be automatically unified (and strong) at the same scale, the inverse radius of compositeness. If we exclude gravity from the picture, this approach reduces to the one of Ref. 2. My presentation here will follow closely the papers written by Amati and myself on the subject[3].

This unconventional unification scheme leads to the usual framework in a "low energy" regime to be carefully defined later. Indeed, as we move down from the unification mass, the effective low energy interactions will be the ones dictated by gauge invariance and the principle of equivalence, by the presence of the corresponding dynamically generated gauge bosons and by dimensional arguments. Conventional gauge theories and gravity appear then as phenomenological low energy approximations. Away from that regime, our theory will differ from the conventional ones not even allowing, for instance, an unambiguous definition of a space-time metric. This last point clearly indicates the difference between our approach and those which try to generate quantum gravity from matter fluctuations in a background metric field[4]. Let us remark that, as discussed in Ref. 2, the vector gauge theories appearing in the low energy regime have the property of going towards unification at higher energies in a crescendo. This sort of asymptotically non-free GUT has been proposed and shown to be satisfactory on pure phenomenological grounds in Refs. 5 and 6. For us this sort of unification is an automatic consequence of gauge field compositeness.

In the preceding discussion I have taken the attitude that the gauge symmetries observed at low energies ($\lesssim$ 100 GeV) are those relevant for our unification procedure. One could take the alternative attitude that the gauge symmetry that unifies with gravity at M_{Planck} is the GUT interaction itself. In this case GUT would appear as an effective "low energy" theory valid at $E \approx M_{GUT} << M_{P\ell}$ which could further break at lower energies down to $SU(3) \otimes SU(2) \otimes U(1)$ following one of the conventional GUT schemes. This scheme would necessitate several GUT families above M_{GUT} in order to reach a small unified gauge coupling at M_{GUT}. This idea is perhaps less appealing in our view, but shows the possibility of the co-existence of two very different unification mechanisms.

The outline of the talk is as follows. I shall start by constructing actions invariant under a set of gauge transformations on the basis of only fundamental fermions. To avoid internal indices (colour, flavour, etc.) I shall limit the treatment to Lorentz and U(1) gauge invariances (i.e., gravitation and QED), (see Ref. 2 for inclusion of other gauge symmetries). After having written the invariant action, I shall discuss a strategy to analyze it through the introduction of a suitable set of auxiliary fields. Next I shall discuss the vacuum properties, associated with homogeneous classical solutions. They are shown to generate a scale Λ through a spontaneous breaking of local Lorentz and general co-ordinate transformations. This scale, which appears as a vacuum property, represents the only dimensional quantity of the theory.

Subsequently, I shall outline the computation of the quadratic fluctuations around the previous stationary field configuration up to second order derivatives. This will allow us to recognize induced mass and kinetic terms of the eigenmodes. A light sector (massless bosons and arbitrarily light fermions) is identified, together with a heavy sector (masses of order Λ). If the heavy eigenmodes are not excited - as expected when all energies are much smaller than Λ - the remaining light modes are shown to represent a spinor, a photon and a graviton described by an effective action which is the usual Einstein-Dirac-Maxwell action with the Newton constant and the electric charge given in terms of Λ. This relation can be described by saying that Λ is to be identified with both the Planck mass and the Landau pole position. At the end, I shall comment briefly on how this theory differs from the conventional one in the short distance domain where one can hope it solves some of the well-known difficulties of quantum gravity.

2. We begin by writing, in terms of a single spinor field ψ, an action invariant under local U(1) and Lorentz transformations. In their infinitesimal form these read*):

$$\psi(x) \to (1 - i\alpha(x))\psi(x) \; ; \quad U(1) \tag{1a}$$

$$\psi(x) \to (1 - \frac{i}{4}\, \sigma_{\beta\gamma}\omega^{\beta\gamma}(x))\psi(x) \; ; \quad O(3,1) \tag{1b}$$

Besides, we shall write the action as a space-time integral of a scalar density thus enforcing invariance under general co-ordinate transformations $x_\mu \to x'_\mu(x)$. Our first talk will be to construct a covariant derivative. Defining

$$D_\mu = \partial_\mu + A_\mu + \sigma^{\beta\gamma}A_{\mu,\beta\gamma} \quad , \tag{2}$$

$$\begin{aligned} A_\mu &= \frac{1}{8}\, \mathrm{Tr}(H^{-1}\overleftarrow{H}_\mu - \overrightarrow{H}_\mu H^{-1}) \\ A_{\mu,\beta\gamma} &= \frac{1}{16}\, \mathrm{Tr}(\sigma_{\beta\gamma}H^{-1}\overleftarrow{H}_\mu - \sigma_{\beta\gamma}\overrightarrow{H}_\mu H^{-1}) \end{aligned} \tag{3}$$

*) We adopt the normalizations and conventions of J.D. Bjorken and S.D. Drell, Relativistic Quantum Fields, McGraw-Hill (1965). The usual contraction of Lorentz indices through $\eta_{\alpha\beta} = (1,-1,-1,-1)$ is tacitly understood.

and the Dirac index matrices

$$H = \psi\bar{\psi} \ , \ \overleftrightarrow{H}_\mu = \psi \overleftrightarrow{\partial}_\mu \bar{\psi} \tag{4}$$

it is easy to check that D_μ is a good covariant derivative under the transformations (1). This means that $D_\mu \ \psi(x)$ transforms as $\psi(x)$ in (1).

As a consequence, the composite operator

$$W_{\alpha\mu}(x) = \bar{\psi}(x)\frac{i}{2}\left[\gamma_\alpha, D_\mu\right]\psi(x) \equiv \frac{i}{2}\ \bar{\psi}(x)\gamma_\alpha \vec{D}_\mu \psi(x) - \frac{i}{2}\ \bar{\psi}(x)\overleftarrow{D}_\mu \gamma_\alpha \psi(x) \tag{5}$$

transforms as a usual vierbein, i.e., as a set of four four-vectors (one for each index α) under general co-ordinate transformations and, for each μ, as a vector under local Lorentz transformations. Therefore $\det W_{\alpha\mu}$, where the determinant refers to the 4×4 matrix of indices $\alpha\mu$, is a scalar density. This can still be multiplied by a scalar invariant which, due to the lack of a space-time metric tensor to saturate space-time indices, can only be an arbitrary scalar function $V(H)$. We are thus led to define

$$A_f = \int d^4x(\det W_{\alpha\mu})V(H) \tag{6}$$

as the most general relativistic and U(1) invariant action of our single spinor system.

Let us notice that even if non-polynomial in ψ and $\bar{\psi}$ [due to H^{-1} in Eqs. (3)] A_f is polynomial of fourth order in the derivatives and only first order in the derivative with respect to any space-time co-ordinate. The four derivatives compensate the dimension of d^4x so that ψ is dimensionless and so is H. Thus $V(H)$ and therefore A_f are free of any dimensional parameters. We shall see that the arbitrarity of $V(H)$ will not jeopardize the possibility of analyzing the consequences of (6). $V(H)$ will indeed enter into the condition for the minimum of A_f in terms of H. As we shall see, the existence of a minimum will be important while its location (arbitrary for arbitrary V) will be irrelevant. Further details of V will influence the spectrum of the theory but in a way which will not affect its basic physical aspects.

We shall be interested in extending the single spinor case discussed up to now by introducing possible replicas, i.e., different fields ψ_i transforming analogously under the local transformation (1). In so doing we may assign a different weight K_i to the contribution of the ith replica to the definition (4) of H that enters into the definition (2), (3) of the covariant derivative. This introduces therefore a set of dimensionless parameters K_i into the invariant action (6) where now

$$W_{\alpha\mu} = \frac{i}{2} \sum_i \bar{\psi}_i [\gamma_\alpha, D_\mu] \psi_i \tag{7}$$

D_μ being given by (3) in terms of

$$H = \sum_i K_i \psi_i \bar{\psi}_i \ , \quad \overleftrightarrow{H}_\mu = \sum_i K_i \psi_i \overleftrightarrow{\partial}_\mu \bar{\psi}_i \tag{8}$$

We shall actually use this extension in two ways. One will be to use N equal weights (K_i=1, i=1, ..., N) in order to define a 1/N expansion. The other will be to enforce a Pauli-Villars type of regularization.

The quantum theory we are considering implies the functional integration over the fundamental fields ψ_i, $\bar{\psi}_j$ of the exponential of the action A_f of Eq. (6) to which source terms $\bar{j}_i\psi_i + j_i\bar{\psi}_i$ are tacitly added. This is obviously a complex task because of the high non-linearity of A_f. Our strategy is to introduce as many auxiliary bosonic fields as needed in order to render A bilinear in ψ_i, $\bar{\psi}_i$ and to perform next the fermion functional integral. This leads us to a new action which, though equivalent to the original one, is written only in terms of the auxiliary fields.

For this purpose we introduce four functional Lagrange multipliers $\xi^{\mu\alpha}$, $\lambda^{\mu,\beta\gamma}$, λ^μ and ϕ needed to enforce the definitions of $W_{\alpha\mu}$, $A_{\mu,\beta\gamma}$, A_μ and H as given in Eqs. (3), (7) and (8)*). The action written in terms of the original fermions <u>and</u> of the auxiliary bosonic fields, reads:

$$\begin{aligned} A_{f,b} = \int d^4x \, \det W\{ & [V(H) - \xi^{\mu\alpha} W_{\alpha\mu} - \lambda^{\mu,\beta\gamma} A_{\mu,\beta\gamma} - \lambda^\mu A_\mu + \mathrm{Tr}(H\phi)] + \\ & + \sum_i \bar{\psi}_i (\xi^{\mu\alpha} \tfrac{i}{2} [\gamma_\alpha, (\partial_\mu + A_\mu + \sigma_{\beta\gamma} A_{\mu,\beta\gamma})] - K_i \phi + \\ & + \frac{K_i}{4} H^{-1} \overleftrightarrow{\partial}_\mu \lambda^\mu + \frac{K_i}{16} (\sigma_{\beta\gamma} H^{-1} \overleftarrow{\partial}_\mu - H^{-1} \sigma_{\beta\gamma} \overrightarrow{\partial}_\mu) \lambda^{\mu,\beta\gamma}) \psi_i \} \end{aligned} \tag{9}$$

The integration over ψ_i, $\bar{\psi}_i$ is now formally straightforward. However, in order to give meaning to the tr log obtained in that way, an ultra-violet regularization procedure must be introduced. It is crucial to do this while respecting all the invariances of the theory. We shall adopt a Pauli-Villars (PV) prescription by choosing replicas characterized by parameters K_i in Eq. (8) and by PV parameters C_i satisfying

$$\sum_i C_i K_i^{2n} = 0, \quad n = 0, 1, \ldots \tag{10}$$

---

*) Notice that ϕ and H are dimensionless matrices in Dirac space, that $W_{\alpha\beta}$, $A_{\mu,\beta\gamma}$, A_μ have dimensions of a mass and that $\xi^{\mu\alpha}$, $\lambda^{\mu,\beta\gamma}$ and λ^μ have dimension m^{-1}.

extending the usual PV conditions. Recall that, as usual, negative C_i represent PV ghosts. We shall discuss later the physical relevance of the regularization parameters.

Let me stress that in order to regularize the integration over the fermion fields I had to introduce into the theory a set of dimensionless parameters satisfying a set of conditions. The important matter, nevertheless, is that all invariances of the theory have been preserved. In particular, our regularization has not broken conformal invariance unlike what inevitably happens when mass and kinetic terms in the action have the same degree of homogeneity in ψ, $\bar{\psi}$.

The condition (10) allows one to give a meaning to the formal expression arising from integration over the fermion fields and hence to the effective bosonic action A_b defined by

$$\exp A_b = \int \prod_i \mathcal{D}\psi_i \mathcal{D}\bar{\psi}_i \exp A_{f,b} \tag{11}$$

Using (9) for $A_{f,b}$ we immediately obtain

$$\begin{aligned} A_b = \int d^4x \det W \Big[& V(H) - \xi^{\mu\alpha} W_{\alpha\mu} - \lambda^{\mu,\beta\gamma} A_{\mu,\beta\gamma} - \lambda^{\mu} A_{\mu} + \mathrm{Tr}(H\phi) \Big] + \\ & + \sum_i C_i \,\mathrm{tr}\log \{ \xi^{\mu\alpha} \tfrac{i}{2} [\gamma_\alpha, (\partial_\mu + A_\mu + \sigma_{\beta\gamma} A_{\mu,\beta\gamma})] - K_i \phi + \\ & + \tfrac{K_i}{4} H^{-1} \overleftrightarrow{\partial}_\mu \lambda^\mu + \tfrac{K_i}{16} (\sigma_{\beta\gamma} H^{-1} \overleftarrow{\partial}_\mu - H^{-1} \sigma_{\beta\gamma} \overrightarrow{\partial}_\mu) \lambda^{\mu,\beta\gamma} \} \end{aligned} \tag{12}$$

We have therefore translated our initial action, written in terms of only fermions, into the action (12) written in terms of only bosons. This language will be better suited for analyzing the spectrum of the theory. Before doing that, let us stress that among the variety of boson fields we are left with, none has the properties of a metric field. We have two fields $W_{\alpha\mu}$ and $\xi^{\mu\alpha}$ which could be associated with lower and upper index vierbeins and could be used to define space-time metric tensors such as $W_{\alpha\mu}W_{\alpha\nu}$ and $\xi^{\mu\alpha}\xi^{\nu\alpha}$. However, these two tensors cannot be identified with $g_{\mu\nu}$ and $g^{\mu\nu}$ unless one is the inverse of the other. We thus see that the possibility of defining a metric needs the identification of W with ξ^{-1} up to a proportionality constant. This is not of course the case in our theory since W and ξ are independent fields. We shall be able nevertheless to determine an asymptotic kinematical regime in which that identification will be correct, thus allowing the definition of a space-time metric as an approximate concept.

3. We shall attempt the analysis of our action (12) written in terms of auxiliary bosonic fields by a semi-classical treatment, which consists of finding stationary points of the action and expanding around them. We note that this procedure becomes justified as a 1/N expansion if N identical replicas of each fermion plus regulator families are considered. A factor N then multiplies the tr log term of (12) playing the role of the $1/\hbar$ factor of the other terms in the action.

Let us now discuss the stationarity condition which implies the vanishing of the first derivatives of A_b with respect to all auxiliary fields. The only first order functional derivative not involving the tr log in (11) is

$$\frac{\delta A_b}{\delta W_{\alpha\mu}} = \det W\{\xi^{\mu\alpha} - (W^{-1})^{\mu\alpha}(V(H) - \xi^{\mu\alpha}W_{\alpha\mu} - \lambda^{\mu}A_{\mu} - \lambda^{\mu,\beta\gamma}A_{\mu,\beta\gamma} + \mathrm{Tr}(\phi H))\} \tag{13}$$

Its vanishing gives a proportionality between $\langle\xi\rangle$ and $\langle W\rangle^{-1}$ so that neither $\langle\xi\rangle$ nor $\langle W\rangle$ can be zero, implying a spontaneous breaking of local Lorentz invariance. In order to discuss further the stationarity conditions we encounter the difficulty of having to evaluate derivatives of the tr log operator. This operator is a functional of the classical field configurations. Therefore the classical equations will themselves depend on the class of functions among which one looks for solutions. We will be interested in a vacuum that preserves global Poincaré invariance. We therefore will look for classical homogeneous solutions (i.e., spacetime independent) in the absence of external sources. In this case the evaluation of the tr log expression and its derivatives is simple.

Let us first discuss the stationarity condition on H. The presence of H^{-1} in A_b implies that $H \sim 0$ configurations should be strongly damped. Indeed, the vanishing of $\delta A_b/\delta H$ equates $V'(H)$ with an expression that, depending on the fields, becomes singular for $H \to 0$. We assume that this equation, which has a well-defined meaning for $H \neq 0$, may be satisfied by some value of H. This may eventually imply some condition on V(H) which we assume to be met. By taking into account Lorentz invariance we thus write

$$\langle H\rangle = v\,\mathbb{1} \tag{14}$$

and as we shall see, the actual value of v as long as it is nonzero will be immaterial.

Continuing our analysis, it is easy to see that the four equations

$$\frac{\delta A_b}{\delta\lambda^\mu} = \frac{\delta A_b}{\delta A_\mu} = \frac{\delta A_b}{\delta\lambda^{\mu,\beta\gamma}} = \frac{A_b}{\delta A_{\mu,\beta\gamma}} = 0 \tag{15}$$

are solved by

$$<\lambda^\mu> = <A_\mu> = <\lambda^{\mu,\alpha\beta}> = <A_{\mu,\alpha\beta}> = 0 \tag{16}$$

in accordance with global Lorentz invariance. The remaining two stationary conditions give

$$\frac{\delta A_b}{\delta\phi} = 0 \Longrightarrow v \det <W> = -\frac{1}{4}\sum_i C_i \int\frac{d^4p}{(2\pi)^4}\,\mathrm{Tr}\,\frac{K_i}{<\xi>^{\mu\alpha}\gamma_\alpha p_\mu - K_i<\phi>} =$$

$$= -\frac{1}{\det<\xi>}\,\frac{1}{u}\sum_i C_i \int\frac{d^4r}{(2\pi)^4}\,\frac{K_i^2u^2}{r^2-K_i^2u^2} \tag{17}$$

$$\frac{\delta A_b}{\delta\xi^{\mu\alpha}} = 0 \Longrightarrow <W_{\alpha\mu}> \det <W> = -\sum_i C_i \int\frac{d^4p}{(2\pi)^4}\,\mathrm{Tr}\,\frac{\gamma_\alpha p_\mu}{<\xi>^{\nu\beta}\gamma_\beta p_\nu - K_i<\phi>} =$$

$$= \frac{(-1)}{\det<\xi>}<\xi^{-1}>_{\alpha\mu}\sum_i C_i \int\frac{d^4r}{(2\pi)^4}\,\frac{r^2}{r^2-K_i^2u^2} \tag{18}$$

where

$$r^\alpha \equiv <\xi>^{\mu\alpha}p_\mu \tag{19}$$

and

$$<\phi> = u\,\mathbb{1} \tag{20}$$

as required by global Lorentz invariance.

By using the condition $\Sigma_i C_i = 0$ we recognize the equality of the last two integrals of Eqs. (17) and (18). Introducing now the dimensional proportionality constant η by

$$<W>_{\alpha\mu} = \eta<\xi^{-1}>_{\alpha\mu} \tag{21}$$

equations (14), (17) and (18) imply

$$uv = \eta = V(v)$$

$$I_4 \equiv \sum_i C_i \int\frac{d^4r}{(2\pi)^4}\,\frac{r^2}{r^2-h_i^2} = \eta^5\ , \quad h_i \equiv K_i u \tag{22}$$

where, we remark, the h_i satisfy the same PV conditions (10) as the K_i.

Since W and ξ have dimensions, their non-zero constant v.e.v. implies a vacuum generated scale Λ which, taking into account Eq. (21), can be defined without loss of generality by

$$\langle\xi\rangle^{\mu\alpha} = \Lambda^{-1}\,\eta_{\mu\alpha}\,, \quad \langle W\rangle_{\alpha\mu} = \eta\Lambda\cdot\eta_{\mu\alpha}$$

$$\eta_{\mu\alpha} \equiv \mathrm{diag.}\,(1,-1,-1,-1) \tag{23}$$

Equations (23) show explicitly that our vacuum breaks (spontaneously) local Lorentz as well as general co-ordinate transformation invariances leaving a diagonal global Lorentz invariance. The spontaneous breaking of scale invariance generates the scale Λ as a vacuum property[7]. This is the only dimensional scale of the theory, so that all masses (as well as other dimensional quantities) will be proportional to it. In the case of fermions, the masses which appear as poles in the propagator are given by

$$m_i = h_i\Lambda \tag{24}$$

We may, of course, choose $h_0 \ll h_i$, $i \neq 0$ in order to have the original fermion arbitrarily light. This mass is protected by a chiral invariance for the light fermion in the limit $K_0 \to 0$ ($h_0 \to 0$). Indeed, as is evident from the form (9) of the action, K_i measures the strength of the γ_5 non-invariant terms. Furthermore, for $K_0 \to 0$, the light fermion is excluded from the condensate (14).

Let us call h the smallest h_i ($i\neq 0$) so that $M = h\Lambda$ represents the mass of the lightest regulator (or set of regulators). M acts as a cut-off in our theory and will appear explicitly in Green's functions. M cannot be sent to infinity since an evaluation of I_4 gives $I_4 \sim h^4$, hence:

$$V(v)^5 = I_4 \sim h^4 = \left(\frac{M}{\Lambda}\right)^4 \tag{25}$$

Let us discuss for a moment the meaning of Eq. (25). In order for the homogeneous flat-space solution*) to exist, the parameters h_i and V(v) have to be related by Eq. (22). This can be seen either as a condition on the regulator mass M in terms of a given V(v) and Λ or, once M/Λ is arbitrarily chosen, as a fine tuning of an additional constant contribution to V(H). This is not surprising if one realizes that both V and the PV parameters are part of the regularization needed to give meaning to the theory. In this connection let us remark that we will be able to identify some heavy fields (as, for instance, $\lambda^{\mu,\alpha\beta}$), the integration over which would induce an additional contribution to the effective potential

*) This will also imply the absence of an induced cosmological term as discussed later.

V(H) which depends on the PV parameters and is singular for $H \to 0$. One may thus consider V(H) of Eq. (24) as a driving term. We notice, however, that had we considered the regularization mass M as the basic scale of the theory, Eq. (25) would determine, without fine tuning, Λ and hence the vacuum scale of Eq. (23). This could appear to follow from having explicitly broken scale invariance through the introduction of M. Nevertheless, the just stated equivalence with the spontaneous breaking of scale invariance shows that our regularization procedure represents a soft breaking.

The fact that our breaking is really of a spontaneous nature will be substantiated later when we will show a typical generation of massive O(3,1) gauge bosons through the eating of Goldstone particles associated with broken generators.

4. We shall now compute the quadratic part of the action (12) in the shifted bosonic fields. Furthermore, we shall expand this quadratic action up to second order in the field momenta q_i. The q independent terms, once diagonalized, will allow us to determine masses and, in particular, to identify the massless modes. Terms quadratic in the momenta will provide induced kinetic terms for the massless fields and will lead to the identification of the induced Newton and fine structure constants.

We shall denote by a tilded field its fluctuation around its v.e.v., lowering for convenience all its upper indices by the flat η tensor

a) <u>q Independent Terms</u>: since the field $W_{\alpha\mu}$ does not appear inside the tr log there is no dependence of A_b on the derivatives of W. One then finds, at all momenta

$$\frac{\delta A_b}{\delta \tilde{W}_{\alpha\mu} \delta \tilde{W}_{\beta\nu}} = -\eta^3 \Lambda^2 (\eta_{\mu\alpha}\eta_{\nu\beta} + \eta_{\mu\beta}\eta_{\nu\alpha}) \tag{26}$$

$$\frac{\delta A_b}{\delta \tilde{W}_{\alpha\mu} \delta \tilde{\xi}_{\nu\beta}} = -\eta^4 \Lambda^4 (\eta_{\mu\alpha}\eta_{\nu\beta} + \eta_{\alpha\beta}\eta_{\mu\nu}) \tag{27}$$

the derivatives being of course evaluated for vanishing fluctuations.

The dependence of A_b on ξ is more complicated. One finds:

$$\frac{\delta A_b}{\delta \tilde{\xi}_{\mu\alpha} \delta \tilde{\xi}_{\nu\beta}} = -\Lambda^6 \, T_{\alpha\beta,\mu\nu}(q) \tag{28}$$

where $T_{\alpha\beta,\mu\nu}(q)$ is given by a one-loop Feynman diagram i.e.:

$$T_{\alpha\beta,\mu\nu}(q) = \sum_i C_i \int \frac{d^4p}{(2\pi)^4} \frac{p_\mu p_\nu}{D_q D_{-q}} \mathrm{Tr}\left[\gamma_\alpha(\not p+\frac{\not q}{2}+m_i)\gamma_\beta(\not p-\frac{\not q}{2}+m_i)\right] \tag{29}$$

$$D_{\pm q} = (p \pm \frac{q}{2})^2 - m_i^2$$

with m_i given in Eq. (24). At q = 0 one finds

$$T_{\alpha\beta,\mu\nu}(0) = \sum_i C_i \int \frac{d^4p}{(2\pi)^4} \frac{4p_\mu p_\nu \eta_{\alpha\beta}(m_i^2-p^2) + 8p_\mu p_\nu p_\alpha p_\beta}{(P^2-m_i^2)^2} =$$

$$= -\eta_{\alpha\beta}\eta_{\mu\nu}I_4 + \frac{1}{3}(\eta_{\mu\nu}\eta_{\alpha\beta}+\eta_{\mu\alpha}\eta_{\nu\beta}+\eta_{\mu\beta}\eta_{\nu\alpha})\cdot\sum_i C_i \int \frac{d^4p}{(2\pi)^4} \frac{p^4}{(p^2-m_i^2)} \tag{30}$$

with I_4 defined in Eq. (22). With our PV regularization the last integral in (30) is equal to $3I_4$ so that:

$$\left.\frac{\delta A_b}{\delta\tilde{\xi}_{\mu\alpha}\delta\tilde{\xi}_{\nu\beta}}\right|_{q=0} = -I_4\Lambda^6(\eta_{\mu\alpha}\eta_{\nu\beta} + \eta_{\alpha\nu}\eta_{\beta\mu}) = -\eta^5\Lambda^6(\eta_{\mu\alpha}\eta_{\nu\beta} + \eta_{\alpha\nu}\eta_{\beta\mu}) \tag{31}$$

where Eq. (22) has been used.

The mass matrix formed by the quadratic terms (26), (27) and (31) can be written as

$$-\Lambda^4\eta^5(\Lambda\tilde{\xi}_{\alpha\mu} + \frac{1}{\Lambda\eta}\tilde{W}_{\alpha\mu})(\Lambda\tilde{\xi}_{\beta\nu} + \frac{1}{\Lambda\eta}\tilde{W}_{\beta\nu})(\delta_{\alpha\mu}\delta_{\beta\nu} + \delta_{\alpha\nu}\delta_{\beta\mu}) \tag{32}$$

from which one can regognize the two eigenmodes

$$\tilde{V}_{\alpha\mu} = \frac{1}{2}(-\Lambda\tilde{\xi}_{\alpha\mu} + \frac{1}{\Lambda\eta}\tilde{W}_{\alpha\mu}) \; ; \quad \tilde{U}_{\alpha\mu} = \frac{1}{2}(\Lambda\tilde{\xi}_{\alpha\mu} + \frac{1}{\Lambda\eta}\tilde{W}_{\alpha\mu}) \tag{33}$$

the first of which is massless. The mode $\tilde{U}$ is a heavy mode, i.e., a field with mass of $O(\Lambda)$. Later on we shall be able to see that, in a suitable low energy approximation, $\tilde{V}$ or Eq. (33) is the fluctuation of a vierbein field V satisfying the equations of the usual Einstein theory.

Let us now turn to the q independent quadratic terms in the connections $A_{\mu,\beta\gamma}$, A_μ and in their corresponding Lagrange multipliers $\lambda^{\mu,\beta\gamma}$, λ^μ.

Straightforward calculations give

$$\frac{\delta A_b}{\delta A_{\mu,\alpha\beta}\delta A_{\nu,\gamma\delta}} = 8\Lambda^2 I_2 O_{\mu\nu}^{\alpha\beta,\gamma\delta} \tag{34}$$

$$\frac{\delta A_b}{\delta\lambda_{\mu,\alpha\beta}\delta A_{\nu,\gamma\delta}} = \frac{1}{2}\Lambda^4\eta^4\tilde{O}^{\alpha\beta,\gamma\delta}_{\mu\nu} \qquad (35)$$

$$\frac{\delta A_b}{\delta\lambda_{\mu,\alpha\beta}\delta\lambda_{\nu,\gamma\delta}} = \Lambda^6\hat{O}^{\alpha\beta,\gamma\delta}_{\mu\nu} \qquad (36)$$

where I_2 is a regularized (and otherwise quadratically divergent) integral whose evaluation gives:

$$I_2 = c\eta^{5/2} = c'(M^2/\Lambda^2) \qquad (37)$$

with c and c' numerical constants (see Ref. 3). The tensors appearing in Eqs. (34)-(36) consist of products of η's and are antisymmetric in $\alpha \leftrightarrow \beta$, $\gamma \leftrightarrow \delta$.

Turning now to the A_μ, λ^μ sector, the situation is the same as the one discussed in Ref. 2. The two-by-two mass matrix has only one non-zero entry corresponding to a mass of order Λ for the λ^μ field. The absence of a mass term for A_μ or of a $\lambda^\mu A_\mu$ mixing is of course a consequence of U(1) gauge invariance (the case of $A_{\mu,\beta\gamma}$ is more subtle in this respect, as discussed below).

Finally, the two scalar fields ϕ and M can be shown to be also heavy. We do not give here the explicit expression but just remark that the mass of M depends on the details of V(M) [e.g., on V"(v)].

b) q Dependent Terms: we start with terms linear in the momenta. These mix tensors differing by one index, in particular $A_{\mu,\beta\gamma}$ and $\lambda^{\mu,\beta\gamma}$ with $\xi^{\nu\alpha}$. We find

$$\frac{\delta A_b}{\delta\lambda_{\mu,\alpha\beta}\delta\xi_{\gamma\delta}} = \frac{1}{8}\eta^4\Lambda^5 q_\nu\tilde{O}^{\alpha\beta,\gamma\delta}_{\mu\nu} \qquad (38)$$

$$\frac{\delta A_b}{\delta A_{\mu,\alpha\beta}\delta\xi_{\gamma\delta}} = 2\Lambda^3 I_2 q_\nu O^{\alpha\beta,\gamma\delta}_{\mu\nu} \qquad (39)$$

where the tensors O and $\tilde{O}$ of Eqs. (34) and (35), being antisymmetric in $\gamma \rightarrow \delta$, pick up the antisymmetric part of ξ. We shall come back to the above results after discussion of quadratic terms in q for the W,ξ sector which, as mentioned above, contain only ξ. Defining

$$\xi^{\mu\alpha}_{s,a} = \frac{1}{2}(\xi^{\mu\alpha} \pm \xi^{\alpha\mu}) \qquad (40)$$

we find up to second order in q:

$$\frac{\delta A_b}{\delta\tilde{\xi}_s\delta\tilde{\xi}_a} = 0 \; ; \quad \frac{\delta A_b}{\delta\tilde{\xi}^s_{\mu\alpha}\delta\tilde{\xi}^s_{\nu\beta}} = O^{(+)}_{\mu\alpha,\nu\beta} \; ; \quad \frac{\delta A_b}{\delta\tilde{\xi}^a_{\mu\alpha}\delta\tilde{\xi}^a_{\nu\beta}} = O^{(-)}_{\mu\alpha,\nu\beta} \tag{41}$$

where

$$O^{(+)}_{\mu\alpha,\nu\beta} = \frac{1}{6}\Lambda^4 I_2\Big[q^2(\eta_{\alpha\beta}\eta_{\mu\nu} + \eta_{\alpha\nu}\eta_{\beta\mu} - 2\eta_{\alpha\mu}\eta_{\beta\nu}) - $$
$$- q_\alpha q_\beta\eta_{\mu\nu} - q_\mu q_\beta\eta_{\alpha\nu} - q_\mu q_\nu\eta_{\alpha\beta} - q_\alpha q_\nu\eta_{\beta\mu} + 2q_\beta q_\nu\eta_{\alpha\mu} + \tag{42}$$
$$+ 2q_\alpha q_\mu\eta_{\beta\nu}\Big]$$

$$O^{(-)}_{\mu\alpha,\nu\beta} = \frac{1}{2}\Lambda^4 I_2\Big[q_\gamma q_\delta O^{\mu\alpha,\nu\beta}_{\gamma\delta}\Big] \tag{43}$$

Recalling that $\tilde{\xi}_s$ does not appear in Eqs. (38) and (39) we see that the full kinetic term in $\tilde{\xi}_s$ is given by $O^{(+)}$, implying a contribution to A_b of the form:

$$-\frac{1}{6}\Lambda^4 I_2\Big[\Box(\tilde{\xi}^{\mu\alpha}_s\tilde{\xi}^s_{\mu\alpha} - \tilde{\xi}^\mu_{s,\mu}\tilde{\xi}^\alpha_{s,\alpha}) - 2\partial_\alpha\partial_\beta\tilde{\xi}^{\mu\alpha}_s\tilde{\xi}^\beta_{s,\mu} +$$
$$+ 2\partial^\alpha\partial^\mu\tilde{\xi}^s_{\mu\alpha}\tilde{\xi}^\nu_{s,\nu}\Big] \tag{44}$$

The quadratic pieces that contain ξ_a [Eqs. (38), (39), (43)] , can be combined with those containing $A_{\mu,\gamma\delta}$ and $\lambda^{\mu,\gamma\delta}$ to give the following contributions to A_b:

$$4\Lambda^2 I_2 B_{\mu,\alpha\beta}B_{\nu,\gamma\delta}O^{\alpha\beta,\gamma\delta}_{\mu\nu} + \frac{1}{2}\Lambda^4\eta^4\lambda_{\mu,\alpha\beta}B_{\nu,\gamma\delta}\tilde{O}^{\alpha\beta,\gamma\delta}_{\mu\nu} \tag{45}$$

where

$$B_{\mu,\alpha\beta} = A_{\mu,\alpha\beta} + \frac{\Lambda}{4}q_\mu\tilde{\xi}^a_{\alpha\beta} \tag{46}$$

The fact that $A_{\mu,\alpha\beta}$ and $\tilde{\xi}^a_{\mu\alpha}$ appear only in the combination $B_{\mu,\alpha\beta}$ of Eq. (46) is due to the O(3,1) gauge invariance of the theory. $A_{\mu,\alpha\beta}$, being a connection, transforms inhomogeneously under the transformation (2.16):

$$A_{\mu,\alpha\beta} \to A'_{\mu,\alpha\beta} + \frac{1}{4}(\sigma^{\gamma\delta})_{\alpha\beta}\partial_\mu\omega_{\gamma\delta}(x) \tag{47}$$

Naïvely this would seem to forbid terms in the action containing $A_{\mu,\alpha\beta}$ without derivatives as is the case for the electromagnetic field A_μ. On the other hand, in our case we can construct another field transforming as a connection by taking derivatives of $\tilde{\xi}^a_{\alpha\beta}$.

Thus the combination $B_{\mu,\alpha\beta}$ of Eq. (46) is again a good tensor. The first term in Eq. (46) shows that $B_{\mu,\alpha\beta}$ is a massive field which has "eaten up" à la Higgs the six would-be Goldstone bosons $\xi^a_{\alpha\beta}$ of the spontaneously broken O(3,1) invariance. This fact confirms the validity of our statement that the flat-metric expectation values of W and ξ represent a spontaneous rather than explicit breaking of the symmetries of the original action, dilatation symmetry included.

Turning finally to the q dependent terms involving the λ^μ and A_μ electromagnetic fields we find again, as in Ref. 3, that they involve A_μ only through the field strength tensor $F_{\mu\nu}$. This ensures that A_μ is an exactly massless field contributing to the action with a term

$$-\frac{1}{4} I_0 F^2_{\mu\nu} \quad ; \quad I_0 \Big|_{q=0} = \frac{1}{12\pi^2} \log(M^2/m^2) + O(1) \tag{48}$$

I_0 here is the usual QED vacuum polarization integral whose dependence on q is well known.

Summarizing the results obtained so far, we have identified in our spectrum the following light particles:

i) one or several light fermions (according to the number of parameters K_i set to be much smaller than one) protected by chiral invariance from obtaining masses of $O(\Lambda)$;

ii) one massless spin one particle associated with the exact U(1) gauge symmetry, i.e., a photon. In a similar way, one would obtain eight gluons from requiring local SU(3) invariance;

iii) a massless excitation with the structure of the vierbein, associated with general relativistic invariance.

Besides these, we have obtained a number of "heavy" excitations, i.e., fields with mass of $O(\Lambda)$. As long as these heavy degrees of freedom carry energies $E_i \ll \Lambda$, they will not be excited away from their v.e.v's. In this limit we have, to a very good approximation, the following relations

$$\eta \Lambda \tilde{\xi} = -\frac{1}{\Lambda}\tilde{W} \Longrightarrow \xi = \eta W^{-1} \tag{49}$$

$$H = v\, \mathbb{1} \quad , \quad \phi = u\, \mathbb{1} \tag{50}$$

$$B^{\alpha\beta}_\mu = 0 \Longrightarrow A^{\alpha\beta}_\mu = \frac{1}{4} \xi^{-1}_{\alpha\nu}\, \partial_\mu \xi^{\nu\beta} \tag{51}$$

$$\lambda^{\mu} = \lambda^{\mu,\alpha\beta} = 0 \tag{52}$$

In this regime Eq. (49) allows, as discussed before, the definition of a metric after identification of $\Lambda\xi$ ($W/\eta\Lambda$) with the usual vierbein fields with upper (lower) indices. In that case we recognize in Eq. (51) the usual definition of the Lorentz connection in terms of the vierbein. Looking at the momentum-dependent part of the fluctuations, we then see that Eq. (44) is nothing but the quadratic expansion around the flat vierbein solution of

$$\frac{1}{6}\Lambda^2 I_2 \sqrt{-g}\, R \tag{53}$$

with g and R the usual metric determinant and scalar curvature expressed in terms of the vierbein.

At fourth order in the derivatives, a term[8] (log M/m) $(R^2-3R_{\mu\nu}R^{\mu\nu})$ appears. Higher order terms compatible with general relativistic invariance will also be generated, their dimensions being compensated by appropriate powers of M^{-1}. All these R^2 and higher order terms are therefore negligible in the low energy regime under discussion.

We notice at this point the absence of an induced $\sqrt{-g}$ cosmological term that, if present, would have appeared with an M^4 scale factor. This is an expected consequence of the fact that in our approach the flat metric is a classical solution and that a cosmological term would contain linear vierbein fluctuations.

Hence, in the low energy regime under discussion, we recover the Einstein-Dirac action for gravity plus spin 1/2 matter, with a vanishing cosmological constant and an induced Newton constant given by

$$(16\pi G_N)^{-1} = \frac{1}{6}\Lambda^2 I_2 = \frac{1}{6}c' M^2 \ ; \quad c' = \text{numerical constant} \tag{54}$$

Equation (54) shows that the Newton constant is related to the vacuum generated scale Λ by precisely the coefficient that makes it equal to the momentum cut-off M.

As already said, the quadratic terms in λ^{μ} and A_{μ} provide the usual kinetic term $F^2_{\mu\nu}$. As in Ref. 3, in the low energy regime where we can set $\lambda^{\mu} = 0$, we find nothing but the ordinary QED action with an effective fine structure constant given by the inverse of the quantity I_0 of Eq. (48). For $m^2 \ll q^2 \ll M^2$ one finds:

$$\alpha(q^2) = 3\pi(\log(M^2/q^2) + O(1))^{-1} \tag{55}$$

In the case of N light fermions of U(1) charges e_i we would have obtained for the ith fermion the effective electromagnetic coupling

$$\alpha_i(q^2) = 3\pi\, e_i^2 \left(\sum_{j=1}^{N} e_j^2 \log(M^2/q^2) + O(1)\right)^{-1} = \\ = 3\pi\, e_i^2 \left(\sum_{j=1}^{N} e_j^2 \log\left(\frac{1}{G_N N q^2}\right) + O(1)\right)^{-1} \tag{56}$$

where we have used Eq. (54) with the appropriate factor coming from the multiplicity of fermions. Equation (56) expresses the fact that, up to some factors O(N), the Landau pole of QED and the Planck mass have to coincide in our approach. A similar statement can be made for other unbroken gauge symmetries (as for instance QCD) provided that enough fermionic thresholds make them eventually asymptotically non-free[2].

5. The regime in which we have seen ordinary gauge theories arise as effective interactions is the one in which all fields carry, by decree, small momenta. This regime, however, does not necessarily coincide with that of the full quantum theory at low energy. It is only so at the tree level or if high momentum quantum fluctuations (as appearing through ultra-violet behaviours of loops) are strongly damped. We do not know yet if our approach will lead to to such behaviour.

We can only give a few qualitative hints of why such behaviour could be milder than in the conventional theory, by underlining the modifications implied by our theory on a conventional calculation of gravity effects at short distances.

Sticking to the bosonic language that proved useful in recognizing the low-lying spectrum, we find differences with the usual treatment of gauge and gravity theories at various levels.

i) The rich heavy sector (masses of order M, the Planck mass) we have found will contribute as much as the light one as soon as the momenta they transfer approach M. Cancellations may of course occur. We recall nevertheless that among these heavy particles with well-defined masses and couplings, we find also PV ghosts. These represent a nuisance even if the meaning of tree-unitarity at a level in which not even a space-time metric can be defined is far from clear. It is possible that a supersymmetric extension of our model could alleviate this problem.

ii) Both the light and heavy sectors have quadratic terms in the fields with higher derivatives rescaled, of course, by powers of 1/M. These higher derivatives may be resummed and lead to propagators which are calculable functions of q^2. Only for $q^2 \ll M^2$ do they coincide, for the light sector, with the conventional propagators. A calculation at this level to test the short-distance behaviour of the theory is in progress. A behaviour milder than the conventional one would already be a gratifying signal for our alternative approach to gravity.

iii) Besides the aforementioned contributions that count as soon as $q^2 \sim M^2$, we also meet an infinite series of induced many field couplings. They would, of course, contribute to any given process at higher and higher loop levels and therefore should be depressed by higher and higher powers of 1/N. It is nevertheless very unclear whether or not we have any right to expect an ultra-violet behaviour which is uniform in 1/N.

All the high frequency modifications to the conventional theory, i.e., extra particles, higher derivative interactions and extra couplings, are there to recall the basic fact that the graviton and gauge bosons were composite objects. At short distances we should indeed see the effects of their structure through some sort of form factor which softens their contribution as one approaches momentum transfers of order M.

These considerations show perhaps that the language of auxiliary fields, so convenient for studying the spectrum and the low energy structure of the theory, is not the correct language for studying the short-distance behaviour. The complexity of this language here is to remind us of its duality with respect to the language of components which, in our theory, is the one of the original fermions. This is perhaps better suited to short distances and our hope for a mild behaviour in this limit can be recomforted by the high derivative structure of our original action which should penalize very high frequencies. In this regime we should also lose the scale Λ, provided by the vacuum condensate, and recuperate therefore all the local symmetries of the theory. This could suggest a deep ultraviolet behaviour which might be not worse than that of ordinary renormalizable theories. All these arguments are only hints, but they nevertheless support our hope that the pregeometric gravity we propose could be a valid alternative for overcoming the pathologies of the conventional theory.

REFERENCES

1. For a review on GUT cf. for instance J. Ellis in Scottish Universities Summer School, ed. K.C. Bowler and D.G. Sutherland, (1981), p. 201.
2. D. Amati, R. Barbieri, A.C. Davis and G. Veneziano, Phys. Lett. 102B:408 (1981);
3. D. Amati and G. Veneziano, Phys. Lett. 105B:358 (1981) and CERN preprint TH.3197 (1981).

4. A.D. Sakharov, Dokl. Akad. Nauk. SSSR 177:70 (1967); [Soviet Physics Dokl. 12:1040 (1968)];
Ya.B. Zel'dovich, Zh. ETF. Pis. Ref. 6:922 (1967); [JETP Lett. 6:345 (1967)];
O. Klein, Phys. Scr. 9:69 (1974);
K. Akama, Y. Chikashige, T. Matsuki and H. Terazawa, Progr. Teor. Phys. 60:868 (1978);
K. Akama, Progr. Theor. Phys. 60:1900 (1978);
S.L. Adler, Phys. Rev. Lett. 44:1567 (1980), Phys. Lett. 95B:241 (1980);
B. Hasslacher and E. Mottola, Phys. Lett. 95B:237 (1980);
A. Zee, Phys. Rev. D23:858 (1981);
In a recent paper, to be published in Rev. Mod. Phys. S.L. Adler reviews these approaches to induced gravity and extends the analysis of renormalizable field theories with dynamical scale invariance breaking to the treatment of a quantized metric field.

5. L. Maiani, G. Parisi and R. Petronzio, Nucl. Phys. B136:115 (1978);
N. Cabibbo, L. Maiani, G. Parisi and R. Petronzio, Nucl. Phys. B158:295 (1979).

6. N. Cabibbo and G. Farrar, to be published.

7. We recognize here an analogy with the approach of V. De Alfaro, S. Fubini and G. Furlan [Phys. Lett. 97B:67 (1980)] where a scale invariance breaking mass parameter also appears through the v.e.v. of a vierbein (or metric) field. Their approach differs however from our since its starting point is the conventional Einstein action with a fundamental metric (or vierbein) field.

8. Cf. K. Akama et al., Ref. 4.

STRONG CP VIOLATION AND AXIONS

C. Jarlskog

Department of Physics, University of Bergen
Bergen, Norway

1. Introduction

Once upon a time there was a serious problem in physics. Experiments did not agree with the theory: energy-momentum and angular momentum conservation laws seemed to be violated in beta decay. In 1930 Pauli suggested a clever cure: a "light penetrating particle" which he called the neutron but which later on got the name "neutrino". He wrote to a friend, "I've done a terrible thing today, something which no theoretical physicist should ever do. I have suggested something that can never be verified experimentally". A few years later, Pauli, in a jolly mood, had remarked [1] that perhaps he should have called it (the neutrino) the hypothon. Now some fifty years later, there is again a serious problem in physics, however this time not due to any new experimental discovery, but because the theory has changed. The problem is that, to the best of our knowledge, quantum chromodynamics respects neither parity nor time reversal invariance. The cure this time is again "a light penetrating particle", as we shall discuss in the following. This time the "hypothon" has many names: the axion, Achion, higglet, etc. In some of its versions, the "hypothon" of our time is expected to "never be verified experimentally". That's why it is sometimes called the "invisible axion", or "phantom axion", etc.

The axionic solution of the strong CP-violation is quite an exciting possibility. Its ingredients span over a large domain of theoretical ideas. The problem involves the vacuum structure of QCD, the so-called θ-vacuua, topological charges and instantons. The essential elements of the solution are the axial anomaly, chiral symmetry and its spontaneous breakdown, Goldstone bosons, current algebra, Higgs scalars, gauge theories and more recently grand unification.

The axion is supposed to be a Goldstone boson, born in the Higgs potential, in the process of the spontaneous breakdown of a global chiral symmetry (the Peccei-Quinn symmetry). It acquires, in general, a tiny mass due to the anomaly.

The observation of the axion, with expected properties, would constitute a beautiful confirmation of gauge theories. We must not forget that so far we haven't seen any Higgs particles or gauge bosons (except the photon).

2. The Effective Lagrangian and Chiral Symmetry

QCD has a complicated vacuum structure [2] and because of that it does not automatically respect parity or time reversal invariance. The effective Lagrangian (density) in QCD is given by (we use the Feynman metric and apologize for mismatch in Lorentz indices)

$$\mathcal{L}_{eff}(\theta) = \mathcal{L}(G) + \mathcal{L}(G,f) + \hat{\mathcal{L}}(\theta) \,, \tag{2.1}$$

$$\mathcal{L}(G) = -\frac{1}{4}\, G^j_{\mu\nu} G^j_{\mu\nu} =$$ 

$$\mathcal{L}(G,f) = \bar{f}(i\gamma_\mu D_\mu - m)f =$$  (2.2)

$$\hat{\mathcal{L}}(\theta) = \theta\, a\,, \quad a \equiv \frac{g^2}{32\pi^2}\, G^j_{\mu\nu}\tilde{G}^j_{\mu\nu} \,, \quad \tilde{G}^j_{\mu\nu} = \frac{1}{2}\varepsilon_{\mu\nu\rho\sigma} G^j_{\rho\sigma} \tag{2.3}$$

Here G is the gluon field, D denotes the covariant derivative, f stands for a quark with mass m (the generalization to several quarks is trivial) and a is the anomaly. Finally θ is a real constant. It is the presence of the last term in (2.1) which constitutes the problem to be solved, viz., $a \sim \bar{\bar{E}}\cdot\bar{\bar{B}} + \ldots$ (where E(B) is the colour electric (magnetic) field) is odd under both parity and time reversal. There would be no problem if θ happened to be zero. But there is no à priori reason why θ should vanish.

The above situation is drastically changed if the quark mass, in (2.2), happens to be zero. Then the QCD Lagangian (2.1) is formally invariant under the chiral global phase transformation (abbr. chiral rotation)

$$f \to \exp(i\alpha\gamma_5)\cdot f \tag{2.4}$$

where α is a real constant, $\bar{f}\gamma D f \to \bar{f}\gamma D f$. So we would expect that the Lagrangian, Eq. (2.1), should not change under (2.4) but we would be wrong, because of the anomaly [3]. We may calculate the change in the Lagrangian from Mme. Noether's theorem.

$$\delta\mathcal{L} = \partial^\mu \left(\frac{\delta\mathcal{L}}{\delta(\partial^\mu f)}\right) = -\alpha\, \partial^\mu(\bar{f}\gamma_\mu\gamma_5 f) \sim \partial^\mu A_\mu$$
$$\partial^\mu(\bar{f}\gamma_\mu\gamma_5 f) = 2a, \tag{2.5}$$
$$\delta\mathcal{L} = -2a\alpha .$$

Here A_μ is the axial current and a is the anomaly, eq. (2.3). The net effect of the chiral rotation is thus

$$\mathcal{L}_{eff}(\theta) \to \mathcal{L}_{eff}(\theta-2\alpha) . \tag{2.6}$$

By choosing $\alpha=\theta/2$ the troublesome term is eliminated, in other words θ is rotated away. The only problem is that there is no massless[4)] quark in nature, and the mass term changes violently under chiral rotations,

$$\mathcal{L}_{mass} = -m\bar{f}f \to -m\bar{f}\exp(2i\alpha\gamma_5)f . \tag{2.7}$$

Therefore the θ-parameter should be there. The question is then how big is it? In order to answer this question, we may, by a suitable chiral rotation, eliminate the $\hat{\mathcal{L}}(\theta)$ term in (2.1) at the expense of obtaining a P and T violating mass term. For several flavours, we choose to do independent chiral rotations on each, according to

$$f_j \to \exp(i\alpha_j\cdot\gamma_5)f_j \quad , \quad j=1-n \tag{2.8}$$

Then the Lagrangian in (2.1), generalized to several flavours, transforms according to

$$\mathcal{L} \to \mathcal{L}(G) + \bar{f}_j(i\gamma D)f_j - \bar{f}_j\, m_j \exp(2i\alpha_j\gamma_5)f_j + $$
$$+ (\theta - 2\sum_j \alpha_j)a \tag{2.9}$$

Here α_j are constants, n is the number of flavours. We now choose

$$m_j\,\alpha_j = K \quad , \quad K=\text{constant}$$
$$2\,\Sigma\alpha_j = \theta \tag{2.10}$$

Then

$$K = (\theta/2)/(\Sigma\, m_j^{-1}) . \tag{2.11}$$

Assuming θ to be small (which is justified a postriori), we may expand the exponential in (2.9) and neglect terms of order θ^2,

$$\mathcal{L} = \mathcal{L}_{QCD}(\theta=0) + \mathcal{L}_{Bad} ,$$
$$\mathcal{L}_{Bad} = -i\theta\left(\sum_j 1/m_j\right)^{-1} \sum_K \bar{f}_K\gamma_5 f_K \tag{2.12}$$

where the "Bad" term violates P and T. For three flavours it is given by

$$\mathcal{L}_{Bad} = -i\theta \frac{m_u m_d m_s}{m_u m_d + m_d m_s + m_u m_s} (\bar{u}\gamma_5 u + \bar{d}\gamma_5 d + \bar{S}\gamma_5 S) \quad (2.13)$$

Note that this unwanted term vanishes if any of the quark masses goes to zero. The point of the choice in (2.10) is that the thus obtained is SU(3) symmetric and, therefore, it does not disturb the vacuum state, so it may be treated as a small perturbation. $\mathcal{L}_{Bad}$ results in a nonvanishing electric dipole moment for the neutron. The calculation[5] of this quantity and comparison with experiment yields

$$\theta \lesssim 10^{-9} . \quad (2.14)$$

Why is θ so small? Is the effective θ actually zero? If so, which physical principle forces it to vanish? An answer to these questions was given by [6] Peccei and Quinn, as we shall now discuss.

3. The Peccei-Quinn (PQ) Symmetry

The point is that one wishes to be able to perform a chiral rotation in order to rotate θ to zero. The only obstacle is the mass term which is not invariant. But if the mass is due to a Higgs boson then the latter may restore the invariance. For a single flavour we have

$$\mathcal{L}_{mass} = c\, \bar{f}_L\, \phi f_R + h.c. \quad (3.1)$$

$$f \to \exp(i\alpha\gamma_5)\cdot f$$

$$\mathcal{L}_{mass} \to c\, \bar{f}_L\, \phi \exp(2i\alpha\gamma_5) f_R + h.c. = c\, \bar{f}_L \phi \exp(2i\alpha)\cdot f_R + h.c. \quad (3.2)$$

Eq. (3.2) suggests that the chiral rotation should be generalized to involve also the ϕ,

$$U(1)_{PQ} \; : \; f \to \exp(i\alpha\gamma_5)f \;, \phi \to \exp(-2i\alpha)\phi \; . \quad (3.3)$$

Then all that happens under this generalized rotation is that $\theta \to \theta - 2\alpha$ and so θ may be rotated away. Of course we must add in kinetic and potential terms for the scalar field θ into our Lagrangian

$$\mathcal{L}(\varphi) = (\partial^\mu \phi^+)(\partial_\mu \phi) - V(\phi) \; , \quad (3.4)$$

where $V(\varphi)$ is assumed to respect the chiral symmetry of Eq. (3.3). Now the Noether current involves also the ϕ-field

$$\delta \mathcal{L} = -\alpha\, \partial^\mu [\bar{f}\gamma_\mu \gamma_5 f + 2i(\partial_\mu \phi^+ \cdot \phi - \phi^+ \partial_\mu \phi)] \quad (3.5)$$

When the scalar field acquires a vacuum expectation value ($\varphi \to \varphi + v$, $cv=-m$) the (PQ)-symmetry breaks spontaneously. The Higgs potential produces [7] a massless Goldstone boson, associated with this breakdown. This particle is the axion. Note that here we are dealing with a global symmetry, so there is no gauge boson which can swallow the axion. Actually renormalizability in gauge theories requires that there should be no anomalies in the currents associated with the gauge bosons. The removal of θ, on the other hand, demands an anomalous current. Therefore the symmetry whose breakdown produces the axion is different from (the global part of) those which give the longitudinal degrees of polarization of the gauge bosons. The above arguments were given for just one flavour. Now we must generalize to the real world. With our gauge theory prejudice we would write, in general

$$\mathcal{L} = \mathcal{L}(V) + i\bar{f}\gamma D f + \theta a + (D^{\mu}\Phi)^{+}(D_{\mu}\Phi) - V(\Phi) + \mathcal{L}(\phi, f) \qquad (3.6)$$

where V stands for the gauge bosons, $\mathcal{L}(V)$ denotes their kinetic and self-interaction terms and $f(\phi)$ stands for the fermions (Higgs bosons). The most natural choice of the underlying symmetry in (3.6) is the standard electroweak SU(2) U(1) multiplied with the colour SU(3). With the standard assignments of the electroweak model, we have

$$\mathcal{L}(\Phi, f) = c(\bar{u}, \bar{d})_L \Phi d_R + c'(\bar{u}, \bar{d})_L \tilde{\Phi} u_R + h.c. \qquad (3.7)$$

where Φ is the standard Higgs doublet, $\tilde{\phi} = i\sigma_2\phi^*$ and the c's are constants. We have suppressed the sum over flavours as well as the primes on u and d fields, which indicate that they are not mass eigenstates. Now we can easily convince ourselves that

$$\delta\mathcal{L} = [\alpha(u_L) - \alpha(u_R) + \alpha(d_L) - \alpha(d_R)]\, a, \qquad (3.8)$$

vanishes in this case, so θ cann't be rotated away. This is not strange; the U(1) symmetry of (3.7) is the weak hypercharge which doesn't give an anomalous current. Thus if we insist on the standard model, we must take a more complex Higgs sector. The most straight-forward extension is to consider two Higgs doublets. The choice of doublets is convenient because they preserve the successful mass relation $M_W = \cos\theta_W M_z$. With two doublets ϕ and χ we have

$$\mathcal{L}(\Phi, f) = c\bar{D}_L \Phi d_R + c'\bar{D}_L \chi u_R + r\bar{D}_L \tilde{\chi} d_R + r'\bar{D}_L \tilde{\Phi} u_R + h.c. \qquad (3.9)$$

Here D denotes the doublet; c, c', r and r' are constants. Again requiring U(1) invariance yields $\alpha(\phi) = \alpha(\tilde{\chi}) = -\alpha(\chi)$ and θ is not shifted. Furthermore as in (3.9), the quarks with a given charge get their masses from more than one doublet.there are flavour-changing neutral currents. In order to avoid these difficulties one requires (e.g. by imposing discrete symmetries) that r=r'=0.

The Higgs potential $V(\phi,\chi)$ has five independent symmetries, viz. $SU(2)\times U(1)\times U(1)$, where the last symmetry refers to PQ. Four of these symmetries are broken spontaneously as only U(1)-electromagnetic is assumed to remain unbroken. We shall not go into how all this can be arranged, for a certain range of the parameters in the potential. In order to isolate the axion we look into the term $|D\phi|^2$, in (3.6), which reads

$$\mathcal{L}(\phi,V) = \sum \left|(\partial - ig\bar{T}\cdot\bar{W} - ig'yB)(\phi + v(\phi))\right|^2 \tag{3.10}$$

where the sum goes over the ϕ and χ multiplets introduced in (3.9); T and y are the appropriate (matrix representation of) generators and v is the vacuum expectation value. Note that ϕ and χ correspond to 8 states, 4 with charges ± 1 and 4 neutrals. From (3.10) we may easily isolate the Goldstone bosons G^j, j=1-3, which are absorbed by the gauge bosons

$$G^j \sim \sum_{\phi,\chi} \left[v^\dagger(\phi)T^j\phi - \text{h.c.}\right] \quad , \; j=1-3 \; , \tag{3.11}$$

where

$$\phi = \frac{1}{\sqrt{2}}\begin{pmatrix}\phi_3 + i\phi_4 \\ \phi_1 + v_\phi + i\phi_2\end{pmatrix} \quad , \qquad \chi = \frac{1}{\sqrt{2}}\begin{pmatrix}\chi_1 + v_\chi + i\chi_2 \\ \chi_3 + i\chi_4\end{pmatrix} \tag{3.12}$$

By adjusting the phases of ϕ and χ the v's may be chosen to be real. Then from (3.11), (3.12) and what we learned in the standard model, we conclude

$$\begin{pmatrix}G^3 \\ h\end{pmatrix} = \begin{pmatrix}\cos\theta & -\sin\theta \\ \sin\theta & \cos\theta\end{pmatrix}\begin{pmatrix}\chi_2 \\ \phi_2\end{pmatrix} \; , \qquad \sin\theta = \frac{v_\phi}{\sqrt{v_\phi^2 + v_\chi^2}} \; , \tag{3.13}$$

where h is the axion or higglet. The Higgs multiplets may now be rewritten as

$$\phi = \frac{1}{\sqrt{2}}\exp(i\phi_2/v_\phi)\begin{pmatrix}0 \\ v_\phi\end{pmatrix} = \frac{1}{\sqrt{2}}\exp\left(iG^3\sigma_3/v + ih/(xv)\right)\begin{pmatrix}0 \\ v_\phi\end{pmatrix}$$

$$\rightarrow \frac{1}{\sqrt{2}}\exp\left(i\frac{h}{xv}\right)\begin{pmatrix}0 \\ v_\phi\end{pmatrix} \tag{3.14}$$

and similarly

$$\chi \rightarrow \frac{1}{\sqrt{2}}\exp(ihx/v)\begin{pmatrix}v_\chi \\ 0\end{pmatrix} \; , \quad v = (v_\phi^2 + v_\chi^2)^{1/2} , \quad x = v_\phi/v_\chi \tag{3.15}$$

In (3.14) and (3.15) we have only written out the axionic components; the Goldstones G are supposed to have been gauged away. The axion receives a kinetic term from (3.10) and it couples to fermions through (3.9). The piece of the Lagrangian in (3.6) which involves the axion reads

$$\mathcal{L}_h = \tfrac{1}{2}(\partial h)^2 - \Big[m_{dj}\, \bar{d}_{jL} \exp(i \frac{h}{xv}) d_{jR} + $$
$$+ m_{uj}\, \bar{u}_{jL} \exp(i \frac{xh}{v}) u_{jR} + \text{h.c.} \Big] \qquad (3.16)$$

where we have diagonalized the fermion mass matrix. The chiral symmetry of (3.16) is easily obtained from

$$d \to \exp(i\alpha(d)\gamma_5)d, \; u \to \exp(i\alpha(u)\gamma_5)u, \; h \to h + \delta h,$$

$$\text{i.e. } \alpha(u) \equiv \alpha \cdot x, \quad \alpha(d) = \alpha/x \;, \; \delta h = -2v\alpha \;, \qquad (3.17)$$

where α is a constant. Note that

$$D_L \to \exp(-i(x+\frac{1}{x})\alpha/2)\exp(-i(x-\frac{1}{x})\alpha\sigma_3/2)D_L \qquad (3.18)$$

i.e., there is also a rotation involved here. However, since the σ_3 term does not lead to an anomalous current, only the phase transformation is essential. The shift in the Lagrangian due to (3.17) is

$$\delta\mathcal{L} = -2\alpha(x+\frac{1}{x})\cdot N \cdot a \qquad (3.19)$$

where N is the number of doublets. The Noether current associated with the above PQ-symmetry reads[8] (see eq. 2.5)

$$A_\mu^{PQ} = v\,\partial_\mu h + \frac{x}{2}\, \bar{u}_j \gamma_\mu \gamma_5 u_j + \frac{1}{2x}\, \bar{d}_j \gamma_\mu \gamma_5 d_j \;,$$
$$\partial^\mu A_\mu^{PQ} = (x+\frac{1}{x})\cdot N \cdot a \qquad (3.20)$$

A comment about the leptons is in order now. The leptons also get their masses from Higgses, in the standard model. If these Higgses are the ones defined in (3.12) then the axion will couple to leptons as well. It is not so clear why the PQ-symmetry, which is a property of strong interactions, should affect the leptons. Anyway, the lepton-axion couplings are largely arbitrary[8,9] in the standard model. Unfortunately this arbitrariness affects the predictions for the axion decay modes and lifetime. From this point of view the grand unified theories do a better job.

Having obtained the (PQ)-current, Eq.(3.20), Bardeen and Tye argue that this current, which is obtained by formal manipulations, does not describe the axion. The axion current should be anomaly free and divergenceless as m_u or m_d go to zero, the latter because teh axion is intimately associated with the chiral symmetry and in the chiral limit it should become a true Goldstone boson. These requirements are fulfilled by replacing

$$A_\mu^{PQ} \to A_\mu^h = A_\mu^{PQ} - (x+\frac{1}{x})(N/2)\Big[\frac{1}{1+Z}\,\bar{u}\gamma_\mu\gamma_5 u + \frac{Z}{1+Z}\,\bar{d}\gamma_\mu\gamma_5 d\Big],$$
$$Z \equiv m_u v_u/(m_d v_d) \qquad (3.21)$$

where the ν's are constants to be determined later. Now $\partial A^h \sim m_u m_d \to 0$ as $m_u \to 0$ or $m_d \to 0$. The isospin current, which describes the neutral pion is given by

$$A^\pi_\mu = \frac{1}{2}(\bar{u}\gamma_\mu\gamma_5 u - \bar{d}\gamma_\mu\gamma_5 d). \tag{3.22}$$

One may use current algebra techniques to calculate the pion and axion masses [8]. From (3.21) and (3.20) it is straight-forward to compute

$$\begin{aligned} &\langle 0|-i[Q^\pi, D^\pi]|0\rangle = -(m_u \nu_u + m_d \nu_d) \\ &\langle 0|-i[Q^\pi, D^h]|0\rangle = 0 \\ &\langle 0|-i[Q^h, D^h]|0\rangle = -N^2(x+\tfrac{1}{x})^2 Z(1+Z)^{-2}(m_u\nu_u + m_d\nu_d), \end{aligned} \tag{3.23}$$

where Q is the axial charge and $D = \partial^\mu A$, and $\nu_u = \langle 0|\bar{u}u|0\rangle$, $\nu_d = \langle 0|\bar{d}d|0\rangle$. Current algebra gives the LHS of eq. (3.23),

$$\begin{aligned} &f_\pi^2 m_\pi^2 = (m_u + m_d)\nu, \quad \nu = \nu_u \approx \nu_d \\ &m_h^2 = \sqrt{2}\, G_F f_\pi^2 m_\pi^2 N^2 (x+\tfrac{1}{x})^2 Z(1+Z)^{-2}. \end{aligned} \tag{3.24}$$

Here we have used that the axion decay constant, from (3.21) and (3.20), is given by

$$\begin{aligned} &\langle 0|A^h_\mu|h(q)\rangle = i f_h q_\mu, \quad f_h = v, \\ &g^2 v^2/4 = M_W^2, \quad g^2/(8M_W^2) = G_F/\sqrt{2}, \quad v^2 = (\sqrt{2}\, G_F)^{-1}. \end{aligned} \tag{3.25}$$

The axion decay constant is huge, v ~ 250 GeV and yet "harmless". The axion mass, on the other hand, is tiny: $m_h \sim f_\pi m_\pi / v$. From (3.24) using $x+1/x>2$ and $N>2$ one has $m_h>100$ keV. Also (3.21) and (3.22) imply that axion current has a small mixing (of order f_π/v) with the (isovector) pion-current and the isoscalar η-current). Roughly speaking, the axion behaves, with a probability of $(f_\pi/v)^2 \sim 10^{-7}$ as a strongly interacting particle. For example,

$$g^2_{hNN} = g^2_{\pi NN}(f_\pi/v)^2 \approx 10^{-7}\, g^2_{\pi NN}, \tag{3.26}$$

where g denotes the coupling constant obtained by applying the Goldberger-Treiman relation. Also the amplitude for the axion decay into two photons, determined from its pionic component, is given by

$$\begin{aligned} &\frac{A(h\to\gamma\gamma)}{A(\pi\to\gamma\gamma)} = \frac{f_\pi}{v} N(x+\tfrac{1}{x})\cdot Z(1+Z)^{-1} \\ &\tau(h\to\gamma\gamma) \sim \tau(\pi\to\gamma\gamma)\cdot\left(\frac{m_\pi}{m_h}\right)^5. \end{aligned} \tag{3.27}$$

Numerically,

$$\tau(h \to \gamma\gamma) = \frac{0{,}4\ \mathrm{Sec}}{Z} \cdot (100/m_h)^5,$$

where m_h is in keV and $Z = m_u/m_d \approx 0.56$.

In summary, the general properties of the axion are as follows:

$$\begin{cases} f_h \sim v \\ m_h \sim f_\pi m_\pi / v \\ \tau(h \to \gamma\gamma) \sim \tau(\pi^0 \to \gamma\gamma)\,(m_\pi/m_h)^5 \end{cases}$$

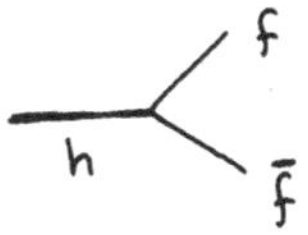

$$\sim (m_f/v)\,\bar{f}\gamma_5 f \cdot h \qquad (3.28)$$

$$g_{hNN} \sim (f_\pi/v)\, g_{\pi NN}$$

$$A(h \to \pi) \sim \mathcal{O}(f_\pi/v).$$

4. The Invisible Axion

A light penetrating particle decaying into two photons, but not to electron-positron pair, has been reported, in a series of recent publications[11)], by Faissner and his collaborators. The mass of the particle is found to be (250 ± 100) keV.

Beam dump experiments so far found no evidence[12)] for the axion. It is clearly very important to further improve the limits.

While some of our experimental colleagues have been working hard to find the axion, the theorists have discovered a distressing possibility: the "invisible" axion. The point is that, if the chiral symmetry breaking which produces the axion has a large vacuum expectation value associated with it, for example $v \sim v(\mathrm{GUT}) \sim 10^{14}$ GeV then from (3-28) we expect that $m_h \sim 10^{-7}$ eV, $\tau(h \to \gamma\gamma) \sim 10^{50}$ yrs, and the axion couplings to fermions as well as axion-meson mixings will all be extremely small.

The recipe for making an invisible axion spelled out by Dine, Fischler and Srednicki[13] is rather simple. We introduce new scalars H which neither couple to fermions nor to W and B. The simplest possibility is to assume that they behave as singlets with respect to the electroweak gauge group. The vacuum expectation value of H is therefore not restricted by the mass of the electroweak gauge bosons (see eq. 3-25). Furthermore, H is assumed to feel the chiral PQ-symmetry, i.e.,

$$\begin{aligned} f &\to \exp(i\alpha(f)\gamma_5) f \quad , \; f = u, d \\ \phi &\to \exp(i\alpha(\phi))\phi , \\ H &\to \exp(i\alpha(H)) H , \end{aligned} \tag{4-1}$$

where ϕ denotes other scalars in the theory and the α's are all multiples of a single real parameter. The Langrangian is requested to be formally invariant under the PQ-symmetry. Again one must see to it that the mass terms are such that δL is nonzero and θ can be shifted. The spontaneous symmetry breaking will in general lead to $H \to H + v(H)$, $\phi \to \phi + v(\phi)$, where v denotes the vacuum expectation value, and there will be a massless Goldstone boson, associated with the breakdown of PQ-symmetry, in the Higgs potential $V(H,\phi)$. The axial current of the PQ-symmetry will now be of the form

$$\begin{aligned} A_\mu^{(PQ)} \sim\; & v(H)\cdot\alpha(H)\partial_\mu h(H) + v(\phi)\,\alpha(\phi)\partial_\mu h(\phi) + \\ & (\cdots)\,\bar{u}\gamma_\mu\gamma_5 u + (\cdots)\,\bar{d}\gamma_\mu\gamma_5 d , \end{aligned} \tag{4-2}$$

where $h(x)$ is a neutral scalar field from the x multiplet, just as in the case of two doublets where we got a linear combination of ϕ_2 and χ_2, see eqs. (3-13) and (3-20). The H and thus also h(H) do not couple to fermions (by assumption). Furthermore, one must assume that $|v(H)| \gg |v(\phi)|$, for example $v(H) \sim 10^{14}$ Gev and $v(\phi) \sim 10^2$ Gev is a typical result in the grand unified models. Then the axion field h, from (4-2), is

$$h = h(H) + \mathcal{O}(v(\phi)/v(H))\cdot h(\phi) + \mathcal{O}(f_\pi/v(H))\cdot\pi + \cdots \tag{4-3}$$

Thus we have an axion with the properties given in Eq. (3-28) but where $v \sim v(H)$ is extremely large. This is the invisible axion.

An example of the models in which the axion is invisible is the SU(5) axion of Wise, Georgi and Glashow[14]. The Higgses are a complex 24, which we denote by H and two fives, ϕ and χ. So there are $2\cdot(24+5+5) = 68$ states. The fermion content of the theory is as usual, the right- (left-) handed fermions are in a five (ten) of SU(5),

$$\mathcal{L}(f,\phi,\chi) = c_1 \bar{T}_L^c T_L \phi + c_2 \bar{T}_L F_R \chi + h.c.,$$

where F(T) stands for the five (ten) dimensional fermion representation, and the superscript c denotes charge conjugation. The reason for introducing two fives of Higgs is as discussed after Eq. (3-9); H has to be complex in order to take part in chiral rotations. One assumes that v(H) is much larger than v(ϕ) and v(χ). Then v(H) breaks SU(5) into SU(3)xSU(2)xU(1); v(H) = v(GUT). The further breaking down is due to v(ϕ) and v(χ) which are of the order of 100 GeV, as usual. The authors find $m_h \sim 10^{-8}$ eV and $\tau(h\to\gamma\gamma) \sim 10^{56}$ yrs.

Further examples of axions in grand unified models are found in the literature[15].

One might wonder whether such super penetrating light particle would not modify gravity at intermediate distances, $m \sim 10^{-8}$ eV implies $\lambda \sim 10$ meters. Would there be an effective Yukawa potential of the form $\frac{1}{r} \cdot \exp(-r/\lambda)$ due to axion exchange? In perturbation theory there is no scalar coupling $\bar{f}fh$, as is seen from the analysis of Bardeen and Tye[8] (see eq. (3-16)), so there is no such long range force. In the presence of instanton effects there could be[16] a small induced scalar coupling $\bar{f}fh$ of the order $f_\pi \cdot \theta / v(GUT)$.

The axion also disappears from the Higgs potential, because there it behaves like a Goldstone boson, which can be rotated away.

The grand unified axion models are so far very extravagant in their Higgs sector. It is hard to imagine that all these scalars were judiciously created just to solve the θ-problem. The grand unified axion, if it exists, plays a fundamental role in nature and yet leaves no marks or footprints. Evidently, there is much to be learned and understood yet.

REFERENCES

1. This recollection is due to Lamek Hulthén.
2. G. 't Hooft, Phys. Rev. Lett. 37, 8 (1976); Phys. Rev. D14, 3432 (1976).

3. J.S. Bell and R. Jackiw, Nuovo Cimento 60A, 49 (1969); S. Adler, Phys. Rev. 177, 2426 (1969).
4. S. Weinberg, in A Festschrigt for I.I. Rabi, Ed. L. Motz (New York Academy of Sciences, N.Y., 1977), and references therein.
5. R.J. Crewther, P. Di Vecchia, G. Veneziano and E. Witten, Phys. Lett. 88B, 123 (1979); V. Baluni, Phys. Rev. D19, 2227 (1979). N.F. Ramsey, Phys. Reports 43C, 409 (1978).
6. R.D. Peccei and H.R. Quinn, Phys. Rev. Lett. 38, 1440 (1977); Phys. Rev. D16, 1791 (1977).
7. S. Weinberg, Phys. Rev. Lett. 40, 223 (1978); F. Wilczek, ibid 40, 279 (1978).
8. W.A. Bardeen and S.-H.H. Tye, Phys. Lett. 74B, 229 (1978); S. Weinberg, Ref. 7.
9. J.-M. Frère, M.B. Gavela and J.A.M. Vermaseren, Phys. Lett. 103B, 129 (1981).
10. T.W. Donnelly et al., Phys. Rev. D18, 16?7 (1978).
11. H. Faissner et al., Phys. Lett. 96B, 2?1 (198?); ibid 103B, 238 (1981); Preprint PITHA 81/32, to be published in the Proceedings of the International Neutrino Conference 1981.
12. L. Lo Secco et al., Phys. Lett. 102B, 209 (1981).
13. M. Dine, W. Fischler and M. Srednicki, Princton Preprint (1981)
14. M.B. Wise, H. Georgi and S.L. Glashow, Preprint HUTP-81/A019 (1981).
15. R. Barbieri, D.V. Nanopoulos and D. Wyler, Preprint TH.3119-CERN (1981); J. Ellis, M.K. Gaillard, D.V. Nanopoulos and S. Rudaz, Preprint TH.3127-CERN (1981).
16. R. Barbieri, R.N. Mohapatra, D.V. Nanopoulos and D. Wyler, Preprint TH.3105-CERN (1981).

NON-LINEAR MECHANICS OF A STRING IN A VISCOUS NOISY ENVIRONMENT*

William G.Faris
Department of Mathematics - University of Arizona - Tucson, Arizona 85721 - USA

Giovanni Jona-Lasinio
Istituto di Fisica "G.Marconi"
Università di Roma - Roma Italy

Stochastic partial differential equations constitute a relatively new subject. There have been a number of interesting papers in the last few years, but the field is still in its infancy if compared with the highly developed theory of stochastic ordinary differential equations. In this paper we report on a study of a non linear heat equation in a finite interval of space subject to a white noise forcing term. Due to the non linearity the equation , without the forcing term, exhibits several equilibrium configurations two of which are stable, actually asymptotically stable. The solution of the complete forced equation is a stochastic process in space and time with continuous sample paths whose behaviour we study in the limit of small noise. We obtain lower and upper bounds for the probability of large fluctuations and then apply our estimates to the calculation of the transition probability between the stable configurations (tunneling).

The equation has the following form:

$$\frac{\partial u(x,t)}{\partial t} = - \frac{\delta S(u)}{\delta u(x,t)} + \varepsilon\alpha(x,t) \qquad (1)$$

*This report is based on the work "Large fluctuations for a Non-Linear Heat Equation with Noise" (to appear) written while the authors were visitors at the I.H.E.S. - Bures-sur-Yvette (France).

where S(u) is the action functional

$$S(u) = \int_0^L \left[\frac{1}{2}\left(\frac{\partial u}{\partial x}\right)^2 + V(u)\right]dx \tag{2}$$

and $\alpha(x,t)$ is a two-dimensional gaussian white noise characterized by

$$E\,(\alpha(x,t)\ \alpha(x',t')) = \delta(x - x')\ \ \delta(t - t') \tag{3}$$

where E denotes the expectation.

From (2) it follows that (1) is the non-linear heat equation

$$\frac{\partial u}{\partial t} = \frac{\partial^2 u}{\partial x^2} - V'(u) + \varepsilon\alpha \tag{4}$$

We assume Dirichlet boundary conditions on the interval $[0,L]$ i.e. $u(0,t) = u(L,t) = 0$ for all t. In the concrete example considered here V(u) is the function

$$V(u) = \frac{\lambda}{4}\ u^4 - \frac{\mu}{2}\ u^2 \qquad \lambda,\ \mu > 0 \tag{5}$$

Our model problem is then directly related to the quantum mechanical double well anharmonic oscillator and makes contact with some recent attempts to describe quantum mechanics in terms of equilibrium states of non-linear stochastic differential equations in infinite dimensions. It is easy in fact to see that the formal equilibrium density of Eq. (4) is

$$\exp\ \left(-\frac{2}{\varepsilon^2}\ S(u)\right) \tag{6}$$

which is just the formal density in the functional integral description of quantum mechanics at imaginary times. However another interpretation is possible for Eq.(4). We may think of it as describing the motion of an elastic string in a high viscosity noisy environment (Smoluchowski approximation). This interpretation is actually a useful reference for intuition.

The picture that emerges from the analysis of Eq(4)$_1$ is the following. For sufficiently large values of $\mu^{\frac{1}{2}}$ L the deterministic string defined by the equation for $\varepsilon = 0$ will have two stable equilibrium position $\pm u_1$ corresponding to the two minima of the potential V(u) at $u = \pm (\mu/\lambda)^{\frac{1}{2}}$ as in Fig. 1.

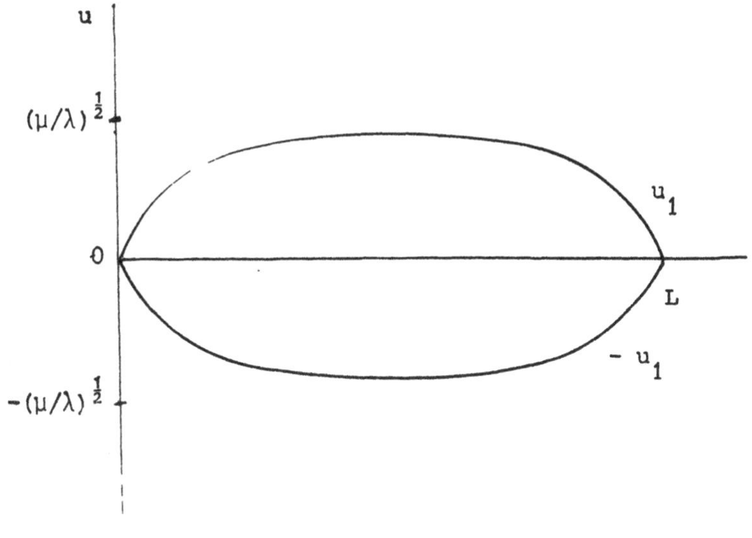

Fig.1 - The stable configurations.

These are absolute minima of S(u) that we shall call the equilibrium action. There are also a certain number (depending on the value of $\mu^{\frac{1}{2}}$ L) of instanton like or multi instanton like unstable equilibria . Fig. 2 shows one of these.

All these solutions are critical points of the equilibrium action S(u). The instantons or multi instanton solutions are saddle points. When the noise $\varepsilon\alpha$ is introduced the string will most likely perform small fluctuations near the stable configurations, but from time to time a particularly lucky fluctuation will

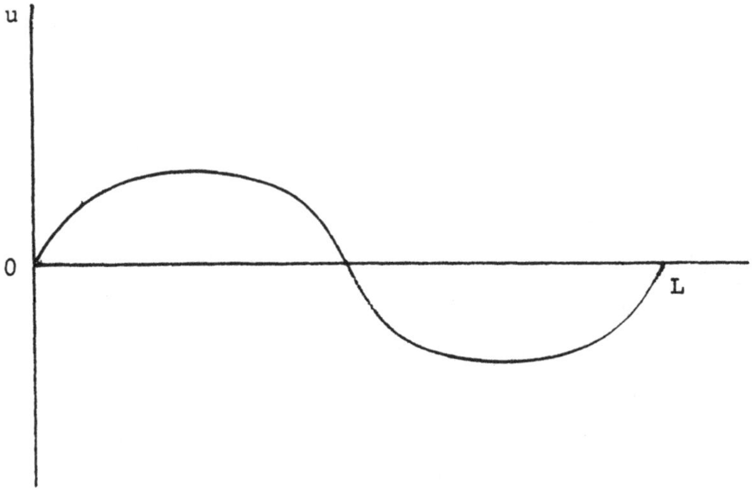

Fig. 2 - An instanton-like solution.

take it from one stable position to the other. We may call this event tunneling. The corresponding probability will be very small and its rough magnitude will be independent of the length of the finite time interval considered. We show that this magnitude is of the order of

$$\exp\left[-\frac{2}{\varepsilon^2} (S(u_2) - S(u_1)) \right] \tag{7}$$

where u_2 is the instanton like solution sketched above and u_1 is a stable equilibrium solution. We must emphasize that this picture is strictly connected with the circumstance that the space interval $[0,L]$ is finite.

We now give a brief outline of our strategy. After converting Eq.(4) into an integral equation using the solution of the linear part, we prove exixtence and uniqueness of global solutions, continuous with probability one. The ergodicity of the process and the existence of a unique equilibrium measure also follow. We then construct the large fluctuation theory for the process. This involves a functional I(u) that we call the action functional of the process (not to be confused with the equilibrium action S).

$$I(u) = \frac{1}{2} \int_0^T \int_0^L \left(\frac{\partial u}{\partial t} - \frac{\partial^2 u}{\partial x^2} + V'(u)\right)^2 dxdt \qquad (8)$$

If B is a Borel set in the space of continuous functions u satisfying Dirichlet boundary conditions and an initial condition $u(x,0) = u_0(x)$, we may also define

$$I(B) = \inf_{u \in B} I(u) \qquad (9)$$

The following estimates are then true
If B is open

$$- I(B) \leqslant \liminf_{\varepsilon \to 0} \varepsilon^2 \ln P(u \in B) \qquad (10)$$

If B is closed

$$\limsup_{\varepsilon \to 0} \varepsilon^2 \ln P(u \in B) \leqslant - I(B) \qquad (11)$$

where $P(u \in B)$ is the probability that $u \in B$.

We define next the event

$$A = \{ u(x,t): u(x,0) = u_0(x),\ u(x,T) \varepsilon Y \} \tag{12}$$

which will be a tunneling according to our previous definition, if we take u_0 in a small neighborhood N of $-u_1$ and the final u(x,T) in a small neighborhood Y of u_1. The topology of uniform convergence is understood.

The estimates needed to apply the lower and upper bounds (10) and (11) on the probability are upper and lower bounds of the functional I. The upper bound states that there is a neighborhood N of $-u_1$ so that for every $\rho > 0$, there is a $T < \infty$ such that

$$I(A) \leqslant 2(S(u_2) - S(u_1)) + \rho \tag{13}$$

This bound is proven by computing I(u) with a suitable trial function in A. The lower bound states that there is a neighborhood Y of u_1, so that for every $\rho > 0$ and every compact set of initial conditions K, there is a neighborhood N of $-u_1$ such that if $u_0 \varepsilon K \cap N$

$$2(S(u_2) - S(u_1)) - \rho \leqslant I(\bar{A}) \tag{14}$$

where $\bar{A}$ is the closure of A.

This bound is more difficult since one has to consider all possible tunneling trajectories. This requires a topological argument that involves the critical point structure of the equilibrium action S. This is a point where the infinite dimensional theory differs from the finite dimensional one for ordinary stochastic differential equations.

In fact S is not continuous in the space of continuous functions and it is remarkable that bounds like (13) and (14) hold.

We may summarize our results in the following

<u>Theorem</u> - There exists a neighborhood Y of u_1 such that for all $\rho > 0$ and all compact K, there exist a neighborhood N of $-u_1$ and a $T > 0$ so that for all u_0 in $K \cap N$ and all ε sufficiently small

$$- 2(S(u_2) - S(u_1)) - \rho \leqslant \varepsilon^2 \ln P(u \varepsilon A) \leqslant -2 (S(u_2) - S(u_1)) + \rho$$

A is the tunneling event (12).

This suggests the picture that in the limit of small noise, tunneling takes place through an instanton configuration.

In conclusion we would like to emphasize that equations similar in structure to the non-linear heat equation considered here arise in different areas of natural sciences. We may mention the phenomenological theories of such different phenomena as the diffusion of a fluid in a porous medium, transport in semiconductors, coupled chemical reactions with possibility of spacial diffusion and population genetics. Also the Navier-Stokes equation may be included in the family. In all these cases, due to the approximate phenomenological character of the equations, it is of interest to test how the solutions change under the effect of small stochastic perturbations. We hope that the methods we have developed will prove useful also in these cases.

REFERENCES

For a detailed bibliography we refer to our paper mentioned in the front page. Here we give a few titles on the mathematical background.

a) Stochastic partial differential equations

S.M.Kozlov: "Some Problems of Stochastic Equations with Partial Derivatives".
Works of the Petroski Seminar 4 (1978) 147 (Russian).
Ya.I.Belopolskaia, Yu.L.Daletzki: "Diffusion Processes in Smooth Banach Spaces and Manifolds". I
Works of the Moscow Math. Soc. 37 (1978) 107 (Russian).
R.Marcus: "Parabolic Ito Equations with Monotone Non-linearities".
J. Functional Analysis 29 (1978) 275.

b) Large Fluctuation Theory for Finite-dimensional Diffusion Processes

S.R.S.Varadhan: "Asymptotic Probabilities and Differential Equations".
Comm. Pure Appl. Math. 19 (1966) 261.
A.D.Ventzel, M.I.Freidlin: "On Small Random Perturbations of Dynamical Systems".
Russian Mathematical Surveys 25 (1970) 1.
A.D.Ventzel, M.I.Freidlin: "Fluctuations in Dynamical Systems under the Action of Small Random Perturbations".
Nauka, Moskva 1979 (Russian) (English translation by Springer-Verlag, to appear).

c) Non-linear Parabolic Equations and Related Stability Problems.

D.Henry: "Geometric Theory of Semilinear Parabolic Equations".
Lecture Notes in Math. 840, Springer, Berlin, 1981.

This book contains also some information on phenomenologies leading to non-linear parabolic equations.

SOME NEW INTEGRABLE MODELS IN FIELD THEORY AND STATISTICAL MECHANICS

Héctor de Vega

Laboratoire de Physique Théorique et Hautes Energies
Université Pierre et Marie Curie - Tour 16 ler étage
4, place Jussieu 75230 Paris Cedex 05 France

The triangular relations [1-3] (also called factorization equations or Yang-Baxter equations)

$$S_{ab}^{k\ell}(\theta)\, S_{kd}^{cm}(\theta+\theta')\, S_{\ell m}^{ef}(\theta') = S_{bd}^{\ell m}(\theta')\, S_{am}^{kf}(\theta+\theta')\, S_{kf}^{ce}(\theta) \qquad (1a)$$

are the clue of the resolution of two-dimensional integrable models in quantum field theory and statistical mechanics [3,4,5]. In (1) the indices run from one to N, $S_{k\ell}^{ij}$ are N^4 functions (in general complex) of the variable θ. The physical interpretation of the $S_{k\ell}^{ij}$ depends on the context where they are considered.

$S_{k\ell}^{ij}(\theta)$ can be interpreted as a two-body S-matrix in two-dimensional space-time. In that case i label the possible internal states of a particle. Namely $|i\,j\rangle$ is the ingoing two-particle state and $|k\ell\rangle$ the outgoing one. θ stands for the relative rapidity if we are in a relativistic theory and it stands for the relative momentum if we consider non-relativistic particles.

The n-body S-matrix can be expressed as a sum of products of two-body S-matrices and eq.(1) precisely guarantees that the factorization is consistent. It is useful to consider $S_{k\ell}^{ij}(\theta)$ as an operator in the tensor product of three one-particles spaces. In this notation (1) reads

$$S^{(12)}(\theta_1-\theta_2)\, S^{(13)}(\theta_1-\theta_3)\, S^{(23)}(\theta_2-\theta_3) =$$
$$= S^{(23)}(\theta_2-\theta_3)\, S^{(13)}(\theta_1-\theta_3)\, S^{(12)}(\theta_1-\theta_2) \qquad (1b)$$

Here

$$S^{(12)}(\theta_1-\theta_2)^{i_1 i_2 i_3}_{j_1 j_2 j_3} = S^{i_1 i_2}_{j_1 j_2}(\theta_1-\theta_2)\,\delta^{i_3}_{j_3} \text{ etc.}$$

Symmetry requirements impose further relations to the S-matrix which are not contained in (1) [3,5]. Namely

1) T-invariance $S = S^T$ or $S^{ij}_{kl}(\theta) = S^{kl}_{ij}(\theta)$ (2)

2) P-invariance $S = PSP$ or $S^{ij}_{kl}(\theta) = S^{ji}_{lk}(\theta)$ (3)

where $P^{ab}_{cd} = \delta^a_d\,\delta^b_c$

3) unitarity $SS^{\dagger} = 1$ or $S^{ij}_{kl}(\theta)\,S^{mn}_{kl}(\theta)^* = \delta^i_m\,\delta^n_j$ (4)

4) real analiticity

$$S^{ij}_{kl}(\theta^*)^* = S^{ij}_{kl}(-\theta) \tag{5}$$

5) internal symmetries. If the S-matrix is to be invariant under a symmetry group G acting on the particle states, we must have

$$S^{ij}_{kl}(\theta) = M^i_{i'}\,M^j_{j'}\,S^{i'j'}_{k'l'}(\theta)\,(M^{-1})^{k'}_k\,(M^{-1})^{l'}_l \tag{6a}$$

for all $M\in G$.

$$\text{or} \qquad S(\theta) = (M\otimes M)\,S\,(M^{-1}\otimes M^{-1}) \tag{6b}$$

we recall that $(A\otimes B)^{ab}_{cd} = A^a_c\,B^b_d$

From eqs. (2), (3) and (4) follows that

$$S^{ij}_{mn}(\theta)\,S^{mn}_{kl}(-\theta) = \delta^i_k\,\delta^j_l \tag{7}$$

This property is often called "unitarity relation".

Crossing symmetry implies

$$S(\eta-\theta) = (1\otimes\sigma)\,S(\theta)\,(\sigma\otimes 1) \tag{8a}$$

Here the matrix σ acts on the one particle states interchanging a particle with its antiparticle (crossing). In the S-matrix context $\eta = i\pi$. In an real basis this equation reads

$$S^{ij}_{kl}(\theta) = S^{il}_{kj}(\eta - \theta) \tag{8b}$$

Let us now discuss S in the context of classical statistical mechanics. In that case one can consider $S^{ij}_{kl}(\theta)$ as the statistical weight of a vertex configuration in a two dimensional lattice (fig 1). Here $i = 1,\ldots, N$ labels the possible states of a link. In this context S must be real and positive.

The transfer matrix connecting two neighboring rows in a NxN square lattice reads

$$T^{l_0 \{\nu'\}}_{l_N \{\nu\}}(\theta) = \sum_{l_1 \cdots l_{N-1}} \prod_{i=0}^{N-1} S^{l_i\, \nu_i'}_{l_{i+1}\, \nu_i}(\theta) \tag{9}$$

The partition function writes in term of the transfer matrix as

$$Z = \mathrm{Tr}\left[\tau(\theta)^N\right] \quad , \quad \tau^{\{\nu'\}}_{\{\nu\}} = \sum_l T^{l\{\nu'\}}_{l\{\nu\}} \tag{10}$$

and Tr means trace over vertical indices. Hence the free energy per site is given in the thermodynamic limit by the largest eigenvalue of τ.

If the local weights S fulfill the triangular relation (1), it follows that

$$[\tau(\theta), \tau(\theta')] = 0 \tag{11}$$

for all θ, θ'

Hence the eigenvectors of $\tau(\theta)$ are θ-independent and one can hope to diagonalize simultaneously $\tau(\theta)$ for all θ.

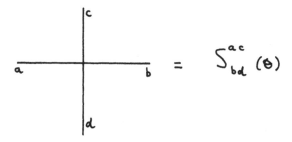

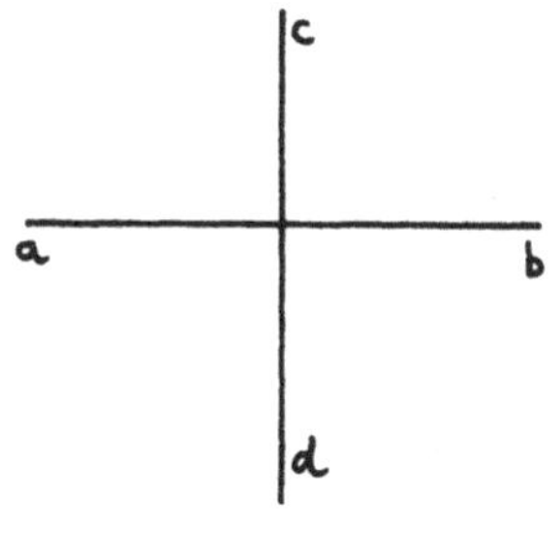

$$= S^{ac}_{bd}(\theta)$$

Figure 1.

A one-dimensional quantum hamiltonian can be derived from $\tau(\theta)$ through

$$H = \frac{d}{d\theta} \operatorname{Log} \tau(\theta) \Big|_{\theta=0} \tag{12}$$

From eqs (11)-(12) follow that H belongs to a family of an infinite number of commuting operators. They are the coefficients of the expansion of $\ln \tau(\theta)$ in powers of θ .

It is important to note that in many cases eqs(1), (2), (4) and (7) allow to compute the largest eigenvalue of $\tau(\theta)$ without constructing the corresponding eigenvector[6-8]. From eqs(7) and (8) follows that

$$S^{ab}_{cd}(\theta)\, S^{ed}_{cf}(\eta+\theta) = \delta^{a}_{e}\, \delta^{b}_{f} \tag{13}$$

[This equation is also useful to derive the quantum Gelfand-Levitan equation, see sections 2 and 3 in Faddeev's lectures.] Eqs (7) and (10) generalize for the transfer matrix as follows

$$\begin{aligned} T(\theta)\, T(-\theta)^{T} &= \mathbf{1} \\ T(\theta)\, T^{t}(\eta+\theta) &= \mathbf{1} \end{aligned} \tag{14}$$

Here

$$\left(T^{T}\right)^{k\{v\}}_{\ell\{w\}} = T^{\ell\{w\}}_{k\{v\}} \qquad\qquad \left(T^{t}\right)^{k\{v\}}_{\ell\{w\}} = T^{\ell\{v\}}_{k\{w\}}$$

One can define another partition function

$$Z' = \operatorname{Tr}\left[T(\theta)^{N}\right] \tag{15}$$

It differs from the former one (eq. (10)) by a slight charge in the boundaries[8]. Hence Z = Z' in the $N\to\infty$ limit. In this limit the largest eigenvalue of $T(\theta)$ must be of the form $\Lambda(\theta)^{N}$ and from eqs (14)

$$\begin{aligned} \Lambda(\theta)\, \Lambda(-\theta) &= 1 \\ \Lambda(\theta)\, \Lambda(\eta+\theta) &= 1 \end{aligned} \tag{16}$$

We can assume analiticity of $\Lambda(\theta)$. This can be justified both in the S-matrix and in the statistical mechanical context. It then follows that

$$\Lambda(\theta) = 1 \quad , \qquad Z = 1 \tag{17}$$

In other words if the statistical weights are normalized such that S verifies unitarity the corresponding partition function equals unity. This means in practice that the computation of Z has been reduced to a manipulation of $N^2 x N^2$ matrix at one vertex (a two-body problem).

FIELD THEORIES ASSOCIATED TO ROOT SYSTEMS OF LIE ALGEBRAS AND GENERALIZED HEISENBERG MODELS

Let $\mathcal{A}$ be a simple Lie Algebra and $\mathcal{H}$ its Cartan subalgebra $\mathcal{H} = \{h_1, \ldots, h_m\}$. The roots $(r_{\alpha_j})^i$ as it is known are the eigenvalues of the linear form $[h_i, \cdot]$ acting in the adjoint representation

$$[h_i, e_{\alpha_j}] = (r_{\alpha_j})^i \, e_{\alpha_j} \quad \text{(no sum over j)} \qquad i = 1, \ldots, m$$

We normalize the eigenvectors e_{α_j} and the h_i such that in the adjoint representation

$$\mathrm{Tr}(h_i h_j) = \delta_{ij} \quad , \quad \mathrm{Tr}(e_{\alpha_i} e_{-\alpha_i}) = 1$$

It is possible to associate to any simple Lie algebra a field theory admitting a Lax representation as follows[9]

$$\left[\partial_x + \partial_t \vec{\varphi}\cdot\vec{h} + \frac{m}{\beta}\sum_{\rho_j\in\Delta} \ell^j\left(\lambda e_j + \frac{e_{-j}}{\lambda}\right)\right]\Psi = 0$$

$$\left[\partial_t + \partial_x \vec{\varphi}\cdot\vec{h} + \frac{m}{\beta}\sum_{\rho_j\in\Delta} \ell^j\left(\lambda e_j - \frac{e_{-j}}{\lambda}\right)\right]\Psi = 0 \tag{18}$$

Here $\ell^j = e^{\beta\vec{\varphi}\cdot\vec{\rho}_j}$, λ is the spectral parameter, Δ is a system of n+1 roots formed by n simple roots plus the sum of all simple roots with opposite sign and $\vec{\varphi} = (\varphi_1, \ldots, \varphi_m)$.

The compatibility condition of eqs (18) give the Euler-Lagrange equations of the lagrangian

$$\mathcal{L} = \frac{1}{2}(\partial_\mu\vec{\varphi})^2 - \frac{m^2}{\beta^2}\sum_{\rho_j\in\Delta} e^{2\beta\vec{\varphi}\cdot\vec{\rho}_j} \tag{19}$$

Let us now consider the quantization of this field theory using the Quantum Inverse Problem Method[4]. This leads as always to ultraviolet divergences that one must regularize. Lattice regularization is particularly appropriated here. In this way the linear system (18) becomes a finite difference system. The choice of discretization is a non-trivial dynamical problem because one needs to preserve integrability on the lattice (At least with enough approximation to take the continuous limit at the end).

We define the local transition operator [10]

$$L_n(\lambda) = u_n - \frac{m}{\beta}\sum_{\Delta} \ell_n^{j}\left(\lambda e_j + \frac{1}{\lambda} e_{-j}\right) \tag{20}$$

so that

$$\psi_{n+1} = L_n(\lambda)\,\psi_n \tag{21}$$

Here

$$u_n = \exp\left(-\int_{x_n-\delta}^{x_n} dx\, \vec{\pi}\cdot\vec{h}\right),\ \ell_n^{j} = \delta\exp\left(\frac{\beta}{\delta}\int_{x_n-\delta}^{x_n} dx\, \vec{\rho}_j\cdot\vec{\varphi}\right),\ \vec{\pi} = \dot{\vec{\varphi}} \tag{22}$$

δ stands for the lattice spacing. In the $\delta \to 0$ limit we recover eq.(18). We have the commutation relations

$$u_n\,\ell_m^{j} = e^{i\beta\,\delta_{nm}\,\vec{h}\cdot\vec{\rho}_j}\,\ell_m^{j}\,u_n \tag{23}$$

We would like to note that L_n is an ultralocal operator [4], namely it depends on quantum operators only at site n. For Lie algebra A_1 it reduces to the L_n of sine-Gordon [11].

The transfer operator is now

$$T_N(\lambda) = \prod_{-N}^{+N} \overleftarrow{L_n(\lambda)} \tag{24}$$

Let us look for a numerical matrix R such that

$$R(\lambda,\mu)\left[T_N(\lambda)\otimes T_N(\mu)\right] = \left[T_N(\mu)\otimes T_N(\lambda)\right] R(\lambda,\mu) \tag{25}$$

Because of ultralocality it is enough to have

$$R(\lambda,\mu)\left[L_n(\lambda)\otimes L_n(\mu)\right] = \left[L_n(\mu)\otimes L_n(\lambda)\right] R(\lambda,\mu) \tag{26}$$

We have explicitly found the R matrix (quantum monodromy matrix) for the Lie Algebra A_n. In that case $\mathcal{L}$ possesses a Z_{n+1} invariance. Thus, we impose eq.(6) to R where $M_{ab} = \delta_{a,b+1}$ $(1 \leq a,b \leq n+1)$

After some work [10] we found that the solution reads

$$\begin{aligned} R_{ii}^{ii} &= w \\ R_{\ell k}^{k\ell} &= w\,\frac{\operatorname{sh}\theta}{\operatorname{sh}(\theta+\gamma)} \qquad k \neq \ell \\ R_{k\ell}^{k\ell} &= w\,\frac{\operatorname{sh}\gamma}{\operatorname{sh}(\theta+\gamma)}\exp\left\{\theta\left[\operatorname{sig}(k-\ell) - 2\,\frac{k-\ell}{n+1}\right]\right\},\ k\neq\ell \end{aligned} \tag{27}$$

where θ, γ are functions of $\lambda-\mu$ and β respectively.

From eq. (26) we get as a consistency condition that R itself satisfies the triangular equations (1). R also verifies the "T-invariance" (2) and the normalization equation (7) if we agree that $w(\theta)\,w(-\theta) = 1$. In the other hand, for n > 1, R is not unitary, neither parity invariant, nor crossing invariant become eqs (4), (8) and (3) do not hold. Hence we cannot interpret R as a physical relativistic S-matrix in the usual way. R has a non-relativistic interpretation as scattering matrix for particles of type 1,2...,n with a complex potential. A minimal solution (in the sense of ref (3)) is [10]

$$w = \frac{\operatorname{sh} \pi(k+\gamma)}{\operatorname{sh} \pi k} \; \frac{\Gamma(1-ik-\gamma)\,\Gamma(\gamma-ik)}{\Gamma(1-ik)\,\Gamma(-ik)} \quad (k = \text{momentum}) \tag{28}$$

This corresponds to a N-body Schrödinger equation with potential

$$V_{\ell j}(x) = \gamma(\gamma-1)\left[\frac{\delta_{\ell j}}{\operatorname{sh}^2 x} - \frac{1-\delta_{\ell j}}{\operatorname{ch}^2\left(x + i\frac{\pi}{2}\Delta_{\ell j}\right)}\right] \tag{29}$$

$$\Delta_{\ell j} \equiv \operatorname{sig}(\ell-j) - \frac{2(\ell-j)}{n+1}$$

In the context of classical statistical mechanics the elements of R can be viewed as the weights of a integrable vertex model (fig 1) . We find in general (n+1)(2n+1) non vanishing weights. For n = 1 we recover the six-vertex model.

As usual a magnetic quantum hamiltonian can be derived from the T matrix associated to our solution R. We write it as

$$H = -\operatorname{sh}\gamma \,\frac{\partial}{\partial\theta} \log \tau(\theta)\Big|_{\theta=0} - 1\, N \cosh\gamma \tag{30}$$

and we find [12]

$$H = -\sum_{j=1}^{N}\left\{\sum_{1\le a\ne b\le n+1} e_j^{ab}\, e_{j+1}^{ba} + \operatorname{ch}\gamma \sum_{a=1}^{n+1} e_j^{aa}\, e_{j+1}^{aa} + \operatorname{sh}\gamma \sum \Delta_{ab}\, e_j^{aa}\, e_{j+1}^{bb}\right\}$$

This hamiltonian describes a chain of (n+1) component spins on a line with nearest neighbour interactions. For n = 1 we recoves the XXZ model. If $\operatorname{Im}\gamma = 0$ or $\operatorname{Im}\gamma = \pi$ H is a hermitian operator. In the first case we have a ferromagnetic interaction and an antiferromagnetic interaction in the second case.

Notice that under parity (exchange of site j with site

$N-j, 1 \leq j \leq N$) the hamiltonian is not invariant (for n>1) since the last term changes sign. We say that H is antiferromagnetic for $\mathrm{Im}\,\gamma = \pi$ because if the spin at the site j has the value r $(1 \leq r \leq n+1)$, the interaction favours the value $r \pm 1$ for the spins at sites $j \pm 1$ $(j \mp 1)$ for $\mathrm{Re}\,\gamma > 0$ $(\mathrm{Re}\,\gamma < 0)$.

In the ferromagnetic case the ground state of H is

$$\| 1 \rangle = | 1 \rangle_1 \otimes | 1 \rangle_2 \otimes \cdots \otimes | 1 \rangle_N \tag{32}$$

In the antiferromagnetic case this reference state $\| 1 \rangle$ is not the ground state $\| \Psi \rangle$. $\| \Psi \rangle$ corresponds to the maximal eigenvalue of $\tau(\theta)$. We obtain it [12] by the QIPM [4] using the commutation relations that follows from eq (25). Namely

$$\begin{aligned} T_{11}(\theta)\, B_a(\theta') &= v(\theta'-\theta)\, B_a(\theta')\, T_{11}(\theta) - p_{a1}(\theta'-\theta)\, B_a(\theta)\, T_{11}(\theta') \\ T_{ij}(\theta)\, B_a(\theta') &= B_a(\theta')\, T_{ij}(\theta) + p_{ia}(\theta-\theta')\, B_i(\theta')\, T_{aj}(\theta) - p_{ij}(\theta-\theta')\, B_i(\theta)\, T_{aj}(\theta') \end{aligned} \tag{33}$$

$$v(\theta) = \frac{\mathrm{sh}(\theta+\gamma)}{\mathrm{sh}\,\theta}\,, \quad p_{ab}(\theta) = \frac{\mathrm{sh}\,\gamma}{\mathrm{sh}\,\theta}\, e^{\Delta_{ab}\theta} \quad (a \neq b)\,, \quad p_{aa}(\theta) = v(\theta) - 1\,.$$

Here, the operators

$$B_i(\theta) \equiv T_{i1}(\theta) \quad ; \quad i = 2, \ldots, n+1$$

create pseudoparticles. We built the eigenvectors of τ by filling the reference state with pseudoparticles.

$$\| X \rangle = \sum_{\sigma \in S_p} X_\sigma\, B_{\sigma(a_1)}(\theta_1) \cdots B_{\sigma(a_p)}(\theta_p)\, \| 1 \rangle \tag{34}$$

It can be shown that there exist coefficients X_σ such that $\| X \rangle$ is an eigenstate of τ. The detailed proof is given in ref (12). The idea is as follows : one applies $\tau(\mu)$ on the vector (34) and uses the permutation relation (33). In that way one gets two kinds of terms : some of them are of the type (34) and others not. There bad terms contain some $B_i(\mu)$. They disappear if X is a common eigenvectors of a set of p new transfer matrices. The key point is that these new transfer matrices act on spins of n components instead of n+1. Hence by iterating n times this procedure the problem is solved. One has at the end n-nested Bethe ansätze. The corresponding eigenvalue equations read

$$\begin{aligned} \delta_{\ell 1}\, N\, \Theta(\lambda_j^1, \gamma) &= 2\pi i\, I_j^{(\ell)} + \sum_{i=1}^{p_{\ell-1}} \Theta(\lambda_i^{\ell-1} - \lambda_j^\ell, \gamma) + \\ &+ \sum_{i=1}^{p_{\ell+1}} \Theta(\lambda_i^{\ell+1} - \lambda_j^\ell, \gamma) - \sum_{i=1}^{p_\ell} \Theta(\lambda_i^\ell - \lambda_j^\ell, 2\gamma) \,. \qquad \begin{array}{l} \ell = 1, \ldots, n \\ j = 1, \ldots, p_\ell \end{array} \end{aligned} \tag{35}$$

Here the $I_j^{(\ell)}$ are integers (or half-integers) that label the eigen-vectors of $\tau(\mu)$ and

$$\Theta(\lambda,\gamma) = -2 \operatorname{arctg}\left[\frac{\operatorname{tg}\lambda}{\operatorname{th}\left(\frac{\gamma}{2}\right)}\right] \tag{36}$$

In the thermodynamic limit the distribution of λ_j^ℓ becomes continuous with a density

$$\rho_m(\lambda) = \lim_{N\to\infty} \frac{1}{N(\lambda_{j+1}^m - \lambda_j^m)}, \quad m=1,\dots,n \tag{37}$$

The transcendental eq.(35) gives in the $N=\infty$ limit a system of linear integral equation

$$-\rho_\ell(\lambda) + \int_{-\pi/2}^{\pi/2} \frac{d\mu}{2\pi} \sum_{m=1}^{n} K_{\ell m}(\lambda-\mu)\rho_m(\mu) = \frac{\delta_{\ell 1}}{2\pi} \frac{\partial}{\partial\lambda}\Theta(\lambda,\gamma) \tag{38}$$

Here

$$K_{\ell m}(\lambda) = \delta_{\ell m} \frac{\partial\Theta(\lambda,2\gamma)}{\partial\lambda} - (\delta_{\ell,m+1} + \delta_{\ell,m-1}) \frac{\partial\Theta}{\partial\lambda}(\lambda,\gamma)$$

The solution reads for the ground state

$$\rho_\ell(\lambda) = \frac{1}{\pi} \sum_{k=-\infty}^{+\infty} e^{2i\lambda k} \frac{\operatorname{sh}\gamma[n+1-\ell]k}{\operatorname{sh}\gamma(n+1)k}$$

This gives for the free energy per site of the (n+1)(2n+1)-vertex model

$$f = -\frac{1}{N}\log\Lambda_{max} = \frac{2n\mu}{n+1} + 2\sum_{k=1}^{\infty} \frac{e^{-k\gamma}}{k} \frac{\operatorname{sh}(n\gamma k)\,\operatorname{sh}(2k\mu)}{\operatorname{sh}[(n+1)\gamma k]} \tag{39}$$

For n = 1 this reduces to the free energy of the six-vertex model [13]. For $\gamma\to 0$ and arbitrary n we recoves the results of ref (14). The ground state energy of the spin hamiltonian (31) follows from eqs (30) and (39)

$$E_0 = \frac{2n\operatorname{sh}\gamma}{n+1} - \operatorname{ch}\gamma + 4\operatorname{sh}\gamma\sum_{k=1}^{\infty} e^{-k\gamma}\frac{\operatorname{sh} n\gamma k}{\operatorname{sh}(n+1)\gamma k}$$

We expect to discuss the structure of excitations elsewhere[15].

REFERENCES

[1] C.N. YANG, Phys. Rev. Lett. 19, 1312 (1967)

[2] R.J. BAXTER, Ann. Phys. 70, 193 and 323 (1972)

[3] A.B. ZAMOLODCHIKOV and Al. B. ZAMOLODCHIKOV, Ann. Phys. 80, 253 (1979)

M. KAROWSKI, Phys. Reports 49C, 229 (1979)

[4] L.D. FADDEEV, lectures in this summer school
L.D. FADDEEV, Soviet Scientific Reviews (section C) 1, 107 (1980)

[5] P.P. KULISH and E.K. SKLIANIN, Tvärminne lectures. J. Hietarinta and C. Montonen eds. Springer Verlag

P.P. KULISH and E.K. SKLIANIN, Zapisky Nauchny Seminarov LOMI 95, 129 (1980).

[6] R.J. BAXTER, Fundamental Problems in Statistical Mechanics, Vol. V p. 109. E.G.O. Cohen editor (1980)

[7] Yu. G. STROGANOV, Phys. Lett. 74A, 116 (1979)
A.B. ZAMOLODCHIKOV, Comm. Math. Phys. 69, 165 (1979)

[8] R. SHANKAR, Yale preprint, YTP-81-21, 1981

[9] A.V. MIKHAILOV, ZhETF Pis 30, 443 (1979) ; JETP Lett. 30, 414 (1980.

A.N. LEZNOV and M.V. SAVELIEV, Lett. Mat. Phys. 3, 489 (1979).

S.A. BULGADAEV, Phys. Lett. 96B, 151 (1980)

[10] O. BABELON, H.J. DE VEGA and C.M. VIALLET, Nucl. Phys. B (FS), B190, 542 (1981).

[11] E.K. SKLIANIN, L.A. TAKHTADZHYAN and L.D. FADDEEV, Teor. Mat. Fiz. 40, 194 (1979) ; Theor. Math. Phys. 4D, 688 (1980)

[12] O. BABELON, H.J. DE VEGA and C.M. VIALLET, Paris preprint, LPTHE 81/09, Nucl. Phys. B (FS) (to be published).

[13] See for example E.A. Lieb and F.Y. Wu in Phase Transitions and Critical Phenomena, vol. 1 ; Ed. by C. Domb and M.S. Green 1972.

[14] B. SUTHERLAND, Phys. Rev. B12, 3795 (1975)
P.P. KULISH and N.Yu. RESHETIKHIN, ZhETF 80, 214 (1981)

[15] O. BABELON, H.J. DE VEGA and C.M. VIALLET (in preparation)

CALCULATING THE LARGE N PHASE TRANSITION IN LATTICE GAUGE THEORIES*

Stuart Samuel

Physics Department
Columbia University
New York, New York 10027

Abstract

Aspects of the large N phase transition in lattice gauge theories are reviewed.

INTRODUCTION

The work on calculating the large N phase transition was done with Fred Green at the Institute for Advanced Study.

The large N limit in continuum perturbation theory (pt), which picks out the planar Feynman diagrams,[1] appears to be an almost insoluble problem. Only in two dimensions has any progress been made.[2] In order to avoid renormalization problems, it is natural to consider large N on the lattice. Then one can attack the large N limit without having to deal simultaneously with ultraviolet infinities. As a consequence of this synthesis, the synthesis of large N and the lattice, new notions have been introduced: the loop space,[3] the lattice large N Schwinger-Dyson equations,[4] and the collective field theory method[5]. Unfortunately, these methods have made little progress in d=4 $U(\infty)$ QCD. What has emerged is the phenomenon of the large N phase transition. In d=2 the lattice gauge theory is solvable for all N including $N=\infty$ for the same reasons as in the continuum: After going to the $A_0=0$ gauge gluonic degrees of freedom decouple, thereby yielding N^2 (in the U(N) case) or N^2-1 (in the SU(N) case) non-interacting massless vector particles, an extremely trivial situation since such particles have no polarization (or physical)

*This research was supported in part by the U.S. Dept. of Energy.

degrees of freedom in d=2. On the lattice the $A_0=0$ gauge corresponds to setting the time-like link variables, the U's, equal to one. A shift in integration variables reduces the model to the so-called one plaquette problem governed by the partition function:

$$Z = (\mathfrak{Z}_0)^{\text{volume}} ,$$

$$\mathfrak{Z}_0 = \int dU \exp\left[\beta N(\mathrm{Tr}\, U + \mathrm{Tr}\, U^\dagger)\right] \qquad (1)$$

U should be thought of as the 1-plaquette variable, that is, the product of the four link variables around an elementary square.

Gross and Witten[6] and Wadia[7] computed Eq.(1) in the $N\rightarrow\infty$ limit and discovered a phase transition as a function of β. The important features are

(a) The critical coupling, $\beta_c \equiv \frac{1}{g_c^2 N}$, is .5.

(b) The transition is third order with the beta function merely undergoing a change in slope of β_c .

(c) The phase transition only occurs in the $N = \infty$ theory; for finite N Eq.(1) is an entire function of β.

(d) Confinement is not lost. Even on the weak coupling (wc) side the Wilson loop possesses the area law. In this sense, the transition is innocuous.

Last year Fred Green and I set out to investigate the large N phase transition phenomenon in higher dimensions. Our goals were to understand, to calculate, and to prove its existence. We were, in fact, able to achieve all these goals.[8,9] Most of this talk will be a discussion of our results. I will, however, mention interesting recent developments (i.e., those since Sept., 1980).

The $N = \infty$ model will be defined as the $N \rightarrow \infty$ limit of the U(N) lattice gauge theory. This is the same as the $N \rightarrow \infty$ limit of the SU(N) model. Our methods, however, involve a limiting procedure which works only in the U(N) case.

Whenever a statistical mechanics system has a phase transition, it is useful to search for an order parameter which distinguishes the phases. After much thought, particularly in four dimensions, we came to suspect that the determinant of the Wilson loop(WL) might be such an order parameter. Recall that on the lattice a WL is a product of matrix link variables around a closed loop. As such, it, too, is a matrix. Usually, one takes the trace. Instead, take the determinant. For the Wilson action (WA) we were able to prove that in all dimensions

$$\langle \det \mathrm{WL} \rangle^{1/N} \underset{N \rightarrow \text{large}}{\sim} \begin{cases} \exp(-\text{area}) & \text{for sc} \\ 1 & \text{for wc,} \end{cases} \qquad (2)$$

(sc = strong coupling; wc = weak coupling). Eq.(2) proves the existence of at least one phase transition. Nothing can be said of its order since the free-energy is not calculated.

At some point in going from sc to wc $< \det WL >^{1/N}$ loses its area law behavior. This is the phase transition point, β_c. As a check, we calculated the determinational order parameter in the d=2 case. The first part of the calculation is simple: Because a determinant of a product of matrices is the product of the determinants, the determinant of a rectangular WL is the product of the determinants of the plaquette variables contained inside the rectangle:

$$< \det WL > = < \prod_{\substack{\text{plaquettes}, p, \\ \text{in } WL}} \det U_p > \qquad (3)$$

Since the model reduces to the 1-plaquette model Eq.(3) becomes

$$<\det WL > = <\det U >^{\text{area}}_{\text{1-plaquette}} . \qquad (4)$$

The expectation in Eq.(4) is calculated using the partition function in Eq.(1). The second stage of the calculation, that of evaluating Eq.(4) in the large N limit, is mathematically complicated because the usual saddle point methods do not work. Instead, Eq.(4) was computed via a sc beta expansion. We had to (i) use large N character expansion techniques developed in earlier work,[10,11] (ii) multiply U(N) group characters of large order, and (iii) know the dimensionalities of representations of the permutation groups of high order. When the dust settled, the first N terms were obtained

$$< \det U > = (\beta N)^N \left[\sum_{\ell=0}^{N} (-1)^\ell \frac{(\beta N)^{2\ell}}{\ell!} \frac{1}{(N+\ell)!} \right] + O(\beta^{2N+2}). \qquad (5)$$

These terms happen to coincide with the first N terms of the Bessel function, $J_N(2\beta N)$. Using uniform asymptotic expansions for Bessel functions of large argument and large order[12]

$$< \det U >^{1/N} \rightarrow \begin{cases} \exp\left[\sqrt{1-4\beta^2} - \log\left(\frac{1+\sqrt{1-4\beta^2}}{2\beta}\right)\right] & \beta < \tfrac{1}{2} \\ 1 \text{ for } \beta > \tfrac{1}{2} . \end{cases} \qquad (6)$$

Eqs.(4) and (6) exemplify Eq.(2). β_c, as calculated this way, is the self-same β_c of .5.

It is convenient to define the U(1) string tension, $\alpha^{U(1)}$, (not to be confused with the usual (trace of WL) string tension) via

$$a^2 \alpha^{U(1)} = - \frac{1}{N} \frac{1}{\text{area}} \log < \det WL > \qquad (7)$$

where a is the lattice spacing. In terms of $\alpha^{U(1)}$, Eq.(2) is

$$\alpha^{U(1)} = \begin{cases} 0 & \text{for wc} \\ \text{positive} & \text{for sc.} \end{cases} \tag{8}$$

For d=2, $\alpha^{U(1)}$ is plotted in Fig.1. Summarizing, $< \det WL >^{1/N}$ is a perfect order parameter in d=2.

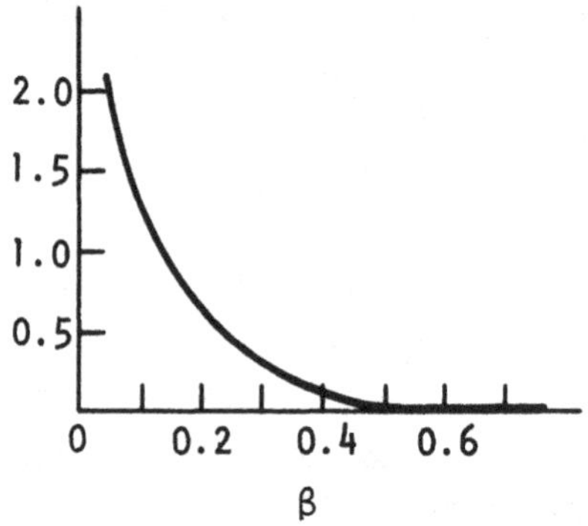

Figure 1. $\alpha^{U(1)}$ in one plaquette model.

The relevance of <det WL> has been checked in two other exactly solvable N = ∞ toy models.[13] Unfortunately, <det WL > and hence β_c have only been approximately calculated (via a sc expansion). For the gauge model on a periodic cube, the results[14] are

$$\beta_c^{\text{approx.}} \approx .32 \ , \quad \beta_c^{\text{exact}} = .33 \tag{9}$$

and for the gauge model on a tetrahedron, the results[14] are

$$\beta_c^{\text{approx.}} \approx .39 \ , \quad \beta_c^{\text{exact}} = \frac{\pi}{8} \approx .393. \tag{10}$$

The agreement in Eqs.(9) and (10) is remarkable considering only a few sc expansion terms were used. < det WL> is a powerful probe in large N matrix U(N) models. In unsoluable systems, it is perhaps the only way to estimate β_c and get a handle on the large N phase transition phenomenon.

At this point, I would like to point out some of its properties: First of all, it is the product of the eigenvalues of the WL matrix, i.e., an N-fold product. Consequently, the Nth root in Eq.(2) is quite natural to assure a smooth large N limit. Secondly, the center of U(N) is U(1), that is, those diagonal matrices proportional to the identity up to a phase. Write the

U(N) WL matrix as a U(1) phase times on SU(N) matrix:

$$WL = \exp\left[i\,\theta \right] \times \left[SU(N)\ \text{matrix} \right] \tag{11}$$

Then

$$\det WL = \exp\left[i\,N\,\theta \right]. \tag{12}$$

Thus, det WL probes the U(N) group via the U(1) center. Roughly speaking

The large N phase transition is closely related to the center of the group fluctuations.	(13)

A detailed discussion of the above statement can be found in Ref.9. Lastly, the mapping $U \to \det U$ is an irreducible one-dimensional representation and character of U(N). Introduce into the theory a particle in this representation. Such a particle is a "colored baryon", i.e., a completely antisymmetric N fold tensor product. In our U(N) theory this colored baryon charge has been gauged. Eq. (2) says that this charge is confined in sc via a string tension proportional to N. In wc this property is lost.

Proving Eq.(2) is actually quite simple. In wc it suffices to use perturbation theory and large N counting. In sc, large sc methods are used. For the details of the proof see Ref.9.

In the SU(N) theory, det WL is identically one and hence a useless operator. To conclude anything about the SU(∞) model, an indirect line of reasoning must be used. The difference between the free energies in the U(N) and SU(N) cases is of order $1/N^2$ in wc and exponentially small in N (for the Wilson action (WA)). Hence, U(N) and SU(N) theories become equal in the large N limit. This can be understood in terms of the difference between the two group manifolds.[8,9] As a consequence, the SU(∞) model must have a phase transition because the U(∞) model does.

$\alpha^{U(1)}$ provides a means for computing β_c. According to Eq.(8), $\alpha^{U(1)}$ is zero in wc and hence need not be calculated there. In sc, $\alpha^{U(1)}$ can be computed order by order in sc series. Each order yields a value of beta where $\alpha^{U(1)}$ vanishes. This sequence of numbers converges to the exact β_c. This was the method used to obtain Eqs.(9) and (10) and was the method we used in the d=4 U(∞) gauge model. Order by order here are the zero's in $\alpha^{U(1)}$:

Order	$\alpha^{U(1)}$ zero	
4	.405	
6	.418	
8	.400	
10	.401	
12	.396	(14)

As you can see, the zero is quite stable. In d=4 there must be a large N phase transition near beta of .40.

Recently, two Monte Carlo investigations have corroborated this result. Actually in d=4, we predicted that the U(N) lattice gauge theory, even for finite N, should have a phase transition. We suggested that the large N transition is the large N limit of these finite N transitions. For U(1), U(2) and U(3) the critical couplings[15], obtained via Monte Carlo methods to an accuracy of about 10% are

$$\beta_c^{U(1)} \approx .50 \quad \beta_c^{U(2)} \approx .41 \quad \beta_c^{U(3)} \approx .38 \tag{15}$$

In d=4 for $N \geq 4$, SU(N) WA gauge theories also have phase transitions. For N=4,5, and 6 the critical couplings are[16]

$$\beta_c^{SU(4)} \approx .31 \quad \beta_c^{SU(5)} \approx .34 \quad \beta_c^{SU(6)} \approx .36 \tag{16}$$

Both Eqs.(15) and (16) are in good agreement with our prediction.

The phase transitions in finite N SU(N) d=4 lattice gauge theories came as a surprise to the theoretical community and was not predicted by us. In fact, even for SU(2) a transition can take place if the WA action is modified to include an adjoint representation term:

$$\text{Action} = \sum_p \left[2\beta_f (\text{Tr}\, U_p + \text{Tr}\, U_p^\dagger) + \beta_A \, \text{Tr}\, U_p \, \text{Tr}\, U_p^\dagger \right] \tag{17}$$

The theoretical analysis of Eq.(17) has been performed in Ref.17. Creutz and Bhanot[18] studied it via Monte Carlo methods and found first order transitions across the lines in Fig.(2). It is suspect-

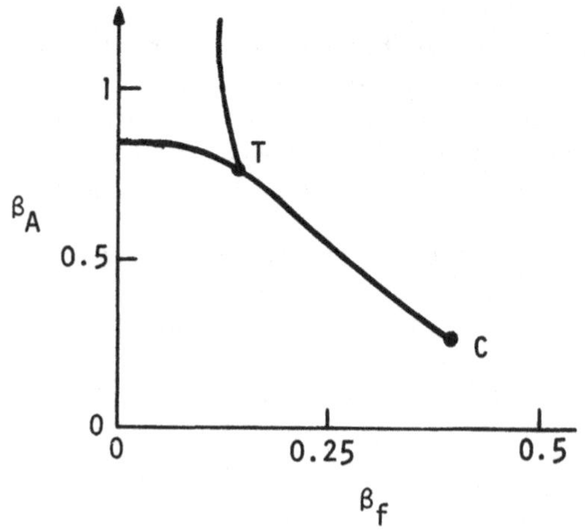

Figure 2.

ed that for $N \geq 4$ the line from T to C crosses the $\beta_A = 0$ horizontal axis thereby yielding the transitions in pure WA theory. Probably by making β_A negative this line can be avoided but as $N \to \infty$ I would conjecture that the point C moves off to infinity in the negative β_A direction thereby making avoidance impossible.

Fred Green and I predicted that monopoles and the center of the group would play a role in the d=4 N=∞ transition. Already in SU(2), the center, Z_2, and monopoles are responsible for the transitions in Fig.(2).[17]

The large N phase transition is an artifact of the lattice. It occurs at intermediate coupling and cannot be seen in the continuum limit which is sensitive only to an infinitesimal region near $g^2N=0$. In d=2 when the continuum limit is taken,[7] the critical coupling, which corresponds to a critical distance, gets pushed off to infinity.

The artificial nature of the N=∞ transition is perhaps more evident in that there exist "good" lattice actions[19] which completely avoid the N=∞ singularity in d=2.[20] One such action, the heat kernel (or non-Abelian Villain) action, is particularly interesting because it appears to resemble the continuum theory more than the WA for a wider range of coupling. In that model in d=2, < det WL> indicates no phase transition,[21] as should be the case. There is no mismatch in N dependence. Instead of Eq.(2),

$$< \det WL > \sim \exp \left[-C(\beta) + O\left(\frac{1}{N^2}\right)\right], \qquad (18)$$

(where C is independent of N) both in wc and sc and in all dimensions. An analysis of this model in d=3 and d=4 was recently carried out.[21] In d=3 the conclusions were not definite but in d=4 there must be a large N phase transition. For finite N C in Eq.(18) is proportional to the perimeter of the WL in wc and proportional to its area in sc. As $N \to \infty$ the critical coupling remains finite and non-zero and is approximately 3. Thus, in d=4 the heat kernel action does not avoid one undesirable feature of the WA.

ACKNOWLEDGEMENTS

I deeply thank K. Pohlmeyer, J. Honerkamp, H. Römer and the Albert-Ludwigs-Universität in Freiburg, Germany for their hospitality and generosity.

REFERENCES

1. G. 't Hooft, Nucl. Phys. B72, (1974) 61.
2. G. 't Hooft, Nucl. Phys. B75, (1974) 461.
3. A. M. Polyakov, Phys. Letts. 82B, (1979) 247; Nucl. Phys. B164 (1980) 171.
4. Y. Nambu, Phys.Letts. 80B, (1979) 372; J.L. Gervais and A. Neveu, Phys. Letts. 80B, (1979) 255; E. Corrigan and B. Hasslacher, Phys. Letts. 81B, (1979) 181; D. Foerster, Phys. Letts. 81B, (1979) 87; T. Eguchi, Phys. Letts. 87B, (1979) 81; D. Weingarten, Phys. Letts. 87B, (1979) 97; Yu M. Makeenko and A.A. Migdal, Phys. Letts. 88B, (1979) 135.
5. A. Jevicki and B. Sakita, Phys. Rev.D22, (1980) 467; Nucl. Phys. B165, (1980) 511.
6. D. J. Gross and E. Witten, Phys. Rev. D21,(1980) 446.
7. S. Wadia, A Study of U(N) Lattice Gauge Theory in 2-Dimensions, University of Chicago Preprint, EFI 79/44 (July, 1979).
8. F. Green and S. Samuel, Phys. Letts. 103B, (1981) 48.
9. F. Green and S. Samuel, Calculating the Large N Phase Transition in Gauge and Matrix Models, to appear in Nucl. Phys. B.
10. F. Green and S. Samuel, Nucl. Phys. B190 [FS3], (1981) 113.
11. S. Samuel, J. Math. Phys. 21, (1980) 2695.
12. See M. Abramowitz and I.A. Stegun, Handbook of Mathematical Functions (National Bureau of Standards)Tenth Edition (Dec.,1974) Formula 9.3.2.
13. F. Green and S. Samuel, Phys. Letts. 103B, (1981) 110.
14. D. Freidan, Comm. Math. Phys. 78 (1981) 353;R. Brower, P. Rossi, and C. Tan, Phys. Rev.D23, (1981) 942; Phys. Rev. D23, (1981) 953.
15. M. Creutz, Phys. Rev. Letts. 43,(1979) 553; B. Lautrup and M. Nauenberg, Phys. Letts. 95B, (1980) 63; K.J. M. Moriarty, Phys. Rev. D24, (1981); M. Creutz and K.J.M. Moriarty, Phase Transitions in U(2) and U(3) Lattice Gauge Theory in Four Dimensions, Brookhaven National Lab. preprint (July,1981).
16. M. Creutz, Phys. Rev. Letts. 46, (1981) 1441; K.J.M. Moriarty, A Phase Transition in SU(4) Four-Dimensional Lattice Gauge Theory, Royal Holloway College preprint (May, 1981); M. Creutz and K.S. M. Moriarity, Phase Transition in SU(N) Lattice Gauge Theory for Large N, Brookhaven Nat. Lab. preprint (June, 1981)
17. I.G. Halliday and A. Schwimmer, Phys. Letts. 101B,(1981) 327; Phys. Letts. 102B, (1981) 337; J. Greensite and B. Lautrup, Phys. Letts. 47, (1981) 9; R. C. Brower, D. A. Kessler, and H. Levine, Monopole Condensation and the Lattice QCD Crossover, HUTP-81/A018 (May, 1981); L. Caneschi, I.G. Halliday, and A. Schwimmer, The Phase Structure of Mixed Lattice Gauge Theories, ICTP/80/81-40 (July, 1981).
18. G. Bhanot and M. Creutz, Variant Actions and Phase Structure in Lattice Gauge Theory,BNL 29640 (May, 1981).

19. J.M. Drouffe, Phys. Rev.D18, (1978) 1174; N.S. Manton, Phys. Letts. B96, (1980) 328.
20. C.B. Lang, P. Salomonson, and B.S. Skagerstam, Nucl. Phys.B190, (1980) 337; P. Menotti and E. Onofri, Nucl. Phys. B190, (1981) 288.
21. S. Samuel, On the Large N Phase Transition in the Heat Kernel Lattice Gauge Theory, CU-TP-203 (Sept., 1981).

EXCITATION SPECTRUM OF FERROMAGNETIC xxz-CHAINS

T. Schneider and E. Stoll

IBM Zurich Research Laboratory
8803 Rüschlikon, Switzerland

INTRODUCTION

The history of xxz-Heisenberg spin chains, defined by the Hamiltonian

$$\mathcal{H} = -J \sum_{\ell=1}^{N} (S_\ell^x S_{\ell+1}^x + S_\ell^y S_{\ell+1}^y + \Delta S_\ell^z S_{\ell+1}^z) , \tag{1}$$

is a rather rich one. It extends over more than fifty years and includes some of the major achievements in mathematical physics.[1-3]

Despite this activity, mainly devoted to the diagonalization of the Hamiltonian in the quantum case,[1-3] and the integrability in the classical counterpart,[4] our understanding of measurable properties and in particular of dynamic form factors (DFF) is much less advanced. In the Schrödinger representation, DFF's are defined by

$$S_{AA}(q,\omega) = \frac{1}{Z} \sum_{\lambda,\lambda'} e^{-\beta\omega_\lambda} |\langle\lambda'|A_q|\lambda\rangle|^2 \, \delta(\omega-\omega_{\lambda'}+\omega_\lambda) , \tag{2}$$

where

$$\mathcal{H}|\lambda\rangle = \omega_\lambda|\lambda\rangle , \quad Z = \sum_\lambda e^{-\beta\omega_\lambda} , \quad \beta = \frac{1}{T} , \tag{3}$$

$$A_q = \frac{1}{\sqrt{N}} \sum_{\ell} e^{iq\ell} A_\ell \ . \tag{4}$$

q denotes the pseudomementum, and ω the energy transfer in the scattering process. T is the temperature, Z the partition function, and A_ℓ an operator of interest. Depending on the choice of A_ℓ, DFF's are seen to probe particular parts of the eigenvalue spectrum, because the accessible spectrum is determined by the matrix elements of A, which are subject to selection rules. Operators A of interest and labeling of the associated DFF are listed in Table 1. $S_{xx}(q,\omega)$, $S_{yy}(q,\omega)$ and $S_{zz}(q,\omega)$ are accessible with neutron-scattering techniques, while $S_{11}(q,\omega)$ can be measured with light-scattering methods.

Before turning to a review of our theoretical understanding of DFF of ferromagnetic xxz-chains as a tool to probe and interprete the excitation spectrum, we briefly review the history and the state of the art.

As shown in Fig. 1, the xxz-chain includes various regimes and special cases for $J > 0$ [Eq. (1)], depending on the value of Δ. For $\Delta > 0$, the ground state is a ferromagnetic (F) and, for $0 \le \Delta < 1$, planar exchange anisotropy, and uniaxial anisotropy for $\Delta > 1$. Special cases are: $\Delta = 0$ the isotropic xx-chain; $\Delta = 1$ the isotropic Heisenberg ferromagnet, and for $\Delta \to \infty$ the Ising ferromagnet. For $\Delta < 0$, the ground state is antiferromagnetic (AF), and planar $-1 < \Delta < 0$, as well as $\Delta < -1$ uniaxial exchange anisotropy is present. $\Delta = -1$ corresponds to the isotropic AF, and $\Delta < -1$ to the uniaxial or Ising-Heisenberg AF.

Since the pioneering work of Bethe in 1931,[3] considerable progress has been made in the diagonalization problem for spin $s = 1/2$. There is the Bethe Ansatz method,[1-3] the relation between the transfer matrix of the two-dimensional six-vertex model and the xxz-

Table 1. Labeling of DFF Associated with Particular Operators A_ℓ.

S^x_ℓ	S^y_ℓ	S^z_ℓ	$S^-_\ell S^-_{\ell+1} + S^+_\ell S^+_{\ell+1}$
$S_{xx}(q,\omega)$	$S_{yy}(q,\omega)$	$S_{zz}(q,\omega)$	$S_{11}(q,\omega)$

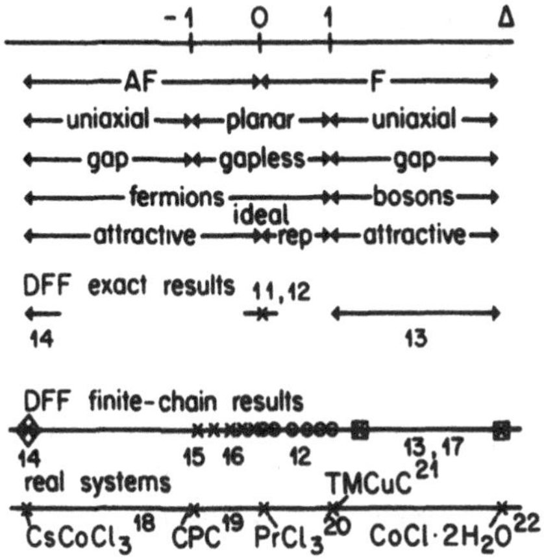

Figure 1. History and state of the art of the evaluation and interpretation of the eigenvalue spectrum and of DFF as explained in the text.

Hamiltonian for $s = 1/2$[5] and the quantum inverse scattering technique.[1,2] These approaches and in particular the work of Johnson, Krinsky and McCoy[5] have led to a rather complete picture of the eigenvalue spectrum. Thus, we know that there are regimes with and without a gap (Fig. 1).

Our understanding of the excited states has been deepened by using the Jordan-Wigner-fermion representation for the $s = 1/2$ spin operators, as introduced by Lieb, Schultz and Mattis[6,7] or by using the Dyson-Maleev representation,[8-10] working best for large s-values. In these representations, the system can be viewed as interacting fermions or bosons, respectively (Fig. 1). For $\Delta = 0$, the fermions are free.[6]

In spite of these achievements, exact results for DFF are available in certain regions and for $T = 0$ only, except for the xx-chain (Δ=0), corresponding to free fermions (Fig. 1).[11-14] This incompleteness reflects the mathematical complexity of the eigenvectors and eigenvalues. Considerable insight has also been gained from finite-chain calculations, where $\mathcal{H}$ is diagonalized numerically.[12-17] There is also an impressive list of quasi-1-d compounds which seem to be well described by the xxz-Hamiltonian. Examples are listed in Fig. 1.

We are now prepared to discuss various results for DFF for $s = 1/2$, with the objective of unravelling the physical origin of the resonance structure.

RESULTS AND DISCUSSION

Ising-Heisenberg Chain

The eigenvalue spectrum of this system, as defined by Hamiltonian (1) for $\Delta \geq 1$ is well known for $s = 1/2$, and consists of magnons, magnon bound-states and magnon continua. Rewriting Hamiltonian (1) in the form

$$\mathcal{H} = -J \sum_{\ell} \frac{1}{g}(S_{\ell}^{x} S_{\ell+1}^{x} + S_{\ell}^{y} S_{\ell+1}^{y}) + S_{\ell}^{z} S_{\ell+1}^{z} , \tag{5}$$

the magnon and n-th magnon bound-state frequencies relative to the ground state are given by[5]

$$\omega_1(q) = J(1 - \frac{1}{g} \cos q) , \tag{6}$$

$$\omega_n(q) = \frac{J}{g} \frac{\sinh\phi}{\sinh n\phi} (\cosh n\phi - \cos q) , \quad \cosh\phi = g , \tag{7}$$

while the two-magnon continuum is bounded by

$$2J(1 + \frac{1}{g} \cos q) = \omega_2^{TC} \geq \omega \geq \omega_2^{BC} = 2J(1 - \frac{1}{g} \cos q) . \tag{8}$$

The resulting dispersion curves up to two flipped spins are shown in Fig. 2 for $1/g = 0.13$ corresponding to an Ising-like system, where large gaps appear at $q = 0$. The simplicity of the ground state, corresponding to aligned up or down spins, makes it possible for the equation of motion of the Green's functions associated with the $S_{AA}(q,\omega)$ to be calculated exactly at $T = 0$.[12,13] The results are

$$S_{xx}(q,\omega) = S_{yy}(q,\omega) = \frac{1}{4} \delta\big(\omega - \omega_1(q)\big) , \tag{9}$$

$$S_{zz}(q,\omega) = 0 , \quad \text{for } g > 1 , \tag{10}$$

$$S_{11}(q,\omega) = \frac{\omega_2(q)}{J}\,\delta\big(\omega-\omega_2(q)\big)\ ,\quad \omega \geq \omega_2^{TC}(q) \tag{11}$$

and

$$S_{11}(q,\omega) = \frac{2}{\pi J}\left(\frac{\cos q/2}{g}\right)^2 \frac{B}{B^2+\left(2\left(\frac{\cos q/2}{g}\right)^2-2+\frac{\omega}{J}\right)^2} \tag{12}$$

for $\omega_2^{BC} \leq \omega \leq \omega_2^{TC}$, where

$$JB = \sqrt{(\omega-\omega_2^{BC})\ (\omega_2^{TC}-\omega)}\ . \tag{13}$$

Accordingly, $S_{xx}(q,\omega) = S_{yy}(q,\omega)$ probes the magnons, while $S_{11}(q,\omega)$, associated with the fluctuations of two adjacent spin deviations (Table 1), probes the two-magnon bound state and the two-magnon continuum.

At finite temperatures, however, the problem becomes much more complicated. In fact, transitions between excited states will give rise to additional contributions. To clarify the detailed origin of this thermally-induced structure, we calculated $S_{xx}(q,\omega)$ and

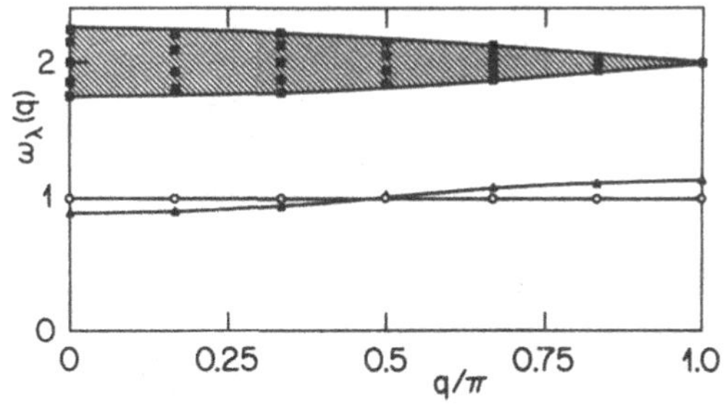

Figure 2. Dispersion curves for $1/g = 0.13$. Solid line with triangles, magnons; solid line with circles, two-magnon bound state. The shaded area is the two-magnon bound continuum. Triangles, circles and squares denote magnons, two-magnon bound states, and two-magnon continuum states, respectively for N = 12, subjected to periodic boundary conditions.

$S_{11}(q,\omega)$ numerically for finite chains subjected to periodic boundary conditions. Clearly, for fixed q, $S(q,\omega)$ will consist of discrete lines of width $\Delta\omega$ whose height is calculated on the basis of Eq. (2) by using the numerically-evaluated eigenvalues and eigenfunctions of Refs. 12 and 13. Figure 3 shows results for $q = \pi/2$ at $T = 0$ and $T/J = 1/3$, for an Ising-like chain ($1/g=0.13$). Even though $T/J = 1/3$ belongs to the low T regime, where the spectrum is still dominated by the magnon creation resonance, interesting thermally-induced structure appears, which on the basis of Eq. (2) can be traced back to particular transitions. The dominant contributions are also listed in Fig. 3. A particularly interesting feature is the appearance of a thermally-induced central peak (CP), associated with transitions involving bound states. Comparison with the corresponding results for $1/g = 0.8$, shown in the lower part of Fig. 3, reveals, however, that the thermally-induced spectrum becomes rather broad by approaching the Heisenberg limit. The effect clearly reflects the more pronounced dispersion and the considerably smaller gaps in the nearly isotropic system. Nevertheless, the physical origin of the thermally-induced but smeared structure turns out to be quite similar to that in the Ising-like case. Next, we turn to the results for $S_{11}(\pi/2,\omega)$ shown in Fig. 4. Comparison with the exact $T = 0$ result for $N = \infty$ clearly reveals that $N = 12$

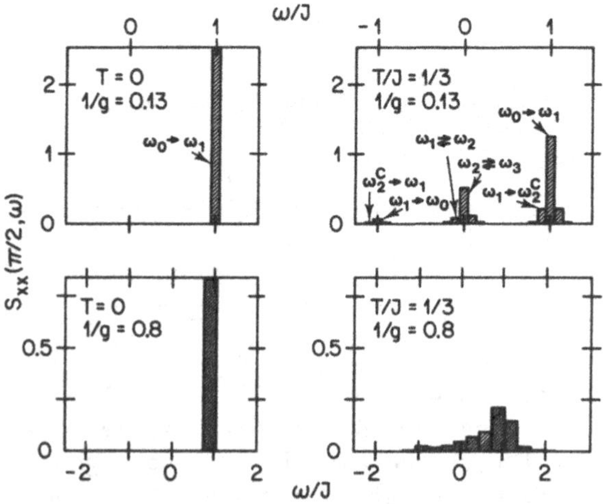

Figure 3. $S_{xx}(q,\omega) = S_{yy}(q,\omega)$ for $N = 12$ at $q = \pi/2$ for $1/g = 0.13$ and 0.8 and $\Delta\omega/J = 0.1$ and 0.3, respectively. ω_n denotes for $n = 0$: ground state; $n = 1$: magnon; $n \geq 2$ n-th magnon bound state. ω_n^c labels the n-th magnon continuum. The arrows denote the transitions.

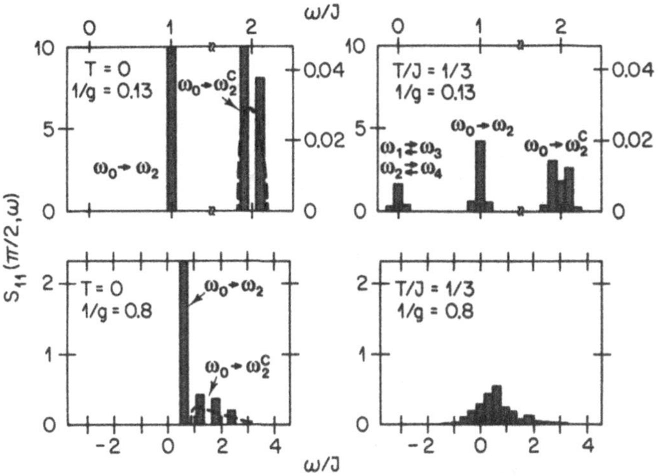

Figure 4. $S_{11}(q,\omega)$ for N = 12 at $q = \pi/2$ for $1/g = 0.13$ and 0.8 and $\Delta\omega/J = 0.1$ and 0.3, respectively. Otherwise, notation as in Fig. 3.

is sufficient to elucidate the physical origin of the resonances associated with the infinite system. At T = 0, we have the expected two-magnon bound state and two-magnon continuum resonance [Eqs. (11) and (12)]. The thermally-induced structure leads to a broadening of the two-magnon bound-state resonance, the corresponding destruction peak and a CP, which again is due to transitions involving bound states. For comparison, we also included the corresponding results for $1/g = 0.8$. There is again a pronounced smearing effect resulting from the more pronounced dispersion, the smaller binding energy of the two-magnon bound state and the smaller gaps in the nearly isotropic system.

Planar-Heisenberg Chain ($0 \le \Delta < 1$)

This model system, defined by Hamiltonian (1), is considerably more complicated, as the exact expression for the ground-state energy per spin for $s = 1/2$ indicates. It is given by[23]

$$\omega_0/J = -\frac{\Delta}{4} - \frac{\sin\mu}{2}\int_{-\infty}^{+\infty} dx\, \frac{\sinh(\pi-\mu)x}{\sinh\pi x\, \cosh\mu x}\,, \qquad (14)$$

where

$$\cos\mu = -\Delta \ . \tag{15}$$

It includes the xx-chain ($\Delta=0$) and the isotropic model ($\Delta=1$) as special cases. In the xx-limit, all properties of interest can be calculated exactly,[11] because the system can be mapped on noninteracting fermions with the aid of the Jordan-Wigner representation of the s = 1/2 spin operator.[6,7] According to Johnson et al.,[5] the lowest eigenvalues are given by

$$\omega_1^{BC} = \frac{J}{2}\frac{\pi\sin\mu}{\mu}\sin q \le \omega \le \omega_1^{TC} = J\frac{\pi\sin\mu}{\mu}\sin q/2 \tag{16}$$

and

$$\omega_n(q) = J\,\pi\frac{\sin\mu}{\mu}\frac{1}{\sin y_n}\sin q/2\,\sqrt{1-\cos^2 q/2\,\cos^2 y_n}\ , \tag{17}$$

where

$$y_n = \frac{n\pi}{2}\left(\frac{\pi}{\mu}-1\right) < \frac{\pi}{2}\ . \tag{18}$$

Equation (16) describes a continuum corresponding to particle-hole pair excitations in the fermion language, and ω_n are "bound states" lying above this continuum. The occurrence of the highest bound state is restricted by Eq. (18).

In terms of the fermion operators, the s = 1/2 Hamiltonian reads

$$\mathcal{H} = -\frac{JN}{4}\Delta + J\sum_k(\Delta-\cos k)\,a_k^+ a_k - \frac{J\Delta}{N}\sum_{k_1+k_2=k_3+k_4}\cos(k_1-k_4)\,\cdot\,a_{k_1}^+ a_{k_2}^+ a_{k_3} a_{k_4}\ , \tag{19}$$

where the third term represents the interaction, vanishing in the xx-limit where $\Delta = 0$. To explore the resonance structure of the dynamic form factor in the presence of the interaction, we evaluated $S_{zz}(q,\omega)$ in the Hartree-Fock approximations (HF) and performed finite-chain calculations.[12] The HF approximation yields for the ground state

$$\omega_0 = -\frac{J}{\pi} + \frac{J}{\pi^2}\Delta\ , \tag{20}$$

for the renormalized particle-hole pair continuum

$$J(1-\frac{2\Delta}{\pi})\ \sin q \le \omega \le 2J(1-\frac{2\Delta}{\pi})\ \sin q/2 = \omega_1^{TC}\ , \tag{21}$$

and for the first "bound state" ($\Delta\to 0$)

$$\frac{\omega_1(q) - \omega_1^{TC}(q)}{\omega_1^{TC}(q)} = 2\Delta^2 \sin^2 q/2\ . \tag{22}$$

These results turn out to be identical with a leading-order expansion in Δ of the corresponding exact results given by Eqs. (14)-(18). Consequently, the planar xxz-chain represents an example where the HF approximation becomes exact, namely, to leading order in the coupling constant Δ. Therefore, it is interesting to discuss the HF expression for $S_{zz}(q,\omega)$. For the PHP contribution, we obtain at $T = 0$,

$$S_{zz}(q,\omega) = \frac{1}{\pi} \frac{\sqrt{d}}{\left(\sqrt{d}+ \frac{f(q,\omega)\Delta}{2\pi^2}\ \ln \frac{\omega_{TC}+\sqrt{d}}{\omega_{TC}-\sqrt{d}}\right)^2 + \frac{f^2(q,\omega)\Delta^2}{4\pi^2}}\ , \tag{23}$$

where

$$\left.\begin{aligned} d &= \left(\omega_1^{TC}(q)\right)^2 - \omega^2 \\ f &= 2\pi \frac{\omega^2-2(1-\frac{2\Delta}{\pi})J^2(1-\cos q)\cos q}{(1-\frac{2\Delta}{\pi})J(1-\cos q)} \end{aligned}\right\} . \tag{24}$$

Above ω_1^{TC}, we find the PHP bound-state resonance

$$S_{zz}(q,\omega) = 2\Delta|\sin q/2|\,\delta\left(\omega-\omega_1(q)\right)\ . \tag{25}$$

Important features are the removal of the square-root singularity at the top of the PHP continuum appearing in Eq. (23) at $\Delta = 0$ (xx-chain) and the occurrence of the PHP bound state above the continuum, reflecting the repulsive nature of the fermion interaction. The HF results for $S_{zz}(q,\omega)$ are illustrated in Fig. 5.

This HF picture, valid to leading order in an expansion in Δ, will be modified in higher orders as follows: From Eqs. (14)-(18)

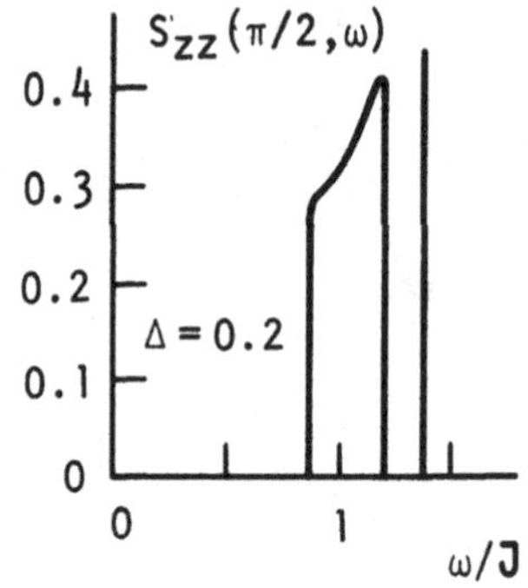

Figure 5. $S_{zz}(\pi/2,\omega)$ according to HF [Eqs. (23) and (25)] at T = 0 and Δ = 0.2.

and (20)-(22), we see that higher orders further renormalize the fermion pseudoparticle, the PHP and their bound-state frequencies. Moreover, as the exact expression for the Green's function associated with $S_{zz}(q,\omega)$ reveals, multiple PHP excitations and bound states thereof (Δ>0) become possible. These excitations are suppressed in HF by construction. Therefore, the weak contributions to $S_{zz}(q,\omega)$ seen in Fig. 6 for Δ = 0.4 and 0.99 must be attributed to multiple-pair excitations. For Δ > 0.5, the second bound state is expected. However, the finite-chain results shown in Fig. 6 for Δ = 0.99 provide no evidence for its appearance in $S_{zz}(q,\omega)$ because there is no weight at about $\omega \approx \omega_2(q)$ [Eq. (17)]. This might be understood by recognizing that only the first-bound state survives in the limit $\Delta \to 1$, because at Δ = 1[13]

$$S_{zz}(q,\omega) = \frac{1}{4}\,\delta\left(\omega - \omega_1(q)\right) , \qquad (26)$$

where $\omega_1(q)$ [Eq. (17)] reduces to the magnon frequency.

The crossover from the PHP bound state to the elementary boson (magnon) also reveals that the so-called des Cloiseaux-Pearson "magnon"[24] with frequency $J(\pi/\mu)\sin\mu\,\sin q$ actually corresponds to the renormalized fermion pseudoparticle or hole excitations given in HF by

$$\omega_k = -J(1 - \frac{2\Delta}{\pi})\cos k \qquad (27)$$

with k replaced by $\pm(\pi/2\ q)$ or $\pm(\pi/2-q)$, respectively.

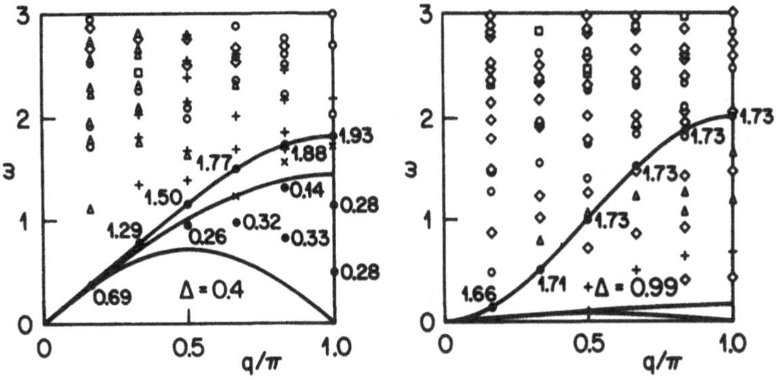

Figure 6. Spectral representation of $S_{zz}(q,\omega)$ at T = 0 for Δ = 0.4 and 0.99, respectively, and N = 12. Full lines denote lower, upper boundary of the PHP continuum [Eq. (16)] and first bound state [Eq. (17)], respectively. Numbers symbols denote the value of $2\pi S_{zz}(q,\omega)$, where × : 10^{-3}; + : 10^{-4}; Δ : 10^{-5}; ○: 10^{-6}; □: 10^{-7} and ◇: $\leq 10^{-8}$.

We hope to have demonstrated that the ferromagnetic s = 1/2 xxz-chain not only represents an interesting model, but that the results reviewed here represent our most complete understanding of DFF associated with nontrivial many-body systems. Moreover, it provides instructive connections between spin systems, interacting fermions and bosons. Finally, the results discussed here include various new aspects: Examples being thermally-induced bound-state effects in terms of central peaks in DFF for Ising-like xxz-chains; the possibility to observe bound states in $S_{zz}(q,\omega)$, accessible by neutron-scattering techniques, in the planar case. Nevertheless, much remains to be done, in particular in the planar regime, where the S-dependence and the modifications introduced by thermal fluctuations have not yet been studied.

We have benefited from illuminating discussions with A. Baratoff, H. Beck, C. P. Enz, U. W. Glaus, H. R. Jauslin and K. A. Müller.

REFERENCES

1 L. A. Takhtadzhan and L. D. Faddeev, Russian Math. Surveys 34:5, 11 (1979).
2 H. B. Thacker, Rev. Mod. Phys. 53:253 (1981).
3 H. A. Bethe, Z. Phys. 71:205 (1931).

4 E. K. Sklyamin (preprint).
5 J. D. Johnson, S. Krinsky, and McCoy, Phys. Rev. A 8:2526 (1973).
6 E. Lieb, T. Schultz, and D. Mattis, Ann. Phys. (NY) 16:407 (1961).
7 A. Luther and I. Peschel, Phys. Rev. B 12:3908 (1975).
8 F. D. Dyson, Phys. Rev. 102:1217, 1230 (1956).
9 S. V. Maleev, Sov. Phys. – JETP 6:776 (1956).
10 T. Schneider, Phys. Rev. B 24:5327 (1981).
11 S. Katsura, T. Horiguchi, and M. Suzuki, Physica 46:67 (1970).
12 T. Schneider and E. Stoll, J. Appl. Phys., (1982), to be pulished.
13 T. Schneider and E. Stoll, Phys. Rev. Lett. 47:377 (1981).
14 N. Ishimura and H. Shiba, Prog. Theor. Phys. 63:743 (1980).
15 G. Müller, H. Beck, and J. C. Bohner, Phys. Rev. Lett. 43:75 (1979).
16 G. Müller, H. Thomas, M. W. Puga, and H. Beck, J. Phys. C 14:3999 (1981).
17 T. Schneider and E. Stoll, to be published.
18 U. Tellenbach and H. Arend, J. Phys. C 10:1311 (1977).
19 Y. Endoh, G. Shirane, R. J. Birgenau, P. M. Richards, and S. L. Holt, Phys. Rev. Lett. 32:170 (1974).
20 J. P. Harrison, J. P. Hessler, and D. R. Taylor, Phys. Rev. B 14:2979 (1976).
21 C. P. Landee and R. D. Willett, Phys. Rev. Lett. 43:463 (1979).
22 J. B. Torrance Jr. and M. Tinkham, Phys. Rev. 187:595 (1969).
23 R. J. Baxter, Ann. Phys. (NY) 70:323 (1974).
24 J. des Cloiseaux and J. J. Pearson, Phys. Rev. 128:2131 (1962).

COMPUTER SIMULATIONS OF A DISCONTINUOUS PHASE TRANSITION AND PERCOLATION CLUSTERS IN THE TWO-DIMENSIONAL ONE-SPIN-FLIP ISING MODEL

E. Stoll and T. Schneider

IBM Zurich Research Laboratory
8803 Rüschlikon, Switzerland

The films "Computer Simulation of a Discontinuous Phase Transition in the Two-Dimensional One-Spin-Flip Ising Model" and "Percolation Clusters" show the temporal evolution of Ising spin configurations. The system is brought to nonequilibrium by an abrupt change of the applied field or temperature, and the approach to the stable equilibrium is shown in the films.

The first film shows the time evolution of the spin states in a system undergoing a discontinuous phase transition. For a detailed description we refer to Ref. 1. By applying a strong positive field, all spins initially point upwards. The field is then switched off at a temperature close but below the critical temperature. The system then evolves towards the equilibrium state. Not all reversed spins are equally distributed over the whole system. Due to the spin-spin interaction, the spins form clusters of up and down separated by walls. As equilibrium is reached, one observes that the size of the clusters fluctuates about a mean value. Due to the rather low density of large clusters, cluster models describe this regime quite well.[2]

By applying a negative field, the cluster size is seen to increase monotonically. Before reaching the critical size, growth becomes slower. This effect is very pronounced at small fields, and the system seems to evolve to a metastable state with finite but long lifetime. Having reached the critical size, the growth rate is accelerated, and new effects occur. Critical clusters touch one another. The cluster border and the associated border energy no longer increase when the clusters grow. Consequently, transition is accelerated until saturation effects close to the new equilibrium occur.

The films illustrate very well that the dynamics of discontinuous transitions is dominated by spin processes occurring close to the cluster borders. One expects, therefore, that first-order transitions depend heavily on the topology of the clusters forming the new phase. The importance of critical clusters indicates that the system size might influence the lifetime of metastable states. In fact, Langer,[3] as well as Penrose and Lebowitz[4] argued that the lifetime vanishes for very large systems.

However, we have shown[5] that the size dependence of the lifetime of metastable states can be subdivided into three regions. For small systems, it is determined by the time needed to form subcritical clusters, giving rise to a vanishing order parameter. In an intermediate region, critical clusters can occur, and the probability of their occurrence increases with the spin number. For large systems, however, this probability no longer depends on the system size, since several critical clusters appear at a time. The lifetime is therefore merely determined by the above-mentioned processes, which link the clusters together. One is then naturally led to the conclusion that the lifetime of metastable states remains finite in the thermodynamic limit, too.

In the percolation film, we studied the conducting probability in the presence of blocking and conducting elements. The transition between blocked and conducting states merely depends on the geometrical distribution of these elements. In our computer simulation, we mapped such a system to a spin-1/2 Ising model. Conducting particles correspond to a spin + 1, and blocking particles to a spin - 1. In random percolation, where geometrical properties are important only, noninteracting spins are assumed. The distribution between conducting and blocking particles is driven by an applied field. For a more detailed discussion, we refer to Ref. 6 and the references therein, where a brief description is given of the basic concepts which arise in percolation theory. The nature of the percolation transition is illustrated by "snapshot" frames from a film of a Monte Carlo simulation of a percolation process with the percolation probability varying from 0 to 1.21 p_c (where p_c is the critical probability). The largest cluster is identified in each frame, and when p is near to p_c this cluster approaches the boundaries. Finally, a short description is given of some more sophisticated percolation problems.

REFERENCES

1 T. Schneider and E. Stoll, *in*: "Anharmonic Lattices, Structural Transitions and Melting," T. Riste, ed., Nordhoff, Leiden (1974), pp. 275-315.
2 E. Stoll, K. Binder, and T. Schneider, *Phys. Rev. B* 6:2777 (1972).
3 J. S. Langer, *Ann. Phys. (NY)* 54:258 (1969).
4 O. Penrose and J. L. Lebowitz, *J. Stat. Phys.* 3:211 (1971).
5 E. Stoll and T. Schneider, *Physica* 86-88B:1419 (1977).
6 C. Domb, E. Stoll, and T. Schneider, *Contemp. Phys.* 21:577 (1980).

POLYACETYLENE: A REAL MATERIAL LINKING CONDENSED MATTER AND FIELD THEORY

David K. Campbell

Theoretical Division and
Center for Nonlinear Studies
Los Alamos National Laboratory
Los Alamos, New Mexico 87545 USA

INTRODUCTION

It sounds at first bizarre, but one of the most active subjects of recent research at the interface between field theory and statistical mechanics concerns a real material that seems more properly to belong to the arcana of organic chemistry than to either field theory <u>or</u> statistical mechanics. The material is polyacetylene ($(CH)_x$), a quasi-one dimensional organic polymer with some very interesting and potentially exotic properties.[1] In this necessarily brief discussion, I will try to give a pedagogical introduction to some of the recent theoretical studies[2-9] of polyacetylene and to convey some of the reasons for the intense excitement in this area. To provide this overview, I will have to survey the work of many individuals, and thus most of what I will present is not my own research. The small part that I have contributed has been done in a close and fruitful collaboration with Alan Bishop and Klaus Fesser.[9,10]

Although I will treat here a number of important theoretical aspects of $(CH)_x$--some background chemistry, a simple continuum model, kink and polaron nonlinear excitations, fractionally charged solitons, confinement of kink-like solutions in <u>cis</u>-$(CH)_x$, and the field theory connection--sadly, I can only refer obliquely to some other equally important features, including the lattice model of $(CH)_x$,[4] the (currently controversial) experimental situation,[1,11] and potential technological applications.[1] I hope the interested reader will delve further into this fascinating subject.

BACKGROUND CHEMISTRY

As the chemical formula $(CH)_x$ suggests, polyacetylene consists of a large grouping of C (carbon)-H (hydrogen) units. The material occurs in two isomeric forms: trans-$(CH)_x$ and cis-$(CH)_x$. Here I shall focus almost exclusively on trans-$(CH)_x$, which is the more stable isomer at room temperature[1] and which has been the major subject of theoretical investigations.[2-9] The schematic chemical structure of trans-$(CH)_x$ is shown in Fig. 1.

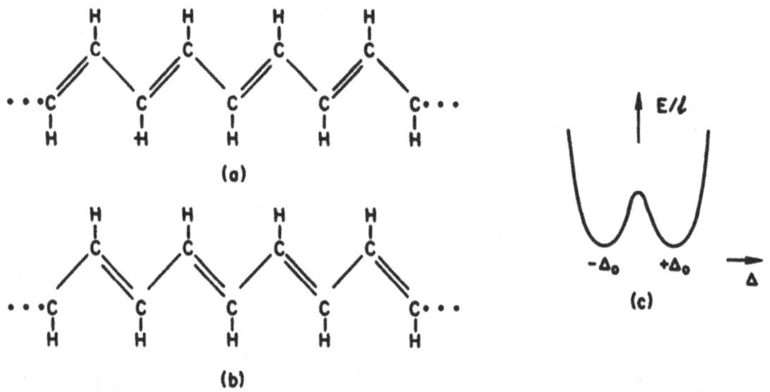

Figure 1: For trans-$(CH)_x$: the two degenerate bond alternation patterns [(a) and (b)]; (c) the energy per unit length versus bond alternation (or band gap).

This idealized (infinite chain) structure summarizes conveniently a number of features contained in the physical model for trans-$(CH)_x$ we will introduce later. First, the structure shows that the compound contains a carbon backbone consisting of alternating single and double bonds. Hence the fundamental unit which repeats contains two carbons and two hydrogens, and the chain is thus said to be "dimerized". Note that the double bonds are physically shorter than the single bonds, and thus the schematic structure indicates that a uniform bond length state is unfavored relative to bond alternation. Second, taken literally as drawn, the structures suggest that polyacetylene is a quasi-one dimensional material, with the only significant physical dimension being along the carbon backbone. Third, the difference between the two structures shown in Fig. 1 is simply whether, for a given carbon atom, the double bond lies to the left or to the right, since all bonds are at the same angle to the backbone axis. For an infinite chain, it seems intuitively clear that this should make no difference to the energy of the system. We shall see below that this is indeed the case and that the ground state of trans-$(CH)_x$, as indicated in Fig. 1c, is two-fold degenerate.

Finally, let me briefly remark on one of the principal reasons for technological interest in $(CH)_x$.[1] Macroscopically, polyacetylene can be prepared as a flexible (silver-colored) semiconducting film. By chemical doping (with AsF_5, Cl, ...) at the level of a few percent, the conductivity of the "plastic sheets" can be changed from 10^{-9} ohm^{-1} cm^{-1} to 10^{+3} ohm^{-1} cm^{-1}, and thus $(CH)_x$ can be transformed by doping from an insulator to a metal. The implications of this for lightweight batteries and electrical and solar cell components may be profound.[1]

THEORETICAL MODELS OF $(CH)_x$

To model microscopically the trans-$(CH)_x$ chain, one must describe the coupled motions of the lattice backbone of C-H units and the single (π-orbital) electron per carbon that,* heuristically speaking, determines where the double bond goes. This can be done either in a (more realistic) lattice formulation[4,7] or in a (more analytically tractable) continuum model.[3,5] Here we focus on the continuum theory and abstract from the lattice model only the important result on the form of the single (π) electron spectrum, shown in Fig. 2.[4]

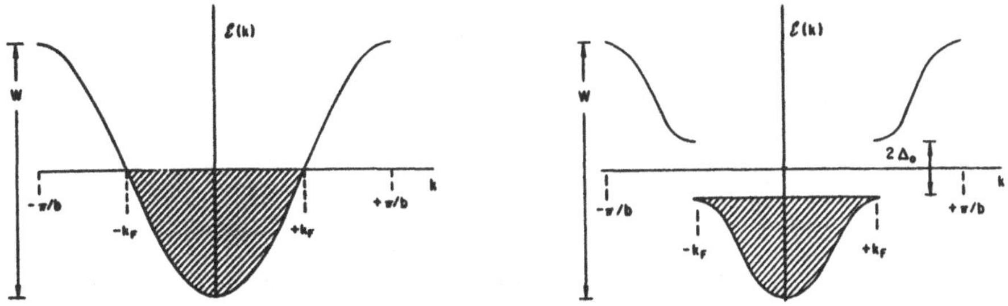

Figure 2: The single (π) electron spectrum in trans-$(CH)_x$ in (a) the hypothetical uniform (undimerized) case and (b) the actual dimerized ground state. The shading indicates filled electron states. W is the full band width (≃ 10 eV) and $2\Delta_0$ is the full band gap (≃ 1.4 eV). Since there is precisely one "active" electron per carbon, the band is half-filled, so $k_F = \pi/2b$, where the lattice spacing b ≃ 1.22 Å.

In the continuum electron phonon model[3,5] of trans-$(CH)_x$ one introduces an order parameter, $\Delta(y)$, (with dimensions of energy)

*One can show that of the four valence electrons per carbon, three form relatively deeply bound orbitals in the $(CH)_x$ polymer and thus can, in this simplified model, be treated as nondynamic.

to describe the phonon (i.e., lattice) motions and a two component electron field $\psi^+ = (U^+, V^+)$ to describe the π-electrons. Keeping only leading terms in the limit of zero lattice spacing ($b \to 0$) and linearizing the electron spectrum about the Fermi surface (see Fig. 2a) leads to an "adiabatic mean field continuum Hamiltonian"

$$H = \int dy \frac{\omega_Q^2}{g^2} \Delta^2(y) + \psi^+(y) \left[-iv_F\sigma_3 \frac{\partial}{\partial y} + \Delta(y)\sigma_1\right]\psi(y) \tag{1}$$

where ω_Q^2/g^2 is the net effective electron-phonon coupling constant, σ_i is the ith Pauli matrix, and v_F is the electron velocity at the Fermi surface (in units with $\hbar = 1$, $v_F = Wb/2$). In deriving (1), the lattice kinetic energy--which would lead to a term proportional to $\dot{\Delta}^2(y)$ in (1)--has been explicity ignored. Obviously this will have no effect on the static excitations we discuss below but it is very significant for on-going studies involving the dynamics of solitons in $(CH)_x$. Variation of H leads to the single particle electron wave function equations

$$\varepsilon_n U_n(y) = -iv_F \frac{\partial}{\partial y} U_n(y) + \Delta(y)V_n(y) \tag{2a}$$

$$\varepsilon_n V_n(y) = +iv_F \frac{\partial}{\partial y} V_n(y) + \Delta(y)U_n(y) \quad , \tag{2b}$$

and the self-consistent gap equation

$$\Delta(y) = g^2(2\omega_Q^2)^{-1} \sum_{n,s} \left[V_n^*(y)U_n(y) + U_n^*(y)V_n(y)\right] \quad . \tag{3}$$

The summation in (3) is over occupied electron states, and the spin label s has been suppressed.

How do Eqs. (2) and (3) express the conclusions of our discussion of the chemical structure of trans-$(CH)_x$? Recall that we anticipated that the configurational energy as a function of (constant) Δ would look like Fig. 1c, so that a uniform bond length--equivalently, a gapless electronic spectrum--state ($\Delta = 0$) is unstable (the "Peierls instability"[12]) and the ground state is two-fold degenerate. Using the explicit plane wave solutions to (2) appropriate for constant Δ, one can show that all these conclusions are indeed correct, that $E(\Delta)$ varies as $\Delta^2 \ln \Delta^2$,[9,13] and that the gap equation determines Δ_0 self-consistently via

$$\Delta_0 = g^2(2\omega_Q^2)^{-1} \frac{1}{2\pi} \int_{-K}^{K} \frac{dk}{\sqrt{k^2 v_F^2 + \Delta_0^2}} \tag{4}$$

Note the crucial result that the cut-off wave vectors, K, is chosen such that the energy at the cut-off corresponds to the correct π-electron band width (W, as shown in Fig. 2) in $(CH)_x$. Thus $W = 2(K^2v_F^2 + \Delta_0^2)^{\frac{1}{2}} \simeq 2Kv_F$, since $\Delta_0 \ll Kv_F$. The need for this cut-off comes solely from the linearization of the spectrum about k_F made in going to the continuum limit; the true spectrum of the discrete model, as shown in Fig. 2, automatically covers a finite range. Solving (4) for $Kv_F \gg \Delta_0$ gives

$$\Delta_0 = W \exp(-\lambda^{-1}) \quad \text{with } \lambda = g^2(\pi v_F \omega_Q^2)^{-1} \tag{5}$$

where the parameters have all been previously introduced (see Fig. 2). The electron spectrum for ($\Delta = \Delta_0$) consists of a valence band $[\varepsilon(k) = -(k^2v_F^2 + \Delta_0^2)^{\frac{1}{2}}]$ and a conduction band $[\varepsilon(k) = +(k^2v_F^2 + \Delta_0^2)^{\frac{1}{2}}]$ separated by a gap of $2\Delta_0$. These sub-bands are the continuum versions of those shown in Fig. 2b for the lattice model.

EXCITATIONS IN trans-$(CH)_x$

One obvious (and linear) excitation from this ground state is a particle-hole pair, in which an electron is promoted from the valence band to the conduction band. The minimum energy for this excitation is $E_{ph}^{min} = 2\Delta_0$. More interesting are the intrinsically nonlinear, soliton-like excitations: "kinks" and "polarons". Thus let us discuss these in some detail. The existence of kink solutions to (2) and (3) is a deep consequence of the two-fold degeneracy of the trans-$(CH)_x$ ground state.[3,6] As sketched in Fig. 3a, the kink soliton solution interpolates between the two degenerate ground states and has the explicit form

$$\Delta_K(y) = \Delta_0 \tanh v_F y/\Delta_0 \quad . \tag{6}$$

There is also an anti-kink ($\bar{K}$) soliton with $\Delta_{\bar{K}}(y) = -\Delta_K(y)$. The associated single electron energy levels consist of extended states in the conduction and valence bands which are modified (essentially "phase shifted") from the plane waves of the ground state and an additional localized "mid-gap" state (at $\varepsilon_0 = 0$, see Fig. 3).

For either K or $\bar{K}$, this "mid-gap" state can be occupied by 0, 1, or 2 electrons, leading to localized excitations with the bizarre spin/charge assignments indicated in Fig. 3. Explicitly one finds that the neutral kink (K^0) has a single electron in the state ε_0; since all states in the valence band remain spin-paired, the spin of K^0 is 1/2![4,5] Similar arguments show that K^+, which has no electrons in ε_0, and K^-, which has two electrons in ε_0, both have spin zero. These results seem to violate conventional

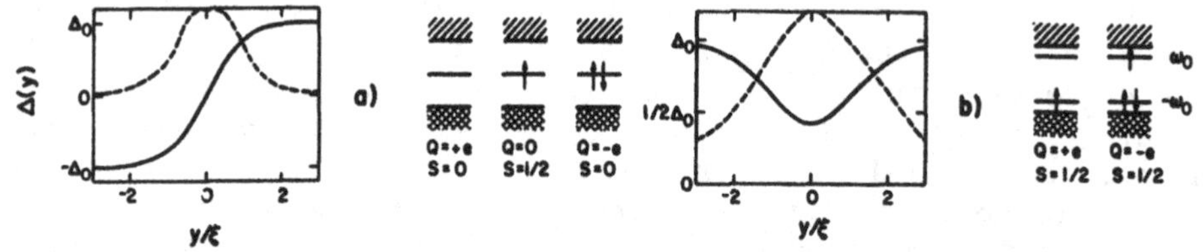

Figure 3: The spatial structure of the gap parameter [Δ(y)] and the associated electronic levels for <u>trans</u>-$(CH)_x$ for (a) the kink soliton, and (b) the polaron. Solid lines in the left-hand drawings indicate Δ(y); dashed lines show electron densities for localized states. The right-hand drawings show the electronic levels in the conduction band (single shading), valence band (crossed shading), and gap states. Q = charge and S = spin.

solid state spin/charge relations which follow quite generally from the fact that the basic charge carrying object, the electron, has spin 1/2 and charge -e. I shall return to this intriguing point and show its relation to "fractional charge" later.

The energy of a single kink of any charge is $E_K = E_{\bar{K}} = 2\Delta_0/\pi$. One vital consequence of the kink soliton's interpolating between the degenerate ground states is that there is a restriction - usually called a "topological constraint"[9,14,15] - on the production of kinks. Specifically, in the infinite chain polymer, kinks can be excited from the ground state only in pairs of kink and anti-kink. The interested reader should convince himself of this by showing that the production of a single kink from the ground state requires overcoming an infinite (for an infinite length polymer) energy barrier. Since kinks must be produced in pairs, the minimum energy state involving kinks which can be excited from the ground state has a single $K\bar{K}$ pair and has energy $E^{min}_{K\bar{K}} = 4\Delta_0/\pi$. This is less than $E^{min}_{ph} = 2\Delta_0$ and thus physical processes which in ordinary materials would produce particle-hole pairs may in $(CH)_x$ produce $K\bar{K}$.

Unlike kinks, polaron excitations represent localized deviations from a single ground state. The form of the polaron solution to (2) and (3) is sketched in Fig. 3b and can be written in the revealing form[8,9] (cf. (6))

$$\Delta_P(y) = \Delta_0 - K_0 v_F\{\tanh[K_0(y+y_0)] - \tanh[K_0(y-y_0)]\} \qquad (7)$$

where $\tanh 2K_0y_0 = K_0v_F/\Delta_0$. For this excitation the single electron states again include extended states in the conduction and valence bands, which states are phase-shifted by the polaron "defect"; the explicit forms of the corresponding wave functions are given in Ref. 9. In addition, two localized electronic states form in the gap at $\varepsilon_\pm = \pm\omega_0$ where $\omega_0 \equiv (\Delta_0^2 - K_0^2v_F^2)^{\frac{1}{2}}$.

Interestingly, the polaron configuration for $\Delta(y)$--Eq. (7)--and the associated electron wave functions (see Ref. 9) satisfy the electron part--Eq. (2)--of the continuum $(CH)_x$ equations for any K_0v_F in the allowed range $0 \leq K_0v_F \leq \Delta_0$. It is the self-consistent gap equation--(3)--that determines the specific value of K_0v_F for an actual solution to the coupled equations. Unsurprisingly, the nature of the solution depends on the occupation numbers--call them n_+ and n_---of the gap levels $\varepsilon_\pm = \pm\omega_0$. Using some aspects of soliton theory,[8,9] one can in effect convert Eq. (3) to an algebraic problem of minimizing with respect to K_0v_F the total energy of the full interacting electron-phonon system. One finds that

$$E^P(n_+, n_-, K_0) = (n_+-n_-+2)\omega_0 + \frac{4}{\pi} K_0v_F - \frac{4}{\pi} \omega_0\tan^{-1}(K_0v_F/\omega_0) \ . \quad (8)$$

With $K_0v_F = \Delta_0\sin\theta$ and $\omega_0 = \Delta_0\cos\theta$, one can show that E^P is minimized for $\theta = \theta(n_+,n_-) = (n_+-n_-+2)\pi/4$. From this result it follows that only for $n_+ = 1, n_- = 2$ (electron polaron) or $n_+ = 0$, $n_- = 1$ (hole polaron) can one have polaron structures. In each case $\theta = \pi/4$ so that $K_0v_F = \Delta_0/\sqrt{2} = \omega_0$ and $E^P = 2\sqrt{2}\,\Delta_0/\pi$. This polaron energy is less than $E_{K\bar{K}}^{min}$ and hence the polaron is the lowest energy excitation available to a single electron added--by doping or electron injection--to a trans-$(CH)_x$ chain. Note that polarons, unlike the kinks, have conventional spin/charge relations: for the electron polaron, $S = \frac{1}{2}$, $Q = -e$, and for the hole polaron, $S = \frac{1}{2}$, $Q = +e$. Using Eqs. (7) and (8) we can also answer the natural question, "What happens if we try to form a "bipolaron" by adding a second electron to a polaron state?" Indeed, for any of electron-electron ($n_+=n_-=2$), hole-hole ($n_+=n_-=0$) or electron-hole ($n_+ = n_- = 1$) configurations, E^P is minimized for $\theta = \pi/2$ so that $K_0v_F = \Delta_0$, $\omega_0 = 0$, and from its definition, $y_0 \to \infty$. Thus the would-be bipolarons actually correspond to an infinitely separated ($y_0 \to \infty$) $K\bar{K}$ pair, as Eq. (7) would suggest. Physically, when one electron is added to a trans-$(CH)_x$ chain, it should form a polaron, whereas when many electrons are added to a single chain, it becomes energetically favorable to form many $K\bar{K}$ pairs. A consequence of this is that, at least in this simplest picture, the response and transport properties of lightly-doped--and average of $\leq$ one electron (or hole) per chain--should be controlled by polarons, whereas more heavily doped samples should reflect the kink dominance.

Let me mention one final point regarding nonlinear excitations in polyacetylene. Up to now we have dealt exclusively with the trans-$(CH)_x$, with a degenerate ground state. In cis-$(CH)_x$, the two chain structures analogous to those shown for trans in Figs. 1a and 1b are not energetically equivalent. Hence there is no degeneracy; the cis-$(CH)_x$ ground state is unique. As a consequence, there should be no kink solitons in cis! Indeed, one can formulate a continuum model[8] similar to Eq. (1) for cis-$(CH)_x$ and show that only polaron (and multi-polaron) solutions exist. One can phrase this result in terms of "confinement of kinks" in cis-$(CH)_x$, and this predicted qualitative difference between the two isomers may be important both in verifying the existence of nonlinear excitations in the real materials and in explaining the experimentally observed differences between the isomers.[1,16]

THE FIELD THEORY CONNECTION

From the perspective of this summer institute one of the most interesting features of the continuum electron-phonon equations for trans-$(CH)_x$ is that they can in fact be shown[9] to be identical to the static, semi-classical equations of the N=2 Gross-Neveu model. To explain this result as briefly as possible, we recall that, for arbitrary N, the Gross-Neveu model can be described by the Lagrangian density[17]

$$\mathcal{L}(x) = \sum_{\alpha=1}^{N} \bar{\Psi}^{\alpha}(x)\left(i\gamma_{\mu}\frac{\partial}{\partial x_{\mu}} - g_{GN}\sigma(x)\right)\Psi^{\alpha}(x) - \tfrac{1}{2}\sigma^{2}(x) . \tag{9}$$

Here $\gamma_{\mu}(\mu=0,1)$ are the Dirac γ matrices in two dimensions; we follow the usual convention with $\gamma_0 = \sigma_3$ and $\gamma_1 = i\gamma_1$. Recall that $\bar{\Psi} = \Psi^{+}\gamma_0$. For each α, Ψ^{+} is a two component Dirac spinor, (Ψ_1, Ψ_2). The internal SU(N) symmetry index $\alpha = \{1,\ldots,N\}$ labels the "particle type". Apart from this internal symmetry, $\mathcal{L}$ posseses a discrete "chiral" symmetry, in that it is invariant under $\sigma\to-\sigma$, $\Psi\to\gamma_5\Psi$, with $\gamma_5 \equiv \gamma_0\gamma_1$. The static semi-classical equations corresponding to (9) can be shown[18] to be a Dirac equation for single fermion wave functions and a self-consistency equation for $\sigma(x)$. Specifically, defining stationary solutions by $\Psi^{\alpha}(x,t) \equiv e^{-i\omega_n t}\bar{\Psi}^{\alpha}(n;x)$, the Dirac equation becomes

$$\left(\omega_n\gamma_0 + i\gamma_1\frac{\partial}{\partial x} - g_{GN}\sigma(x)\right)\Psi^{\alpha}(n;x) = 0 \tag{10}$$

where for simplicity we have suppressed the index α. By making the transformations $\Psi_1 \to (U + V)/2$, $\Psi_2 \to -i(U - V)/2$. $g_{GN}\sigma \to \Delta$, and $y \to k_F x$ and performing some straightforward algebra, (10) can be seen to be precisely equivalent to the electron equations in trans-$(CH)_x$. The correct form of the self-consistency equation for $\sigma(x)$ can be shown to be[18]

$$Z(\Lambda)\sigma(x) = -g_{GN} \sum_{\substack{\alpha,n \\ \omega_n<0}} \bar{\Psi}^{\alpha}(n;x)\Psi^{\alpha}(n;x) \quad . \tag{11}$$

In (11) $Z(\Lambda)$ represents the ultraviolet renormalization--Λ is the ultraviolet momentum cut-off--necessary to cancel the divergence of the infinite (as $\Lambda\to\infty$) sum over $\bar{\Psi}\Psi$. This sum is over all $\Psi^{\alpha}(n;x)$ satisfying (10) with energy less than zero. Clearly the sum over the negative energy sea is analogous to the sum over the valence band in $(CH)_x$. Further, the sum over α in (11) becomes just like the spin sum in $(CH)_x$ if we choose N=2. The explicit form of $Z(\Lambda)$ is[18]

$$Z(\Lambda) = \frac{g_{GN}^2}{\pi} N \int_0^{\Lambda} \frac{dk}{(k^2+m^2)^{\frac{1}{2}}} = \frac{g_{GN}^2}{\pi} N \,\ell n \, \frac{\Lambda + \sqrt{\Lambda^2 + m^2}}{m} \tag{12}$$

where $m \equiv g_{GN}\sigma_0$. Although both the structure and interpretation of (11) are obviously directly analogous to the gap equation in $(CH)_x$, establishing an exact equivalence at first appears to present a problem, for the renormalization factor $Z(\Lambda)$ in (11) seems to have no counterpart in (4). Actually, this reflects merely a (very interesting) difference of interpretation between the solid state and field theory applications. Recall that in the continuum solid state model, because of the linearization of the electron spectrum, we had to cut-off the gap equation integral (4) at some wave vector corresponding to the full band width, W. Thus the cut-off is directly related to a physical quantity, and there is no need for an "infinite" renormalization. Said another way, the relation between Δ_0 and W [Eq. (5)] is just the requirement that the self-consistency condition (4) for Δ_0 be solved with a "renormalization constant" equal to 1! We can achieve a comparable situation in the field theory context by choosing σ_0 such that

$$m \equiv g_{GN}\sigma_0 = 2\Lambda \exp(-\pi/N g_{GN}^2) \quad . \tag{13}$$

Using this result, from (12) it follows that $Z(\Lambda) = 1$. Although this choice might appear unconventional in a field theory analysis, it is in fact completely equivalent to the standard prescription that renormalizes the theory by requiring the second derivative of the effective potential to be unity at an arbitrary subtraction point, σ_s.[9,17]

Choosing constants appropriately, the formal equivalence between the continuum electron phonon equations in trans-$(CH)_x$ and the static, semi-classical Gross-Neveu model is thus established. There remains the important difference in interpretation, however, since both Δ_0 and W are measurable in $(CH)_x$ whereas in the field theory, only m is measurable; Λ is an unphysical cut-off introduced to make the theory finite. Let us pursue this difference, as it illustrates how the $(CH)_x$ gap Eq. (3) reflects three important concepts developed in recent field theory studies and embodied in Eq. (13): "dynamical symmetry breaking", "dimensional transmutation", and "asymptotic freedom". First, as noted above, the original Gross-Neveu Lagrangian possesses a "chiral symmetry" and appears to describe massless fermions. Nonetheless, the actual ground state contains a filled Dirac sea of massive fermions and is not invariant under $\sigma \to -\sigma$, $\Psi \to \gamma_5\Psi$. This is "dynamical spontaneous symmetry breaking."[17,19] Second, for fixed N, the original Lagrangian contains only one parameter: the dimensionless coupling constant, g_{GN}. In the final, renormalized theory, the one physical parameter, as shown in (13), is the dimensional quantity the fermion mass. This is the phenomenon of "dimensional transmutation".[19] Third, (13) correctly reflects the "asymptotic freedom"[19,20] of the Gross-Neveu model, for if one insists on a fixed m--to define the same theory--as $\Lambda \to \infty$, then the coupling constant, g^2_{GN}, must go to zero like one over the logarithm of Λ. It is amusing and instructive to see all three of these concepts emerging, albeit with altered interpretation, in Eq. (5), which describes a real material.

Let me now turn to another aspect of the field theory connection. An alternative description of kink solitons in trans-$(CH)_x$ involves a phenomenological ϕ^4 theory,[6] in which the ground state energy's dependence on the band gap parameter, shown in Fig. 1c, is parameterized as $(\Delta^2 - \Delta_0^2)^2$. This phenomenological approach has been questioned[4,5,13] because the Landau-Ginzburg expansion motivating it is known[5] to fail in $(CH)_x$. Nonetheless, previous semi-classical field theoretic studies can be used as post hoc justifications for this approach. In particular, it is known[9,21] that, when the ϕ^4 theory is coupled to phenomenological fermion wave functions describing only the localized electron states, then (i) the ground state is two-fold degenerate; (ii) there exist kink-like solitons having mid-gap electron states leading to unusual spin/charge assignments;[14] and (iii) there are polaron solutions very similar to those in the continuum electron-phonon model.[21] Thus, at least at the level of static excitations, the phenomenological ϕ^4 theories agree with the more microscopic ones.

In solid state terms, the ϕ^4 potential is imitating the effects of the valence band.

I shall close this lecture by mentioning perhaps the most exotic aspect of certain nonlinear excitations in $(CH)_x$ and related compounds and in field theory models: the possible existence of "fractionally charged" solitons.[4,5,14,15,22,23,24] Several years ago, a field theoretical study[14] of a one-space dimensional ϕ^4 theory coupled to spinless fermions--that is, fermions with only one spin state, instead of the two associated with $s = \frac{1}{2}$ electrons--predicted that the kinks of the theory should have fermion number (i.e., charge) $\pm \frac{1}{2}$ (i.e., charge $= \pm \frac{1}{2}$)! More recently[4] but independently, it was shown that the mid-gap electronic state for a kink in trans-(CH) arose from the removal of $\frac{1}{2}$ state per spin degree of freedom from each of the valence and conduction bands. This implies that, if electrons did not have two possible spin states, the kinks would have charges $\pm e/2$, just as in the field theory model. Although the two spin states obscure this would-be fractional charge in $(CH)_x$, the bizarre spin/charge relations of the kink emerge instead.[4,5,22] Further, one can show that in other quasi-one dimensional real materials, the fractional charge should not be obscured and hence may be observable.[22-24] The crucial question is then whether the state carrying the putative fractional charge is truly an eigenstate of the charge operator or simply a superposition of eigenstates with fractional average value for the charge.[22] In this regard the dedicated reader should refer to the lectures of Faddeev in these proceedings, in which the conceptually closely related problem of the spin of the "fundamental" excitations in the Heisenberg antiferromagnetic chain is treated.

In conclusion, I should stress that much remains to be done. In particular, the major theoretical task is to understand the dynamics of soliton interactions--both within the adiabatic mean field theory we have discussed and in a fully quantum mechanical treatment--in polyacetylene. And above all, the central challenge is to establish--or disprove--definitively the relevance of this intricate and elegant theoretical structure to the real material, $(CH)_x$.

REFERENCES

1. For recent reviews see Physics in One Dimension, eds. J. Bernasconi and T. Schneider (Springer-Verlag, 1981); and A. J. Heeger and A. G. MacDiarmid, p. 353-391 in The Physics and Chemistry of Low Dimensional Solids, ed. L. Alcâcer (Reidel, 1980).
2. A. Kotani, J. Phys. Soc. Japan 42, 408 and 416 (1977).

3. S. A. Brazovskii, JETP Letters 28, 606 (1978); Soviet Phys. JETP 51, 342 (1980).
4. W. P. Su, J. R. Schrieffer, and A. J. Heeger, Phys. Rev. Lett. 42, 1698 (1979); Phys. Rev. B 22, 2099 (1980).
5. H. Takayama, Y. R. Lin-Liu, and K. Maki, Phys. Rev. B 21, 2388 (1980); J. A. Krumhansl, B. Horovitz, and A. J. Heeger, Solid State Commun. 34, 945 (1980); B. Horovitz, Solid State Commun. 34, 61 (1980) and Phys. Rev. Lett. 46, 742 (1981).
6. M. J. Rice, Phys. Lett. 71A, 152 (1979); M. J. Rice and J. Timonen, Phys. Lett. 73A, 368 (1979); E. J. Mele and M. J. Rice, in Chemica Scripta 17, 21 (1981).
7. W. P. Su and J. R. Schrieffer, Proc. Nat. Acad. Sci. 77, 5526 (Physics) (1980).
8. S. Brazovskii and N. Kirova, Pisma ZhETF 33, 6 (1981).
9. D. K. Campbell and A. R. Bishop, Phys. Rev. B 24, 4859 (1981); Nuc. Phys. B (in press).
10. K. Fesser, A. R. Bishop, and D. K. Campbell, in preparation.
11. Y. Tomkiewicz, T. D. Schultz, H. B. Brom, T. C. Clarke, and G. B. Street, Phys. Rev. Lett. 43, 1532 (1979); also Ref. 1.
12. R. E. Peierls, Quantum Theory of Solids (Clarendon Press, Oxford, 1955) p. 108; D. Allender, J. W. Bray, and J. Boudreau, Phys. Rev. B 9, 119 (1974).
13. W. P. Su, S. Kivelson, and J. R. Schrieffer, in Ref. 1.
14. R. Jackiw and C. Rebbi, Phys. Rev. D 13, 3398 (1976).
15. R. Jackiw and J. R. Schrieffer, Nuc. Phys. B 190, 253 (1981).
16. S. Etemad and A. J. Heeger, in Nonlinear Problems: Present and Future, eds. A. R. Bishop, D. K. Campbell, and B. Nicolaenko (North Holland, 1981).
17. D. J. Gross and A. Neveu, Phys. Rev. D 10, 3235 (1974).
18. R. F. Dashen, B. Hasslacher, and A. Neveu, Phys. Rev. D 12, 2443 (1975).
19. S. Coleman and E. Weinberg, Phys. Rev. D 7, 1888 (1973).
20. D. J. Gross and F. Wilzcek, Phys. Rev. Lett. 30, 1343 (1973); Phys. Rev. D 8, 3633 (1973); and H. D. Politzer, Phys. Rev. Lett. 30, 1346 (1973).
21. D. K. Campbell, Phys. Lett. 64B, 187 (1976); D. K. Campbell and Y-T. Liao, Phys. Rev. D 14, 2093 (1976).
22. J. R. Schrieffer in Molecular Crystals and Liquid Crystals (Gordon Breach, 1982).
23. W. P. Su and J. R. Schrieffer, Phys. Rev. Lett. 46, 738 (1981).
24. M. J. Rice and E. J. Mele, Xerox Webster preprint, 1981.

THE HIDDEN FERMIONS IN Z(2) THEORIES

Mark Srednicki

Joseph Henry Laboratories
Princeton University
Princeton, New Jersey 08544

In 1964, Schultz, Mattis, and Lieb (SML) [1] showed that the two dimensional Ising model is equivalent to a system of locally coupled fermions. After a canonical transformation, these fermions become noninteracting, and so SML were able to construct a simple, elegant, and exact solution. Recently, Fradkin, Susskind, and I (FSS) [2] showed that the three dimensional Z(2) gauge theory [3] could also be rewritten in terms of locally coupled fermionic degrees of freedom. Unfortunately, the coupling turned out to be quartic, and so we were unable to solve the theory.

In order to make the change to fermionic variables, both FSS and SML used a Jordan-Wigner transformation [4]. This technique is limited to low dimensions (two for the Ising model and three for the gauge theory) if one demands local couplings only. Nevertheless, it seems intuitively plausible that higher dimensional Z(2) theories are also fermionic theories in disguise. In this talk, I will show that this is correct [5]. Ising models and Z(2) gauge theories, in any number of dimensions, are equivalent to systems of locally coupled fermions. Similar results (for the Ising model) have been obtained by Samuel using the partition function formulation [6]. Groeneveld has also considered related ideas.

I will work with the Hamiltonian formulation throughout. The "time" direction is continuous, and the "space" dimensions are discrete. The spatial lattice is hypercubic unless otherwise indicated. The theory in question is specified by a quantum mechanical Hamiltonian. The Ising model, in any number of dimensions, has as its Hamiltonian [7]

$$H = -\sum_{\text{SITES}} \sigma_1 - \lambda \sum_{\text{LINKS}} \sigma_3 \sigma_3. \tag{1}$$

Here σ_1 and σ_3 are Pauli matrices living on the sites. At any one site, $\{\sigma_1, \sigma_3\} = 0$, $\sigma_1^2 = \sigma_3^2 = 1$. Pauli matrices at different sites commute. A link is a nearest neighbor pair of sites, and λ is a coupling constant. The Z(2) gauge theory, in any number of dimensions, has the Hamiltonian [7]

$$H = -\sum_{\text{LINKS}} \sigma_1 - \lambda \sum_{\text{PLAQ}} \sigma_3 \sigma_3 \sigma_3 \sigma_3. \tag{2}$$

Now the Pauli matrices are associated with links (not sites). "PLAQ" is short for plaquette, or an elementary square of four links.

Let us briefly review the SML solution [1] to the 1 + 1 dimensional Ising model. Each site can be labelled by an integer j, and

$$H = -\sum_j [\sigma_1(j) + \lambda\, \sigma_3(j)\, \sigma_3(j+1)]. \tag{3}$$

To transform to fermionic variables, we can write

$$\psi_1(j) = S(j)\, \sigma_3(j) \tag{4}$$

$$\psi_2(j) = S(j)\, \sigma_2(j) \tag{5}$$

where S(j) is the "string" operator, defined below, and $\sigma_2 = i\, \sigma_1\, \sigma_3$.

$$S(j) = \prod_{k<j} \sigma_1(k) \tag{6}$$

It is easy to show that ψ_1 and ψ_2 are Hermitian fields with anticommutation relations

$$\{\psi_a(j), \psi_b(k)\} = 2\delta_{ab}\, \delta_{jk}. \tag{7}$$

Furthermore, products of ψ's can be related to terms in the Hamiltonian.

$$\psi_1(j)\, \psi_2(j) = -i\, \sigma_1(j) \tag{8}$$

$$\psi_2(j)\, \psi_1(j+1) = -i\, \sigma_3(j)\, \sigma_3(j+1) \tag{9}$$

Thus we have

$$H = -i \sum_j [\psi_1(j)\, \psi_2(j) + \lambda\, \psi_2(j)\, \psi_1(j+1)]. \tag{10}$$

It will prove very useful to have a pictorial representation of the ψ's . Thnk of each ψ operator as an arrow attached to the appropriate site. All ψ_1's point to the left, and all ψ_2's to the right, as in fig. 1. Then each "site" term in H is constructed by multiplying together the two operators at that site. Each "link" term is constructed by multiplying together the two operators whose arrows lie along that link.

It is also possible to write the ψ's in terms of fermion creation and annihilation operators.

$$\psi_1(j) = c_j^+ + c_j \tag{11}$$

$$\psi_2(j) = -i(c_j^+ - c_j) \tag{12}$$

where the c's have the usual anticommutation relations. Thus, the fermionic Hamiltonian is a quadratic form which can be diagonalized by a canonical transformation [8]. This will not be true of any of the other models I will discuss.

Now consider the 2+1 dimensional Ising model. Here, the Jordan-Wigner transformation fails us. There is no way to order the sites such that the string operators cancel for each and every link term. There are always infinitely long loops of string left over, and so the couplings are not local.

Instead, let us reason by analogy with the "arrow" picture developed for 1+1 dimensions. Assign four Hermitian operators (ψ_1, ψ_2, ψ_3, ψ_4) to each site of the two dimensional lattice, each operator associated with an arrow pointing along one of the four links attached to that site (see fig. 2). The ψ's satisfy eq. (7). Now make a guess at a Hamiltonian.

$$H = \sum_{\text{SITES}} \psi_1\psi_2\psi_3\psi_4 + i\lambda \sum_{\text{LINKS}} \psi_{\rightarrow}\psi_{\leftarrow} \; . \tag{13}$$

The notation needs some explanation.

Each site term is constructed by multiplying together, in arbitrary order, the four operators living on that site. This leaves us with a sign ambiguity. However, by looking at the original Hamiltonian, eq. (1), one can see that the sign of each site term

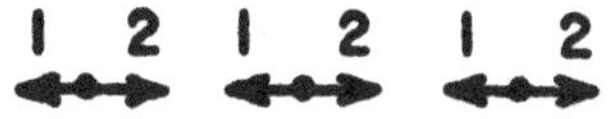

Figure 1.

is irrelevant. Changing the sign of σ_1 only switches the definitions of "spin up" and "spin down" (as measured by that σ_1). Thus, any ordering of the ψ operators in a site term is permitted.

The arrow subscripts in the link terms take on the values 1, 2, 3, 4 in such a way that the two operators whose arrows lie along a particular link are multiplied together, just as in 1+1 dimensions. Again, there is a sign ambiguity. This one is more serious and will be discussed later. The factor i is necessary to make the link terms Hermitian.

How do we check and see if this Hamiltonian is correct? First of all, the various terms must obey appropriate commutation relations. Since the σ_1 at any given site commutes with any other σ_1, all site terms commute with each other. In the fermionic Hamiltonian, eq. (13), each site term is a product of an even number of anticommuting objects. Hence, all site terms commute, as required.

Since all σ_3's commute, all link terms commute with each other. In the fermionic Hamiltonian, link terms are composed of two ψ operators (with different ψ's for different links). Hence, all link terms commute, as required.

A site term and a link term should also commute, unless the site is an endpoint of the link. If this is not the case, the same two terms from the fermionic H will commute for the same reason that site/site and link/link terms commute: each term contains an even number of different ψ's.

If the site is an endpoint of the link, the σ_1 from the site anticommutes with one of the σ_3's from the link; hence, the two terms anticommute. In the fermionic H, one of the four ψ's in the site term will be the same ψ as one of the two ψ's in the link term.

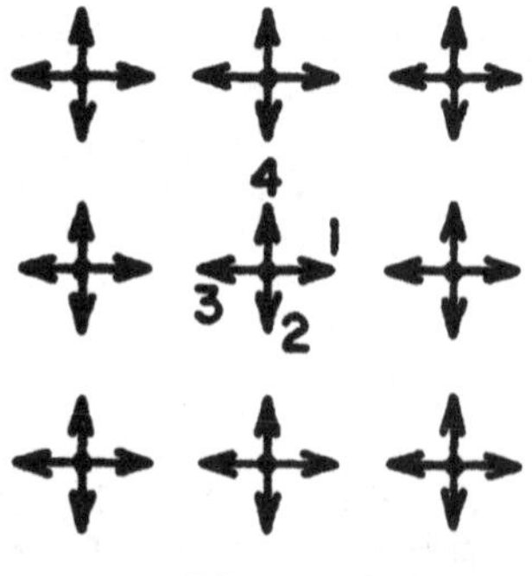

Figure 2.

This ψ commutes with itself ($\psi^2 = 1$). This results in an extra minus sign, as required, when one commutes the site and link terms.

So, the terms in the fermionic H either commute or anticommute in exactly the pattern that is necessary.

Furthermore, since $\sigma_1^2 = \sigma_3^2 = 1$, the square of any term should be unity. Since $\psi^2 = 1$, this condition is also satisfied by the fermionic H.

So far, so good. Now look at the space of states in which the Hamiltonian acts. The original Hamiltonian has a two component spinor on each site, spanned by two basis states. Unfortunately, the fermionic H acts in a Hiblert space with four basis states per site. To see this, recall that each pair of ψ's can be written as a linear comination of fermion creation and annihilation operators [eqs. (11) - (12)]. Thus, there are four basis states per site, which can be labelled by the eigenvalues (0 or 1) of two distinct number operators, c^+c and b^+b. We choose c and c^+ to be associated with ψ_1 and ψ_2, via eqs. (11) - (12), and b and b^+ with ψ_3 and ψ_4.

Things look bad for the fermionic H. It acts in a Hilbert space which is much too large. However, it will prove instructive to examine the dynamical consequences of these extra states. In the end, we will make them go away.

Now consider four links which form a plaquette, and multiply together the associated terms in H. The result is unity; the associated with each site occurs twice, and $\sigma_3^2 = 1$. We can write this symbolically as

$$\Box = 1 \tag{14}$$

What happens when we multiply together the analogous four link terms from the fermionic H? Is eq. (14) satisfied? In this case, $\Box$ equals the product of eight independent Majorans operators (see fig. 3). There is no reason for $\Box$ to equal unity.

Let us study this fermionic plaquette operator, $\Box$. It commutes with the full Hamiltonian, eq.(13). This is true of

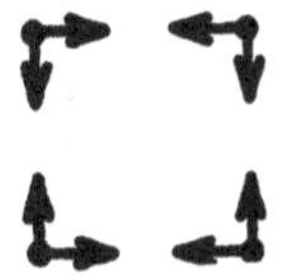

Figure 3.

every $\square$, independent of its location on the lattice. Also, each $\square$ commutes with every other $\square$. Thus it is possible to simultaneously diagonalize H and all the plaquette operators. In other words, the $\square$'s are the generators of a gauge symmetry of the fermionic H. The original H, eq. (1), has no such symmetry.

We can use this extra symmetry to get rid of the unwanted states. Specifically, we require all "physical" states to be gauge invariant. That is,

$$| \phi \rangle = | \phi \rangle \tag{15}$$

where $\square$ is any fermionic plaquette operator, and $|\phi\rangle$ is any physical state. (Since the square of each link term is one, we must have $\square^2 = 1$, and therefore the eigenvalues of $\square$ are +1 and -1.) Thus, in the physical subspace, eq. (14) is satisfied.

We can answer the question of sign ambiguity in the link terms. Start by choosing an arbitrary ordering convention for the two ψ's in each link term of eq. (11). Now, construct each $\square$ operator by multiplying together the appropriate four link terms, using the same ψ ordering convention for each link term that was used in writing down H. Now impose eq. (15). There are no longer any sign problems. If one changes the signs of some of the link terms in the original H, eq. (1), there will always be some $\square$'s with the property $\square = -1$. This is the only way the sign changes can enter the dynamics. We have chosen signs in the fermionic H and the physical subspace such that $\square = +1$ always. Thus there are no sign ambiguities with dynamical consequences.

We have almost proben the equivalence of the two H's. The last thing we must do is show that the full space of states of eq. (1) is contained in the gauge invariant sector of eq. (13). To do so, it is sufficient to formulate a fermionic version of a single σ_3. The σ_3 is used to flip a single spin, and thus create an excited state of the original H (with $\lambda = 0$).

The fermionized $\sigma_3(j)$ is nonlocal. To construct it, multiply together link operators along any continuous path of links with one end at site j and the other end at infinity, as shown in fig. 4. This operator is gauge invariant, as each link operator is gauge invariant. Site j is acted on by a single ψ, which "flips its spin." All other sites are acted on by two ψ's, which leaves them in their ground states (that is, the energy of these when $\lambda = 0$ sites is -1). The location of the string is irrelevant. If one multiplies the string operator by a $\square$ sharing a link with the string, one gets a new string location. Since $\square = 1$, the two strings are equivalent operators in the physical subspace.

Thus, the spectrum of the original H is equivalent to the spectrum of the gauge invariant sector of the fermionic H.

In 2+1 dimensions, the Ising model is dual to the gauge invariant sector of the Z(2) gauge theory [3, 9, 10]. This theory can be fermionized by a standard Jordan-Wigner transformation [2]. The result is eq. (13), with the word "sites" replaced by the word "plaquettes" (a site is dual to a plaquette in two spatial dimensions). We have attempted to understand this fermionization from the point of view of the Ising model, without reference to the dual gauge theory.

We have gained a great deal by doing so. The methods I have used readily generalize to the Ising model and Z(2) gauge theory in any number of dimensions.

Consider the d+1 dimensional Ising model on a hypercubic lattice. To fermionize it, place 2d ψ operators on each site, one for each link attached to that site. The Hamiltonian is

$$H = i^d \sum_{\text{SITES}} \psi_1\psi_2 \cdot\cdot\cdot \psi_{2d} + i\lambda \sum_{\text{LINKS}} \psi_{\rightarrow}\psi_{\leftarrow} \tag{16}$$

The gauge symmetry associated with plaquette operators is still present. Require eq. (15) as before.

Now consider the 3+1 dimensional Z(2) gauge theory. Recall that the degrees of freedom are located on the links rather than

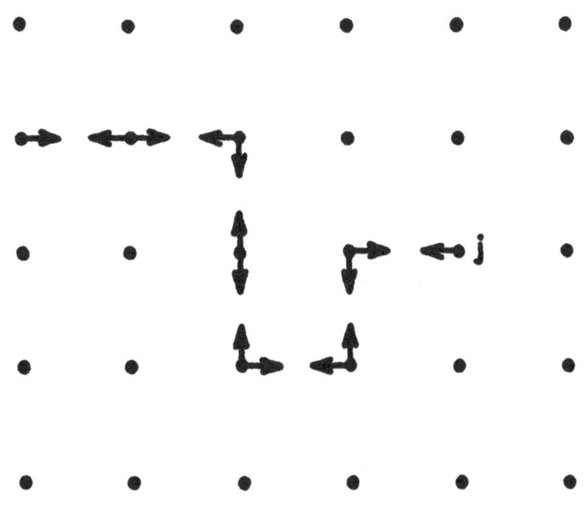

Figure 4.

the sites. Each link is coupled into four plaquettes. To fermionize this system, assign four ψ's to each link, as shown in fig. 5. The link term of the fermionic Hamiltonian is $\psi_1\psi_2\psi_3\psi_4$. The plaquette term is the product of the four ψ's whose arrows point to the center of the plaquette, as shown in fig. 6.

$$H = \sum_{\text{LINKS}} \psi_1\psi_2\psi_3\psi_4 + \lambda \sum_{\text{PLAQ}} \psi\psi\psi\psi \tag{17}$$

The usual gauge symmetry [7] is present in this Hamiltonian. That is, a gauge generator which commutes with H can be constructed by multiplying together the link terms on the six links attached to any given site (I will call these generators *'s). However, we need to find a second kind of gauge symmetry. In the original H, eq. (2), the product of six plaquette terms which form a cube is equal to unity, because the σ_3 on each link occurs twice, and $\sigma_3^2 = 1$. In the fermionic H, the product of six plaquette terms forming a cube has no reason to be unity. The fermionic cube operator (which I will call C) is simply the product of twenty-four ψ's. However $C^2 = 1$, so C has eigenvalues +1 and -1. Furthermore, any C commutes with H, all other C's and (this is important) all the *'s. Thus, we can simultaneously diagonalize H, the *'s, and the C's. For eq. (2) and eq. (16) to define equivalent theories, we require

$$C \mid \phi > = \mid \phi > \tag{18}$$

for any physical state $\mid \phi >$. Usually, one deals with the gauge invariant sector of eq. (2), which means imposing

$$* \mid \phi > = \mid \phi > \tag{19}$$

as well.

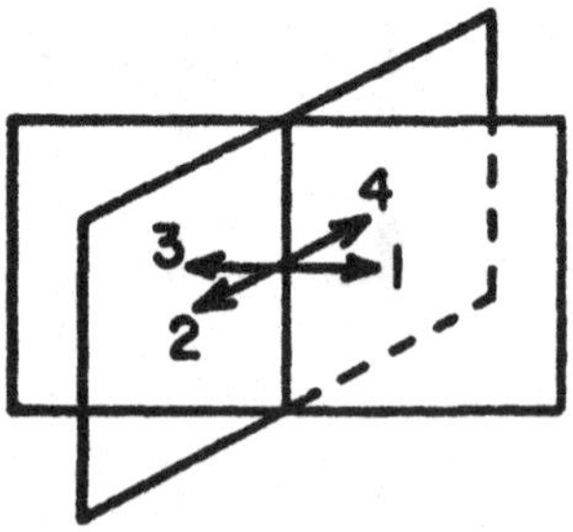

Figure 5.

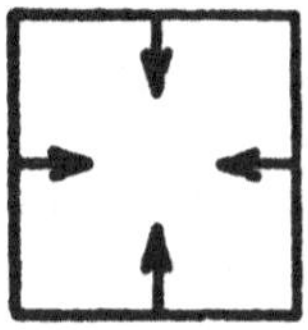

Figure 6.

Once again, all the operator identities satisfied by the terms of the original Hamiltonian are satisfied by the terms of the fermionic Hamiltonian, provided we restrict the Hilbert space to those states satisfying eq. (18). In particular, any closed surface of plaquette operators should equal unity. Since any such surface is a product of C's, this result follows immediately.

There is still a small problem, however. It is possible to show that the gauge condition on cubes, eq. (18), does not restrict the space of states to the point where there is a one-to-one correspondence between the states of the original H and the fermionic H. In fact, there is some more gauge symmetry in eq. (17), associated with "corner" operators defined in fig. 7. The corner operators commute with each other (they can have a single arrow in common). Thus, we cannot diagonalize all of them simultaneously. This means that the ground state satisfying eqs. (18) - (19) is not unique. In contrast, the ground state of the original H which satisfies eq. (19) is unique.

To remove this extra degeneracy of the fermionic H, we must thin the space of states some more. Fortunately, this is trivial. We simply pick any ground state satisfying eqs. (18) - (19) and build excited states on it by applying operators which commute with

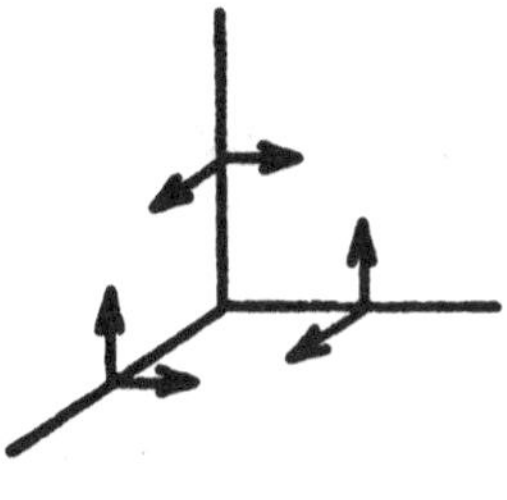

Figure 7.

all of the *'s and all of the C's. Once this is done, the extra states have no dynamical consequences.

There is no need to construct a fermionic σ_3 for the gauge theory, as an individual σ_3 does not commute with all of the *'s.

To extend to higher dimensional gauge theories (on hypercubic lattices), one simply assigns 2(d-1) ψ's to each link of a d dimensional spatial lattice. The Hamiltonian is

$$H = i^{d-1} \sum_{\text{LINKS}} \psi_1\psi_2 \cdot \cdot \cdot \psi_{2(d-1)} + \lambda \sum_{\text{PLAQ}} \psi\psi\psi\psi \tag{20}$$

The plaquette term remains unchanged. We impose C = +1 as we did in 3+1 dimensions. The residual gauge degeneracy discussed above is present, and we must pick a ground state satisfying eqs. (18)-(19). Once this is done, eqs. (18)-(20) define a fermionized d+1 dimensional Z(2) gauge theory.

There is at least one immediate advantage of the "ψ" formalism over the "σ" formalism: duality transformations are completely trivial. Let us consider the very interesting example of the 3+1 dimensional gauge theory. It is known to be self-dual [3, 7]. In three spatial dimensions, a link is dual to a plaquette. To make the dual transformation on (17), we associate the four inward pointing arrows of a plaquette with the four outward pointing arrows of the dual link, as shown in fig. 8. Thus the link (plaquette) terms of eq. (17) map into the plaquette (link) terms of the dual Hamiltonian.

$$H_{\text{DUAL}} = \sum_{\text{PLAQ}} \psi\psi\psi\psi + \lambda \sum_{\text{LINK}} \psi\psi\psi\psi \; . \tag{21}$$

Thus the usual self-duality relation for any energy eigenvalue,

$$E(\lambda) = \lambda\, E(1/\lambda), \tag{22}$$

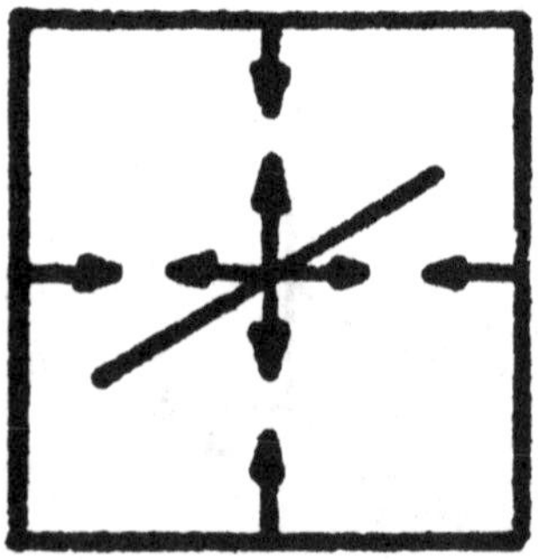

Figure 8.

is obviously true. This one line derivation of self-duality should be compared with the usual derivations [3, 7].

The "ψ" formalism permits extremely simply duality transformations to be made on any Z(2) Hamiltonian. There is no restriction to any gauge invariant sector, as there is in the "σ" formalism. One simply draws the dual lattice, and changes each set of → ← to a ← → set of arrows. The only change made in the Hamiltonian is relabelling the summation signs (i.e., "sites" becomes "plaquettes" in 2+1 dimensions, etc.).

The fermionic formalism does not help us solve the theories in any obvious way. Every unsolved model has products of four or more fermion operators in its Hamiltonian. One might hope, however, that this new formalism will lead to new ideas that allow real progress to be made in determining the quantities of interest (phase diagrams, critical exponents, etc.) in Z(2) theories.

REFERENCES

1. T. Schultz, D. Mattis, and E. Lieb, Rev. Mod. Phys. 36, 856 (1964).
2. E. Fradkin, M. Srednicki, and L. Susskind, Phys. Rev. D21, 2885 (1980).
3. F.J. Wegner, J. Math. Phys. 12, 2259 (1971); R. Balian, J.M. Drouffe, and C. Itzykson, Phys. Rev. D11, 2098 (1975).
4. P. Jordan and E. Wigner, Z. Physik 47, 631 (1928).
5. M. Srednicki, Phys. Rev. D21, 2878 (1980).
6. S. Samuel, J.Math. Phys. 21, 2820 (1980).
7. E. Fradkin and L. Susskind, Phys. Rev. D17, 2637 (1978).
8. P. Pfeuty, Ann. Phys. (N.Y.) 57, 79 (1970).
9. H.A. Kramers and G.H. Wannier, Phys. Rev. 60, 252 (1951).
10. E. Fradkin and S. Raby, Phys. Rev. D20, 2566 (1980).

REMARKS ON ALTERNATIVE LATTICE ACTIONS

AND PHASE STRUCTURE OF LATTICE MODELS

Harald Grosse

Insitut für Theoretische Physik
Universität Wien

ABSTRACT

We state the requirements of lattice gauge theories, remark that Manton's action violates reflection positivity, mention the equivalence of Z_4 and $Z_2 \times Z_2$ theories and discuss how well the SU(3) gauge theory is approximated by taking nonabelian subgroups as a gauge group.

1. FORMULATION

In order to study a quantum field theory one has to introduce cutoffs, and introducing cut offs always destroys symmetries the theory has. If one intends to study a gauge theory one likes to maintain gauge invariance. The only procedure known which allows to stay in the appropriate space time dimension and allows to put an ultraviolet cut off is Wilson's [1] procedure of formulating the theory on a lattice. In this way one obtains gauge invariant Ising models which where introduced first by Wegner [2].

In a classical treatment of a gauge theory one starts with a gauge potential A and a field strength F = dA + [A,A] both being Lie algebra valued forms. Neither A nor F is a gauge invariant quantity but the gauge orbit of A obtained from a field configuration by applying gauge transformations. The gauge orbit can be characterized by the set of parallel transport operators for closed nodeless curves c. These are given as path ordered products

$$g_c = P(\exp i \int_c dz^\mu A_\mu(z)) \tag{1.1}$$

and become the main building blocks of a quantum gauge field theory. The n-point functions are thereby replaced by n-loop functions and g_c should be defined by a suitable normal ordering procedure [3].

In the lattice version one starts from a cubic lattice and assigns to each one dimensional cell (= bond) a group element g_b representing the parallel transport along b. The only requirement on this mapping is that the inverse group element should be assigned to the reversed bond. The set $\{g_b | b \in \Lambda\}$ for bonds out of a volume Λ constitutes a field configuration. Expectation values of local observables [4] (depending on finitely many bonds) are defined by going to the canonical ensemble, summing over all configurations and weighting them with the Boltzmann-factor

$$\langle F \rangle = \lim_{\Lambda \to \infty} \frac{1}{Z_\Lambda} \int \prod_{b \in \Lambda} dg_b \exp \left(\sum_{P \subset \Lambda} S(g_P) \right) F(\{g_b\}) \qquad (1.2)$$

where dg_b denotes the Haar measure of the compact group or averaging in case of a discrete group and Z_Λ, the partition function gives the normalization

$$Z_\Lambda = \int \prod_{b \in \Lambda} dg_b \exp \left(\sum_{P \subset \Lambda} S(g_P) \right) . \qquad (1.3)$$

The total action is written as a sum of plaquette (= two dimensional cell) contributions $S(g_P)$ which should be an approximation to the continuum expression for the action and g_P denotes the parallel transport around the boundary of P. So if $\{b_i | i=1,\ldots,4\}$ form that boundary $g_P = g_{b_1} g_{b_2} g_{b_3} g_{b_4}$. Specifying a theory means fixing the lattice action to which we will come in part 3.

2. REQUIREMENTS

The n-loop functions are given on the lattice by

$$S_n(c_1,\ldots,c_n) = \langle \chi_{i_1}(g_{c_1}) \ldots \chi_{i_n}(c_n) \rangle \qquad (2.1)$$

where χ_{i_j} denotes an irreducible character of the group and c_j denotes closed curves. Constructing a theory would mean trying to establish (after putting counterterms) convergence of these n-loop functions as the lattice constant a goes to zero. This has been done for two dimensional models [3].

Then one would like to check the requirements:

a) regularity property
b) symmetry property
c) euclidean invariance
d) Osterwalder-Schrader (O.S.) positivity
e) cluster property.

The most essential question concerns property d) which implies a positive definite scalar product for the Minkowski theory and the spectral condition and we shall comment on a lattice formulation later on. On the lattice one has to replace c) by

c') lattice translation invariance

and adds in addition

f) formally correct $a \to 0$ limit.

Next one is interested in further properties of the theory, especially confinement properties. Also there are a number of ways one could define confinement for a complete theory, they are at present not useful in practice. Therefore one tries to get information about an effective potential for fields coupled to a gauge field within the pure gauge theory. Starting from the one-loop function one defines

$$V(L) = - \lim_{T\to\infty} \frac{1}{T} \ln |S_1(c)| \tag{2.2}$$

where c denotes a rectangular curve of extension $L \times T$; then one speaks of confinement in the sense of Wilson if

$$V(L) \underset{L\to\infty}{\simeq} \sigma \cdot L \cdot a^2 \tag{2.3}$$

with $\sigma > 0$ being the string tension.

O.S. positivity on the lattice (also often called reflection positivity) is formulated in a way which allows to introduce also fermions as follows:

One divides the volume Λ with the help of a plane ($t = 0$) lying half between two lattice planes into two parts Λ_+ (with $t > 0$) and Λ_- (with $t < 0$); Λ is assumed to be symmetric with respect to this plane. Denote by A_+ and A_- the algebra of observables depending only on bond variables with bonds lying in Λ_+ and Λ_-. Then one asks for an antilinear mapping Θ:

$$A_+ \overset{\Theta}{\to} A_- \ . \tag{2.4}$$

Reflection positivity [4] requires that all expectation values

$$\langle F \cdot \Theta F \rangle \geq 0 \qquad \forall\ F \in A_+ \tag{2.5}$$

have to be positive. To establish (2.5) one goes first to the axial gauge in which all bond variables in time direction are put equal then to the unit element of the group. After introducing periodic boundary conditions one observes that only plaquettes crossing the $t = 0$ plane are relevant since in general

$$\int \prod_{b\in\Lambda} dg_b \ F \cdot \Theta F \geq 0 \qquad \forall\ F \in A_+ \ . \tag{2.6}$$

The contributions to the action can now be splitted into parts

$$\sum_{P\subset\Lambda} S(g_P) = S_+ + \Theta S_+ + \delta S \tag{2.7}$$

where $S_+ \in A_+$, $\Theta S_+ \in A_-$ and δS comes from plaquettes crossing the $t = 0$ plane. Then (2.5) is equivalent to the requirement

$$\int dg\, dh\, F^*(g) \exp\,(S(gh^{-1}))\, F(h) \geq 0 \tag{2.8}$$

which means that exp S(.) should be a function of positive type.

3. CHOOSING AN ACTION

The standard choice for the action per plaquette in (1.2) and (1.3) is Wilson's original proposal

$$S(g) = \beta(\mathrm{Re}\ \chi(g)/\chi(e) - 1)\ , \quad \beta = 1/g_u^2 \tag{3.1}$$

where g_u denotes the unrenormalized coupling constant.

Clearly in (3.1) one singles out one particular representation; it has been argued that a smoother continuum limit will be obtained by taking contributions from different representations. Therefore two alternatives have been discussed:

Manton's [5] proposal amounts to replacing (3.1) by the square of the distance D(g,e) of the shortest geodesics going from the unit element to the group element g:

$$S(g) = \beta\ D^2(g,e) \tag{3.2}$$

while already some times ago Drouffe [6] and Itzykson [7] have proposed and studied the heat kernel action (Villain form). Here one takes the fundamental solution of the heat equation on the compact Lie group and gets

$$\exp S(g) = \sum_\mu \chi_\mu(g)\chi_\mu(e)\exp(-\frac{g_u}{2}\, C(\mu))/\sum_\mu \chi_\mu^2(e)\ \exp(-\frac{g_u}{2}\, C(\mu)) \tag{3.3}$$

where $C(\mu)$ denotes the eigenvalues of the Casimir operator of the group in the representation μ and summation goes over all inequivalent unitary irreducible representations.

Let us try to check the most essential requirement d). Since it is known that exp $\chi(.)$ represents a function of positive type (expand the exponential function) and since all Fourier coefficients of the heat kernel are positive, both (3.1) and (3.3) fulfill reflection positivity. For Wilson's action this argument has first been given in [4].

Here we will point out, however, that Manton's proposal (3.2) violates it [8]. To see that (2.8) is not fulfilled it suffices to show the violation for some subgroups, since one can take functions F concentrated on it; so for SU(n) we may restrict ourselves to a U(1) group. For U(1) one checks easily that the Fourier coefficients change sign for small β depending on whether n is even or odd:

$$I_n(\beta) := \int_{-\infty}^{\infty} \frac{d\alpha}{2\pi} e^{in\alpha} e^{-\beta\alpha^2} \underset{\beta\to 0}{\simeq} (-)^{n+1} \frac{2\beta}{n^2} + O(\beta^2) , \quad n > 0 , \quad (3.4)$$

One could argue that one needs this condition only in the $\beta \to \infty$ limit. The quantities $I_n(\beta)$ of equation (3.4) are indeed positive for large but finite β if $n \leq n_0(\beta)$ where n_0 increases with β, but then it is a delicate question how "large" the violation of (2.8) is.

4. EQUIVALENCE BETWEEN Z_4 AND $Z_2 \times Z_2$ THEORIES

Abelian Z_N theories have been treated due to various reasons. They are simple enough to develop techniques which may be used then for other theories; but also certain (confinement) properties of SU(N) theories are implied by properties of their center Z_N.

It is known that in two dimensions the Z_4 spin model is equivalent to two independent Z_2 models [9] put onto the same lattice. Together with C. Lang and H. Nicolai [10] we studied the analogous problem for the gauge theories. In two dimensions (since all pure gauge theories are solvable) one finds again that Z_4 and $Z_2 \times Z_2$ can be identified. Furthermore both theories are known to be self-dual in four dimensions and their critical temperatures fulfill

$$\beta_{cr}^{(4)} = 2\beta_{cr}^{(2\times 2)} = \ln (1 + \sqrt{2}) . \quad (4.1)$$

We have been able to prove equivalence of both theories for all values of β for which the strong coupling expansion is convergent (which goes beyond β_{cr}); by duality the equivalence holds for all values of β.

The partition function of a Z_N theory is given similar to equ. (1.3) by

$$Z_\Lambda^{(N)}(\beta) = \int \prod_{b\in\Lambda} dg_b \prod_{P\subset\Lambda} \exp (\beta \operatorname{Re} \chi_1^{(N)}(g_P)) \quad (4.2)$$

where dg_b means averaging over the group and the characters of Z_N are given by

$$\chi_p^{(N)}(\exp \frac{2\pi i n}{N}) = \exp \frac{2\pi i n p}{N} \quad (4.3)$$

Using the character expansion for the Boltzmann-factor one gets for $N = 4$

$$\exp (\beta \operatorname{Re} \chi_1^{(4)}(g)) = c_\beta^2 \cdot \{1 + t_\beta(\chi_1(g) + \chi_3(g)) + t_\beta^2 \cdot \chi_2(g)\} \quad (4.4)$$

$$c_\beta = \cosh \frac{\beta}{2} , \qquad t_\beta = \tanh \frac{\beta}{2} .$$

Using (4.4) in (4.2), expanding the product over all plaquettes and using the orthogonality relations for the characters gives terms [11] which can be associated to polyhedral complexes. These can be divided into regular components, which are two dimensional surfaces whose boundary is homeomorphic to a circle. In our case these surfaces carry one of the two possible character functions, either $\chi_1 + \chi_3$ or χ_2.

Putting on the other hand two noninteracting Z_2 theories to the same lattice, adding the actions and following the same procedure as above lead to an expression similar to (4.4)

$$\exp\left(\frac{\beta}{2}(g+\bar{g})\right) = c_\beta^2 \cdot \{1 + t_\beta(g+\bar{g}) + t_\beta^2 \cdot g \cdot \bar{g}\} \ . \qquad (4.5)$$

Putting (4.5) into the expression for the partition function

$$Z_\Lambda^{(2\times2)}\left(\frac{\beta}{2}\right) = \int \prod_{b\in\Lambda} dg_b \, d\bar{g}_b \prod_{P\subset\Lambda} \exp\left(\frac{\beta}{2}(g+\bar{g})\right) \qquad (4.6)$$

and expanding again the product over plaquettes leads like before to contributions associated to two dimensional polyhedral complexes.

After examining the rules for integrating along the boundaries of these surfaces in both theories carefully we have been able to identify contributions coming from one theory with those obtained in the other theory. This leads to the conclusion that for any finite Λ

$$Z_\Lambda^{(4)}(\beta) = Z_\Lambda^{(2\times2)}\left(\frac{\beta}{2}\right) = \left(Z_\Lambda^{(2)}\left(\frac{\beta}{2}\right)\right)^2 \ , \qquad (4.7)$$

an equality which therefore holds in the limit $\Lambda \to \infty$. For the free energy we get

$$F^{(4)}(\beta) = 2F^{(2)}\left(\frac{\beta}{2}\right) \ , \qquad F = \lim_{N\to\infty} \frac{1}{N} \ln Z_\Lambda \qquad (4.8)$$

and a similar relation for the Wilson loop variable $S_1(c)$. Clearly also (4.1) is implied by (4.7).

5. PHASE STRUCTURE FOR THEORIES WITH NONABELIAN SU(3) SUBGROUPS

Recently it has been realized that one can get insight into the phase structure of gauge theories by using Monte Carlo simulations. Moreover by using discrete subgroups one may speed up computer calculations.

A study of the phase structure for U(1) together with Z_n theories has been done in Ref. [12]; the same question for SU(2) has been treated in [13]. Together with H. Kühnelt [14] we looked to theories where one takes a nonabelian subgroup of SU(3) as gauge group.

Surprisingly there exists an infinite number of nonabelian subgroups of SU(3). They have been discussed at the time of the eightfold way [15] and also recently by the Bonn group [16]. There are finitely many crystal-like groups (like in SU(2)), but in addition two sequences of nonabelian groups which are denoted by $\Delta(3n^2)$ and $\Delta(6n^2)$; the number in brackets denotes the order of the group and n is an integer. The group multiplication law can be written down in a compact form and shows that both sequences are semidirect products of abelian groups

$$\Delta(3n^2) = Z_n \times Z_n \circledS Z_3 , \qquad \Delta(6n^2) = Z_n \times Z_n \circledS S_3 \tag{5.1}$$

where the symmetric group of 3 elements is themselves a semidirect product: $S_3 = Z_3 \circledS Z_2$.

We have studied these theories using the Monte Carlo procedure; starting with large β and a completely ordered lattice one performs thermal cycles by changing β in small steps and using the Metropolis method to achieve equilibrium for the new coupling constant. In each sweep through the lattice one asks whether the action is improved by changing the group element and improved values are accepted. Otherwise one changes to the new value with probability $\exp(-\Delta S)$ with $\Delta S = S_{old} - S_{new} > 0$.

The common feature of these calculations are broad hysteresis loops around $\beta \simeq 2.7$ for $\Delta(3n^2)$ with $n \geq 6$ and $\beta \simeq 3.1$ for $\Delta(6n^2)$ with $n \geq 8$. We have compared the results with strong coupling expansions, which break down at the phase transition point. Carefull investigation has shown that for big enough n a second phase transition appears (like in Z_n) which goes out to temperature zero while increasing n.

One might ask how well SU(3) is approximated by these subgroups. Unfortunately this subject is not very well worked out. However there are indications which show that the approximation is poor: These subgroups have irreducible representations with dimensionality at most six; so no octet or decuplet representation occurs. Furthermore, the limiting groups (for $n \to \infty$) are two parameter subgroups consisting of three respectively six disconnected pieces, while the full SU(3) manifold is eight dimensional.

Calculations for the full SU(3) gauge theory have shown a turnover point from the strong coupling to weak coupling behaviour (possibly connected with a roughening transition) around $\beta \simeq 6$. As we have remarked before with the help of discrete groups one approximates the SU(3) results very well up to phase transition point $\beta \simeq 3.1$. Therefore it is not possible in this case (contrary to the SU(2) case) to study the region beyond the turnover point.

One should remark that also the crystal-like groups have been investigated [17]. For the largest group S(1080) $\beta_{cr} \simeq 3.58$ which implies that one is in a similar bad situation as with the $\Delta(6n^2)$ groups.

REFERENCES

1. K.G. Wilson, Phys. Rev. D10 (1974)2445

2. F. Wegner, Jour. Math. Phys. 12 (1971) 2259
3. E. Seiler, Gauge Theories as a Problem of Constructive Quantum Field Theory and Statistical Mechanics (Troisieme Cycle de la Physique en Suisse Romande, 1981)
4. K. Osterwalder and E. Seiler, Ann. Phys. 110 (1978) 440
5. N.S. Manton, Phys. Lett. 96B (1980) 328
6. J.M. Drouffe, Phys. Rev. D18 (1978) 1174
7. J.C. Itzykson, private communication
8. H. Grosse, Seminar given at the Schladming Winter School 1981, to be published in Acta Phys. Austr. Suppl.
9. M. Suzuki, Progr. Theor. Phys. 37 (1967) 770
10. H. Grosse, C.B. Lang and H. Nicolai, Phys. Lett. 98B (1981) 69
11. R. Balian, J.M. Drouffe and J.C. Itzykson, Phys. Rev. D11 (1975) 2104, Err. D19 (1979) 2514
12. M. Creutz, L. Jacobs and C. Rebbi, Phys. Rev. D20 (1979) 1915
13. G. Bhanot and C. Rebbi, CERN preprint TH 2979 (1980)
14. H. Grosse and H. Kühnelt, Phys. Lett. 101B (1981) 77 and Nucl. Phys. (to be published)
15. W.M. Fairbairn, T. Fulton and W.H. Klink, Jour. Math. Phys. 5 (1964) 1038
16. A. Bovier, M. Lüling and D. Wyler, preprint Bonn HE80-21 (to be published in Jour. Math. Phys. 1981)
17. G. Bhanot and C. Rebbi, BNL preprint (1981)

FINITE-SIZE SCALING THEORY

V. Rittenberg

Physikalisches Institut
Bonn University

1. Introduction to Finite-Size Scaling

Fisher's[1] finite-size scaling describes the cross-over from the singular behaviour of thermodynamic quantities at the critical point to the analytic behaviour of a finite system. In the last two years the method was extended to the transfer matrix technique by Nightingale and Blöte[2] and to the Hamiltonian formalism[3] by Hamer and Barber[4-6]. We will be concerned only with these applications.

I first present the method and then proceed to show how it is applied to the specific example of a Z_3 symmetric quantum chain[7].

Let us consider a two-dimensional infinite system which has a second-order phase transition at a temperature T_c with a correlation length ξ behaving like $\xi \sim A(T-T_c)^{-\nu}$ (ν is the critical exponent and A a constant). We now consider a slab of the same material which is infinite in one direction but has only n layers in the other one. The computed (or measured) correlation length of the slab of thickness $L = na$ is $\xi_n = \xi_n(T)$. This function is finite for all T. We take the finite-size scaling limit which consists of increasing the number of layers n and let T approach T_c keeping fixed the product $n(T_c - T)^\nu$. Fisher[1] has argued that in this limit, the ratio L/ξ_n scales:

$$\lim \frac{L}{\xi_n} = f\left(\frac{L}{\xi}\right) \tag{1}$$

where ξ is the correlation length of the infinite system. Eq. (1) can be written in an equivalent way:

$$\lim_{n\to\infty,\, T\to T_c,\, z \text{ fixed}} G(n,T) = \lim \frac{n}{\xi_n} = S(z) \tag{2}$$

where

$$z = n^{1/\nu}(T - T_c) \tag{3}$$

The scaling function S(z) is universal (it does not depend on the type of lattice, etc...) but does depend upon the boundary conditions.

For small z, the function S(z) is sensitive to the boundary conditions. For large z, S(z) is determined by the behaviour of the infinite system.

$$\lim_{z \to \infty} S(z) = A z^{\nu} \tag{4}$$

It was noticed in the case of the Ising model[4] and for Z_3 models[7] that the limit (4) is reached exponentially:

$$\lim_{z \to \infty} S(z) = A z^{\nu} + \mathcal{O}(e^{-cz}) \tag{4'}$$

Based on Eq. (2) one can derive various ways to estimate the critical temperature T_c and exponent ν.

If T_n is the solution of the equation

$$G(n, T_n) = G(n-1, T_n) \tag{5}$$

then

$$\lim_{n \to \infty} T_n = T_c \quad ; \quad \lim_{n \to \infty} G(n, T_n) = S(0) \tag{6}$$

The sequence T_n gives an estimate of the critical point T_c. In many examples[6,7] one has:

$$T_n - T_c = b_1 n^{-c_1} + b_2 n^{-c_2} \tag{7}$$

where c_1, $c_2 \approx 2 \div 3$. An alternative way to determine T_c is based on the observation[7] that for certain boundary conditions S(o) = o and the sequence $\overline{T}_n$ of solutions of the equations

$$G(n, \overline{T}_n) = 0 \tag{8}$$

can approximate T_c.

The critical exponent ν can be estimated using Eqs. (2) and (5). Since

$$\lim_{n \to \infty} \left(\frac{\partial G(n,T)}{\partial T} \right)_{T=T_n} = n^{1/\nu} S'(0) \tag{9}$$

from the sequences

$$\psi_n = \left(\frac{\partial G(n,T)}{\partial T} \right)_{T=T_n} ; \quad g_n(\epsilon) = (n+\epsilon) \frac{(\psi_n - \psi_{n-1})}{\psi_{n-1}} \tag{10}$$

(ε is a parameter which can be introduced to improve the convergence) we have

$$\lim_{n\to\infty} \rho_n(\varepsilon) = \frac{1}{\nu} \tag{11}$$

Instead of using S(z) in the vicinity of the origin one can consider a whole region of z and determine ν_n from the condition that the approximate equality

$$G\left(n, n^{1/\nu_n}(T-T_n)\right) = G\left(n-1, (n-1)^{1/\nu_n}(T-T_n)\right) \tag{12}$$

holds in a certain domain of T.

We consider now another thermodynamic quantity like the specific heat. If for the infinite system, at the critical point, the specific heat diverges like $c_v \sim (T-T_c)^{-\alpha}$, the specific heat for the finite system $c_v(n,T)$ should scale like:

$$\lim_{n\to\infty, T\to T_c, z \text{ fixed}} C_v(n,T) = n^{\alpha/\nu} F(z) \tag{13}$$

where F(z) is again a universal function dependent only on the boundary conditions. From the specific heat we can determine T_c, α and ν as follows. The function $c_v(n,T)$ has a peak at a certain temperature T_n. The sequence T_n gives an estimate of T_c. The value of the specific heat at the maximum $c_v(n,T_n)$ behaves like

$$C_v(n, T_n) \approx n^{\alpha/\nu} F(0) \tag{14}$$

From Eq. (14) one can determine α/ν. Finally we can determine ν from the sequence ν_n:

$$\zeta = \frac{C_v\left(n, n^{1/\nu_n}(T-T_n)\right)}{C_v(n, T_n)} = \frac{C_v\left(n-1, (n-1)^{1/\nu_n}(T-T_{n-1})\right)}{C_v(n-1, T_{n-1})} \tag{15}$$

where the number ζ can be chosen at will. For example 0.9 or 0.5.

At fixed T, the functions $c_v(n,T)$ converge exponentially to the specific heat of the infinite system:

$$c_v(n,T) - C_v(T) \sim e^{-n/\xi} \tag{16}$$

The convergence is poorer in the vicinity of the critical point where the correlation length ξ is large.

The application of finite-size scaling to one-dimensional quantum chains[3]

$$H = -(V + \lambda T) \tag{17}$$

is straightforward. In Eq. (17) V and T are the potential and kinetical energies and the coupling constant λ can be interpreted as the inverse of the temperature. The inverse of the correlation length is the mass gap:

$$m(\lambda) = E_1(\lambda) - E_0(\lambda) \tag{18}$$

where E_0 and E_1 are the ground and first excited state energies. The specific heat per spin is:

$$C_V(\lambda) = -\frac{\lambda^2}{n}\frac{\partial^2 E_0(\lambda)}{\partial \lambda^2} \tag{19}$$

We close this section with a technical comment. As we have seen, finite-size scaling gives us several sequences like T_n or ρ_n from which we have to extract the limit. It turns out that for this purpose the Vanden Broeck-Schwartz approximants[8] are very useful. Consider the sequence P_n which converges to P. The approximants $[n,L]$ ($L = 1,2,..$) are defined as follows:

$$\begin{aligned}&[n,-1]=\infty\ ,\ [n,0]=P_n\ ,\\ &([n,L+1]-[n,L])^{-1}+\alpha_L([n,L-1]-[n,L])^{-1}=\\ &\qquad =([n+1,L]-[n,L])^{-1}+([n-1,L]-[n,L])^{-1}\end{aligned} \tag{20}$$

Here the parameters α_L can be chosen at will. In practice for sequences which converge exponentially (see Eq. (16)) it is convenient to take $\alpha_L=0$ and for sequences which converge like powers (see Eq. (7)) one takes $L=2$, $\alpha=-1$.

When applied to spin systems (like Potts models), finite-size scaling gives estimates for the critical exponent ν which are a bit too high[2,5,6,7] as compared to the theoretical prejudices. An intelligent use by Hamer and Barber of the Vanden Broeck-Schwartz approximants has however settled the Z_3 case where they have found $\nu = 0.8333 \pm 0.0003$ very close to the expected value 5/6. At the same time they have shown that in the Z_5 case, the specific heat does not diverge. This was done using Eq. (16).

In the next section we will show that diagonalizing the Hamiltonian for finite chains allows us not only to estimate the critical exponents but also to discover new phase transitions at lower temperatures which could not be seen through the strong coupling expansion method.

2. Non-Universality in Z_3 Symmetric Quantum Chains

As an application of finite-size scaling we study a 3-states Hamiltonian with Z_3 global symmetry[7]. This Hamiltonian depends on a parameter. We compute the critical exponent ν for several values of this parameter. It turns out that ν varies continuously with the parameter. A similar situation is known to occur in the Baxter[9], Ashkin-Teller[10] and the nn/nnn[11] models.

We consider the one-dimensional quantum chain given by the Hamiltonian

$$\bar{H} = \sum_{\ell=1}^{n} (2\cos\varphi - e^{i\varphi}\sigma_\ell - e^{-i\varphi}\sigma_\ell^+) - \xi\Big[\sum_{\ell=1}^{n-1}(\Gamma_\ell \Gamma_{\ell+1}^+ + \Gamma_\ell^+ \Gamma_{\ell+1}) + B\Big] \tag{21}$$

where the 3x3 matrices σ and Γ are:

$$\sigma = \begin{pmatrix} 1 & 0 & 0 \\ 0 & \omega & 0 \\ 0 & 0 & \omega^2 \end{pmatrix} \; ; \; \Gamma = \begin{pmatrix} 0 & 1 & 0 \\ 0 & 0 & 1 \\ 1 & 0 & 0 \end{pmatrix} \tag{22}$$

$\omega = \exp(\frac{2\pi i}{3})$, ϕ is a parameter $(0 \leq \phi < \frac{\pi}{3})$ and ξ can be interpreted as the inverse of the temperature. $\bar{H}$ has Z_3 as global symmetry and for $\xi \to \infty$ the ground state is 3-fold degenerate. For $\phi=0$, $\bar{H}$ becomes the Potts Hamiltonian which has S_3 symmetry $(S_3 \supset Z_3)$.

The matrix B in Eq. (21) is determined by the choice of the boundary conditions. We have considered two of them:

a) Periodic $(B=B_0)$
In this case we take

$$B_0 = \Gamma_n \Gamma_1^+ + \Gamma_n^+ \Gamma_1 \tag{23}$$

b) "Self-dual" $(B=B_1)$

$$\begin{aligned} &\langle \alpha_1', \alpha_2', \dots, \alpha_n' | B_1 | \alpha_1, \alpha_2, \dots, \alpha_n \rangle = \\ &= \omega^{(\sum_{i=1}^{n}\alpha_i + 1)} \langle \alpha_1', \dots, \alpha_n' | \Gamma_n \Gamma_1^+ | \alpha_1, \dots, \alpha_n \rangle + \\ &+ \omega^{-(\sum_{i=1}^{n}\alpha_i + 1)} \langle \alpha_1', \dots, \alpha_n' | \Gamma_n^+ \Gamma_1 | \alpha_1, \dots, \alpha_n \rangle \end{aligned} \tag{24}$$

The property of the boundary condition B_1 is that even for finite n, in the case $\phi=0$, the mass gap satisfies the self-duality condition

$$m(\lambda) = \lambda \, m\left(\frac{1}{\lambda}\right) \tag{25}$$

We denote the Hamiltonians corresponding to the boundary conditions $B_0(B_1)$ by $\bar{H}_0(\bar{H}_1)$.

It is convenient to redefine $\bar{H}$ and the coupling constant ξ as follows:

$$H = \left(2\sqrt{3}\sin\left(\frac{\pi}{3}-\varphi\right)\right)^{-1}\bar{H} \; ; \; \lambda = \left(\frac{2}{\sqrt{3}}\sin\left(\frac{\pi}{3}-\varphi\right)\right)^{-1}\xi \tag{26}$$

$$H = \sum_{\ell=1}^{n} \Omega_\ell - \frac{\lambda}{3}\Big[\sum_{\ell=1}^{n-1}(\Gamma_\ell \Gamma_{\ell+1}^+ + \Gamma_\ell^+ \Gamma_{\ell+1}) + B\Big] \tag{27}$$

where

$$\Omega=\begin{pmatrix} 0 & 0 & 0 \\ 0 & 1 & 0 \\ 0 & 0 & \rho \end{pmatrix}, \quad \rho=\frac{\sin\left(\frac{\pi}{3}+\varphi\right)}{\sin\left(\frac{\pi}{3}-\varphi\right)} \quad ; \quad 1\leq\rho<\infty \tag{28}$$

Our purpose is to determine the critical points λ_c where the mass gap vanishes and the critical exponents ν for various values of the parameter ρ (or ϕ). For $\rho = 1$ (the Potts case) we have $\lambda_c = 1$ and $\nu = 5/6$. For $\rho \to \infty$, the Hamiltonian (27) reduces to

$$H=\sum_{\ell}\frac{(1-\sigma^z_\ell)}{2}-\frac{\lambda}{6}\sum_{\ell}\left(\sigma^x_\ell\sigma^x_{\ell+1}+\sigma^y_\ell\sigma^y_{\ell+1}\right) \tag{29}$$

where σ^x, σ^y and σ^z are Pauli matrices. The system (29) is integrable[12] and the mass gap vanishes for $\lambda_c = 3/2$ with $\nu = 1$. The phase transition is not a second-order one since for $\lambda > 3/2$ the ground state remains degenerate. It is not a massless phase since the states having the same energy as the vacuum have a non-vanishing momentum. We denote such a transition by (XXM) (XX model in a transverse magnetic field). For the intermediate region ($1<\rho<\infty$) we have performed the standard strong coupling expansion[13] and have obtained the solid curves shown in Figs. 1 and 2. The curves suggest that λ_c and ν vary smoothly between the extreme values: $1\leq\lambda_c\leq 3/2$, $5/6\leq\nu\leq 1$.

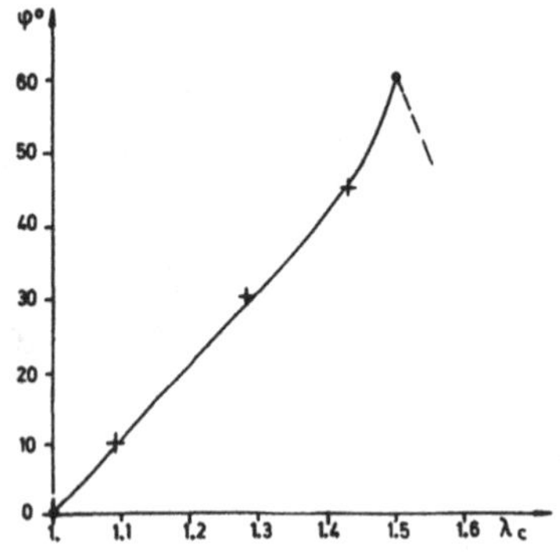

Fig. 1
The critical points λ_c for various angles ϕ. The solid curve is obtained from the strong coupling expansion. The dashed line represents the (XXM) transition. The crosses are the results obtained from finite-size scaling. The dots are exact results.

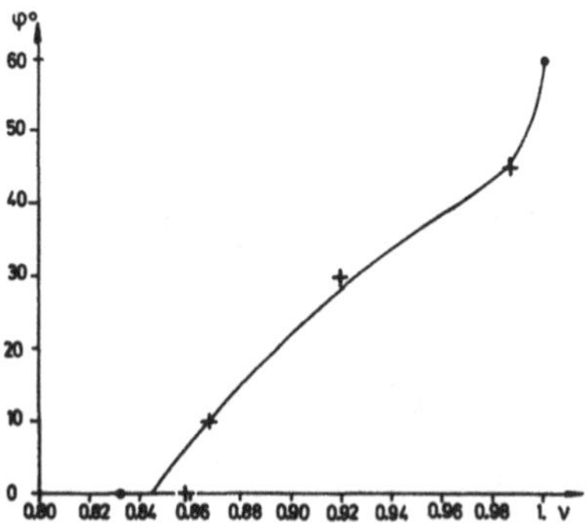

Fig. 2
The critical exponents ν for various angles ϕ. The notations are the same as for Fig. 1.

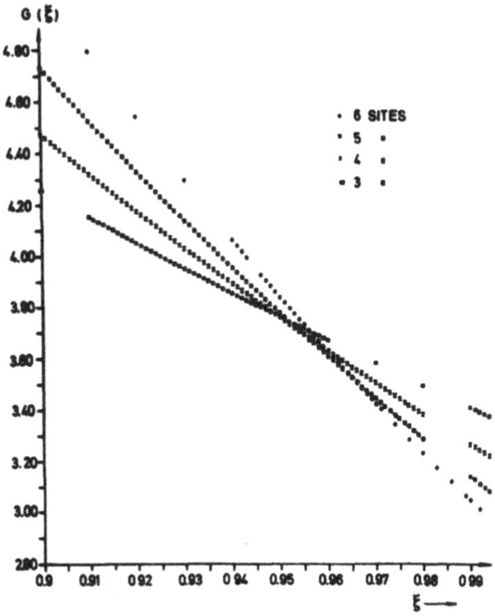

Fig. 3
The mass gap $G(\xi)$ for H_o as a function of ξ, for 3,4, 5 and 6 sites in the case $\phi=10^o$.

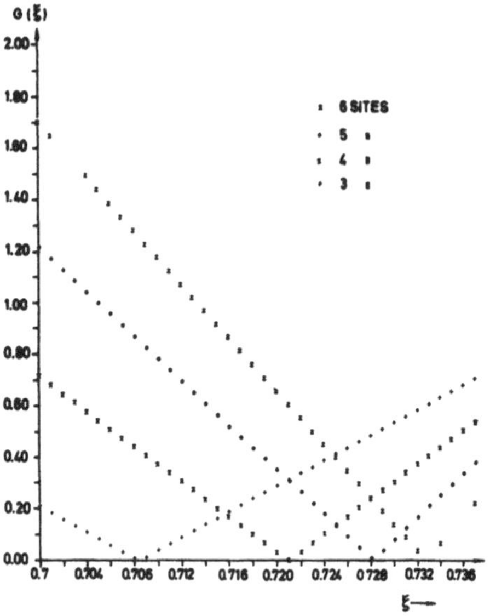

Fig. 4
The mass gap $G(\xi)$ for H_1 as a function of ξ, for 3,4, 5 and 6 sites in the case $\phi=30^o$.

We now apply finite-size scaling to our problem. We have done a sloppy job considering only the cases $n = 3, 4, 5$ and 6 and four angles: $\phi=0^o$ $(\rho=1)$, $\phi=10^o$ $(\rho=1.23)$, $\phi=30^o$ $(\rho=2)$ and $\phi=45^o$ $(\rho=3.73)$. The Hamiltonian was diagonalized by brute force. The two boundary conditions given by Eq. (23) (Hamiltonian H_o) and Eq. (24) (Hamiltonian H_1) have been considered and the quantities

$$G(n,\lambda) = n\, m(n,\lambda)$$

have been computed. Here $m(n,\lambda)$ is the mass gap for the n-sites problem.

First we consider the Hamiltonian H_o. In Fig. 3 we show the functions $G(n,\xi)$ (ξ is related to λ by Eq. (26)) for $\phi=10^o$. The pictures for the other angles are similar. Using Eq. (5) we have determined the critical points. The results are shown in Table 1.

We next consider the Hamiltonian H_1. For Φ=0 (the Potts model), the mass gap functions vanish at λ=1 for all n. The critical point is thus correctly determined even at n=3! For the other angles the situation looks like in Fig. 4 where we have shown the case $\phi=30^{\circ}$. All the mass gap functions vanish for certain values λ_n. We consider (see Eq. (8)) these values as estimates for λ_c. The corresponding values are shown in Table 1 and we see that they converge faster than the estimates obtained from H_o.

Table 1: The critical points λ_c determined from H_1, H_0 (defined in the text) and the specific heat C_v. n denotes the number of sites.

ϕ	0			10°		
n	H_1	H_o	C_v	H_1	H_o	C_v
3	1		0.72	1.069		0.791
		0.983			1.067	
4	1		0.84	1.078		0.949
		0.992			1.075	
5	1		0.91	1.083		1.028
		0.996			1.084	
6	1		0.95	1.085		1.04
ϕ	30°			45°		
n	H_1	H_o	C_v	H_1	H_o	C_v
3	1.23		1.21	1.37		—
		1.20			1.31	
4	1.25		1.23	1.40		—
		1.23			1.36	
5	1.262		1.3	1.42		—
		1.24			1.39	
6	1.269		1.3	1.422		—

In order to determine ν we have used our experience gained from the strong coupling expansion[7] in order to choose the right scaling variable and from now on we consider only H_1.

We first define

$$u = \frac{4\lambda\lambda_n}{(\lambda+\lambda_n)^2} \quad ; \quad z = n^{2/\nu}(1-u) \tag{30}$$

where λ_n are listed in Table 1. We then compute

$$S(n,z)=\left(1+\frac{\lambda}{\lambda_n}\right)^{-1} m(n,u) \qquad (31)$$

where m(n,u) is the mass gap for n-sites and tune ν such that S(n,z) scales (is independent of n). The operation is done using the computer terminal. It turns out that similar to the Ising model, the scaling function S(z) reaches its asymptotic limit exponentially (see Eq. (4')) and thus ln S(z) is a straight line already for small values of z. This simplifies the tuning procedure since it is very easy to check the spreading of the points around a straight line. In Fig. 5 we show ln S(z) as a function of z for $\phi=30^{\circ}$ and the choice $\nu=0.93$. The precision achieved in this way for the estimates of ν is of the order of 2%. Table 2 summarizes our results. They are in very good agreement with the estimates obtained from the strong coupling expansion as seen from Figs. 1 and 2.

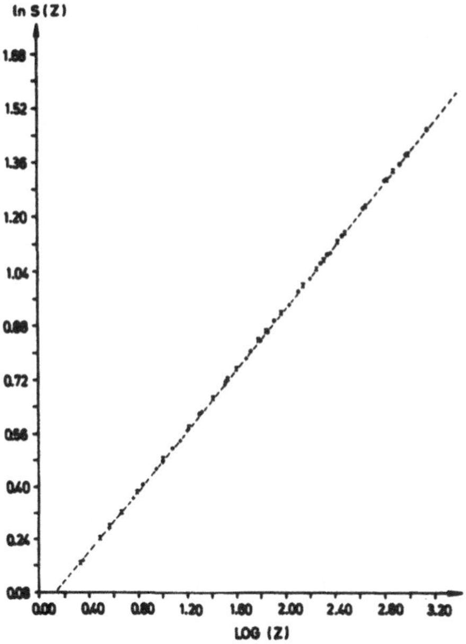

Fig. 5. ln S(z) as a function of z for $\nu = 0.93$, $\phi = 30^{\circ}$.

We now consider the specific heat c_v. This is determined considering the Hamiltonian H_o (periodic boundary conditions), and Eq. (19). From the positions of the maxima we can again estimate λ_c as shown in Table 1. The convergence here is rather poor. We can notice that no values for λ_c have been given at $\phi=45^{\circ}$. It turns out that in this case the energy gap vanishes for a value of λ slightly larger than λ_c. This does not imply that the mass gap vanishes since in our computer program we have not diagonalized the states according to their momenta. Because of the degeneracy of the ground state, the expression (19) loses its sense. This phenomenon should not be surprising since we know that at $\phi = \frac{\pi}{3}$ we do not have a second-order transition but what we have called an (XXM) transition and one should expect a whole line of (XXM) transitions. In Fig. 1 we have drawn tentatively such a line (see the dashed curve). In fact, we have checked that even at smaller angles the energy gap vanishes but it does so at values of λ large enough not to "perturb" the calculation of c_v at λ_c. For example for $\phi=30^o$ the transition occurs at $\lambda=2.4$.

Table 2: The critical points λ_c and indices ν estimated from finite-size scaling.

$\phi = 0$		$\phi = 10^o$		$\phi = 30^o$		$\phi = 45^o$	
λ_c	ν	λ_c	ν	λ_c	ν	λ_c	ν
1	0.858	1.09	0.868	1.280	0.920	1.430	0.974

References

1. M.E. Fisher, Critical Phenomena, in Proc. Enrico Fermi Intern. School of Physics, M.S. Green ed. (Academic Press, N.Y. 1971)
2. M.P. Nightingale and H.W.J. Blöte, Physica 104A, 352 (1980); H.W.J. Blöte, M.P. Nightingale and B. Derrida, J. Phys. A14, L45 (1981)
3. J. Kogut, Rev. Mod. Phys. 51, 689 (1979)
4. C.J. Hamer and M.N. Barber, J. Phys. A13, L169 (1980), 14, 241 (1981)
5. C.J. Hamer and M.N. Barber, J. Phys. A14, 259 (1981); J. Phys. A14, 2009 (1981); M.N. Barber and P.M. Duxbury, J. Phys. A14. L251 (1981)
6. H.H. Roomany, H.W. Wyld and L.E. Holloway, Phys. Rev. D21, 1557 (1980); H.H. Roomany and H.W. Wyld, Phys. Rev. D21, 3341 (1980), Phys. Rev. B23, 1357 (1981)
7. P. Centen, V. Rittenberg and M. Marcu, Preprint Bonn-HE-81-29
8. J.M. Vanden Broeck and L.W. Schwartz, SIAM J. Math. Anal. 10, 658 (1979)
9. R.J. Baxter, Ann. Phys. (N.Y.) 70, 193 (1972)
10. J. Ashkin and E. Teller, Phys. Rev. 64, 178 (1942)
11. M.N. Barber, J. Phys. A12, 679 (1979)
12. Th. Niemeyer, Physica 36, 377 (1967)
13. S. Elizur, R.B. Pearson and J. Shigemitsu, Phys. Rev. D19, 3698 (1979)

ANDERSON TRANSITION AND NONLINEAR σ-MODELS

Franz Wegner

Institut für Theoretische Physik
Ruprecht-Karls-Universität
D-6900 Heidelberg, F. R. Germany

ABSTRACT

The talk covers the behaviour of a quantum mechanic particle moving in a random potential with special emphasis on the critical behaviour near the mobility edge. Symmetry arguments are reviewed which allow the mapping of such a system onto a field theoretical model of interacting matrices. This model yields an expansion of the critical exponents at the mobility edge around the lower critical dimensionality two. Since most of this lecture has already been published as a contribution to the Les Houches institute 1980 "Common Trends in Particle and Condensed Matter Physics" I give here only some new results and refer the reader to reference [1].

NEW RESULTS

Hikami [2] gives the W-(ß-) function in three loop order for the orthogonal and symplectic case of the nonlinear σ-model. As a consequence the critical exponent s for the residual conductivity of a system with spin-independent potential with time-reversal invariance vanishes like $(E - E_c)^s$ with

$$s = 1 + O(d - 2)^3.$$

Here E is the Fermi energy, E_c the mobility edge, and d the dimensionality of the system. Pruisken and Schäfer [3] have mapped the site-diagonally disordered electron system on a system obeying local gauge invariance. This supports strongly the assumption that the mobility edge behaviour is governed by local gauge invariant interactions.

The present author has proven [4] that for a class of tight-binding models with diagonal disorder and short-range one-particle interaction the density of states is positive (non-zero) and finite within the whole band. This class of systems includes models with rectangular, Lorentzian, and Gaussian distribution of the diagonal matrix elements. Systems in which the site-diagonal and the off-diagonal matrix elements are independent random variables with symmetric distribution (on a compact support) have a non-zero density of states everywhere inside the band. (In the case of off-diagonal disorder only, the proof does not work for $E = 0$). Thus, predictions that at the mobility edge the density of states for site-diagonal disorder would diverge or that it would vanish for off-diagonal disorder ($8 - \varepsilon$ expansions) or predictions which allow for both options (inhomogeneous fixed point ensemble [5]) can be ruled out. The homogeneous fixed point ensemble [5], however, is in agreement with these findings.

In a region not too close to the mobility edge the measurements of reference [6] on P doped Si are in good agreement with an exponent $s = 0.55$. However, in comparing with experiments [6,7] one has to keep in mind that the Anderson model does not incorporate interactions between electrons, which modifies the mobility edge behaviour [8,9].

REFERENCES

[1] F. Wegner, Physics Reports 67, 15 (1980).
[2] S. Hikami, Prog. Theor. Phys. 64, 1466 (1980).
[3] A. M. Pruisken, L. Schäfer, Phys. Rev. Lett. 46, 490 (1981).
[4] F. Wegner, Z. Physik B44, 9 (1981).
[5] F. Wegner, Z. Physik B25, 327 (1976).
[6] T. F. Rosenbaum, K. Andres, G. A. Thomas, R. N. Bhatt, Phys. Rev. Lett. 45, 1723 (1980).
[7] B. W. Dodson, W. L. McMillan, J. M. Mochel, Phys. Rev. Lett. 46, 46 (1981).
[8] W. L. McMillan, Phys. Rev. B24, 2739 (1981).
[9] R. Oppermann, Proceedings of the Rome Conference on Disorder and Localization, Rome, May 1981.

FUNCTIONAL INTEGRATION FOR KINKS AND DISORDER VARIABLES

B. Schroer

Institut für Theoretische Physik
Freie Universität Berlin
D-1000 Berlin 33

Kinks or topological solitons have been known for a long time in classical nonlinear field theories. The kink solution of the two-dimensional ϕ^4 model and the one-soliton solution of the Sine-Gordon equation are important and well-known special cases. Arguments based on topological stability assert that particle quantum states corresponding to such classical solutions are expected to exist, and hence a quasiclassical approach for the particle spectrum and the S-matrix appears theoretically justified.[1] A systematic study of this approach with a hindsight from Nuclear Physics suggests the introduction of collective coordinate methods.[2] The many papers published on the semiclassical and collective coordinate approach to the kink problem allow one clearcut conclusion: The formalism becomes complicated and looses its esthetical appeal if used for the construction of field theoretic ("off mass shell") objects such as vacuum expectation values or matrix elements of products of fields. This feature is in fact to be expected in a formalism which treats a Lagrangian field simultaneously with the quantum mechanical degrees of freedom of the kink states.

Following the spirit of local quantum field theory, one should rather aim at the construction of local interpolating kink operators $\mu(x)$. In the ϕ^4 model a kink should correspond to a "half-space" Z_2 symmetry transformation:

(1) $$\mu(x)\,\phi(y) = e^{i\pi\theta(y-x)}\,\phi(y)\,\mu(x)$$

The required scalar transformation property of μ would then imply the validity of this relation for all space-like distances. This is the well-known dual algebra relation first introduced by Kadanoff[3] in the context of the lattice Ising model and later formally generalized to higher dimensional systems by 't Hooft [4]. The classical kink function enters the duality relation in the form of an idealized (structureless) step function. Only in the ordered phase of the ϕ^4 model does one expect the μ to create a new sector of states. The formfactor of ϕ in this new state or in other words the three-point function

$$F_{x_1,x_2}(y) = \langle \mu(x_1)\,\phi(y)\,\mu(x_2) \rangle \tag{2}$$

is the simplest quantum field theoretic object describing the kink structure as y varies from space-like asymptotic left points to space-like asymptotic right ones. The correlation functions for one μ and an arbitrary number of ϕ should vanish in agreement with the selection rule of the Z_2 quantum number carried by μ. For Lagrangians with other multiplicative symmetries the introduction of kink operators with "half-space" commutation relations:

$$\mu(x)\,\phi(y) = \left\{ \begin{array}{ll} \phi(y)\,\mu(x) & , \; y < x \\ \phi^g(y)\,\mu(x) & , \; y > x \end{array} \right\} \tag{3}$$

(where ϕ^g = linearly transformed field) generalizes the Z_2 relation (1). The absence of a broken symmetry phase for two-dimensional systems with a continuous symmetry does not prevent the construction of the μ's. They will however not carry any new selection rules. The existence of invariant condensates i.e. $\langle\phi^2\rangle \neq 0$ is not sufficient for the formation of new sectors.

How can one obtain a realization and a mathematical control of this scenario of local kink operators in the framework of euclidean functional integration? Generalizing a construction given by Kadanoff for the Ising model [3] to continuous field theories one obtains a remarkably beautiful and simple formula [5] for the mixed correlation functions:

$$\langle \mu(x_1)\ldots\mu(x_m)\,\phi^\#(y_1)\ldots\phi^\#(y_n)\rangle = \frac{1}{Z}\int[\mathcal{D}\phi][\mathcal{D}\phi^*]\,\phi^\#(y_1)\ldots\phi^\#(y_n)\exp\{-S_{BA}\} \tag{4}$$

$$\phi^\# = \phi \text{ or } \phi^*$$

Here S_{BA} is the old action S, modified by the introduction of Bohm-Aharonov fluxes at the positions $x_1 \ldots x_m$ coupled minimally to the matter field ϕ. Hence d = 2 kink operators appear in the euclidean (statistical mechanics) domain as some very special gauge theories with "flat connections" ($F_{\mu\nu} = 0$) in a pointed euclidean space from which the points $x_1 \ldots x_m$ have been removed. The derivation of this formula, its general properties, and several applications to simple Lagrangians is the main aim of this lecture.

Before I enter this discussion, I should mention that the problem of kink sectors outside of the quasiclassical and collective coordinate approximation schemes has been investigated notably by J. Fröhlich [6]. The mathematical technique there was the construction of certain automorphisms within the C*-algebra framework. The main difference to the present construction is that by insisting in local fields via functional integrals, we obtain a mathematically more concrete and physically richer formalism in which the renormalization aspects of the new fields are delegated to the selfenergy properties of Bohm-Aharonov fluxes [7] coupled to matter fields. We hope that the few explicit examples will convince the reader of the usefulness and esthetical appeal of this method.

The correctness of (4) can almost be guessed on the basis of formal canonical arguments. For a complex field ϕ with an abelian symmetry (in Minkowski space), the Ansatz

$$(5)\quad \mu(x) = \exp\{2\pi i\alpha \int_{x,C}^{\infty} \varepsilon^{\mu\nu} j_\nu \, d\xi_\mu\} \quad \text{with} \quad j_\mu = i\phi \overleftrightarrow{\partial}_\mu \phi^*$$

would have the desired commutation relation (3) with ϕ and ϕ^* at equal times. However, this does not define a covariant field μ, as can be seen from the functional integral representation of the two-point function

$$(6)\quad \langle \mu(x)\mu^+(y)\rangle = \frac{1}{Z}\int [\mathcal{D}\phi][\mathcal{D}\phi^*] \exp\{-S + 2\pi i\alpha \left(\int_{x,C}^{\infty} - \int_{y,C}^{\infty}\right) \varepsilon^{\mu\nu} j_\nu \, d\xi_\mu\}$$

which is a path-dependent object. (Note the change of reality properties.) But the exponent is a piece of a gauge theory action, and by adding a bilinear term without derivatives one obtains the action

$$(7)\quad \begin{aligned} S_{BA} &= \int d^2x \, \mathcal{L}(\partial_\mu \to D_\mu), \quad D_\mu = \partial_\mu - i\alpha A_\mu \\ A_\mu(z) &= 2\pi \left(\int_{x,C}^{\infty} - \int_{y,C}^{\infty}\right) \varepsilon^{\mu\nu} \delta(z-\xi) \, d\xi_\nu \end{aligned}$$

Now the path-independence is just gauge invariance which hence ensures the the scalarity of the new field. The path C behaves as the fictitious string of a Dirac monopole.

A better derivation of (4) is obtained by generalizing the statistical mechanics argument of Kadanoff for the $d = 2$ lattice Ising model. There the action for the disorder variable μ is obtained from the Ising action by changing the action on a "dual" path C. At those links which are intersected by C one adds a term, e.g. for the two-point function (x,y dual lattice points)

$$\langle \mu(x)\,\mu(y)\rangle = \frac{1}{Z}\sum_{[\sigma]} \exp\{-S_C[\sigma]\}\,, \tag{8}$$

$$S_C[\sigma] = K\sum_{\langle ij\rangle} \sigma_i\sigma_j + A_C[\sigma]\,, \tag{9}$$

where C connects x and y. (C may pass through the periodic boundary or through infinity after the thermodynamic limit has been taken.) The form of A_C is determined by the underlying symmetry $\sigma \rightarrow -\sigma$ and the requirement that C is fictitious. For the case at hand we obtain

$$A_C[\sigma] = -2K\sum_{\langle ij\rangle\in C} \sigma_i\sigma_j\,. \tag{10}$$

The extension to higher point functions is obvious. For mixed correlation functions involving σ's one obtains a sign ambiguity resulting from passing the (locally) fictitious path through the position of a σ.

Consider now a complex field ϕ with Lagrangian

$$\mathcal{L} = \partial_\mu\phi^*\partial_\mu\phi + M^2\phi^*\phi + U(\phi,\phi^*) \tag{11}$$

which has a U(1) or Z_N symmetry. In analogy with (9), the Lagrangian for the two-point function of μ is written as

$$\mathcal{L}_C = \mathcal{L} + \psi_\nu \int_{x,C}^{y} \varepsilon^{\mu\nu}\,\delta(z-\xi)\,d\xi_\mu + \chi_C(\phi,\phi^*)\,. \tag{12}$$

ψ_ν is a local functional of ϕ,ϕ^* and the partial derivatives $\partial_\mu\phi$, $\partial_\mu\phi^*$, and χ_C is a renormalization counterterm which may be path dependent. Let C' be another path and consider the loop $\Gamma = C - C'$. Inside the region S bounded by Γ apply the basic symmetry transformation of the model, i.e.

$$\phi(x) \rightarrow e^{2\pi i\alpha}\phi(x) \text{ inside S}\,,\quad \phi(x) \rightarrow \phi(x) \text{ outside S}\,, \tag{13}$$

where $\alpha = 1/N$ for the case of a Z_N symmetry. Under this transformation, the kinetic part of $\mathcal{L}$ produces a charge living on the boundary Γ:

$$\delta\mathcal{L} = 2\pi i\alpha\;\phi\overleftrightarrow{\partial_\nu}\phi^* \oint_\Gamma \varepsilon^{\mu\nu}\,\delta(z-\xi)\,d\xi_\mu$$

$$+ (2\pi\alpha)^2\,\phi^*\phi \oint_\Gamma\oint_\Gamma \delta(z-\xi)\,\delta(z-\eta)\,d\xi_\mu\,d\eta_\mu$$

Observe next that the Γ-integral term obtained by writing

$$\psi_\nu \int_C \cdots = \psi_\nu \int_{C'} \cdots + \psi_\nu \oint_\Gamma \cdots$$

will compensate against the first term in $\delta\mathcal{L}$ if

(14) $$\psi_\nu = -2\pi\alpha j_\nu = -2\pi i\alpha\, \phi \overleftrightarrow{\partial}_\nu \phi^* .$$

This gives

$$\delta\psi_\nu = -2(2\pi\alpha)^2 \phi^*\phi \oint_\Gamma \varepsilon^{\mu\nu}\, \delta(z-\xi)\, d\xi_\mu .$$

We now choose the counterterm χ_C in such a way that it cancels the C-path contributions to the Γ-integral terms arising from the second term in $\delta\mathcal{L}$ and from $\delta\psi_\nu$; thus

(15) $$\chi_C = (2\pi\alpha)^2 \phi^*\phi \int_C \int_C \delta(z-\xi)\, \delta(z-\eta)\, d\xi_\mu\, d\eta_\mu .$$

With these choices the system closes, i.e. under the transformation (13), $\mathcal{L}_C$ is taken to $\mathcal{L}_{C'}$. Transcription to gauge theoretic language yields (4). The consideration for spinor fields is analogous but somewhat simpler.

It is illustrative to understand how this path-independence works for the construction of Mandelstam's Sine-Gordon soliton operator [8]. In that case the symmetry is additive:

(16) $$\varphi \longrightarrow \varphi + 2\pi n\gamma \quad , \qquad \gamma = \text{const.}$$

This simplifies the situation. With

(17) $$\mathcal{L} = \frac{1}{2}\partial_\mu\varphi\, \partial_\mu\varphi + a \cos \varphi/\gamma$$

we obtain (omitting c-number counterterms)

(18) $$\mathcal{L}_C = \mathcal{L} + \beta\, \partial_\nu\varphi \int_{x,C}^{y} \varepsilon^{\mu\nu}\, \delta(z-\xi)\, d\xi_\mu ,$$

the constant β being quantized in units of the periodicity $2\pi\gamma$ in order to obtain path independence. The corresponding scalar disorder variable μ satisfies the half-space version of the additive symmetry:

(19) $$\mu(x)\,\varphi(y) = \varphi(y)\,\mu(x) + 2\pi\gamma\, \theta(y-x)\, \mu(x)$$

In order to come to the dual relation (3) one defines as the order variable not φ but

$$\sigma(x) = \exp i\alpha\varphi(x) . \tag{20}$$

Because the additive symmetry is always broken, the existence of the kink sector is always assured. The Mandelstam spinor field is obtained for a very special value of α leading to the s = 1/2 spinor transformation property of

$$\begin{pmatrix} \psi_1 \\ \psi_2 \end{pmatrix} \sim \text{l.s.d.} \begin{pmatrix} \sigma\mu \\ \sigma^+\mu \end{pmatrix} . \tag{21}$$

Here l.s.d. denotes the leading short distance operator term (with directionally dependent coefficients). If one interprets φ as a two-dimensional electric potential than the σ - terms in the euclidean functional integral correspond to (imaginary) point charges and the $\mu(x)$ to a dipole layer of charges which ends at x and leads to the same field line picture as that of a magnetic vortex. Hence in statistical mechanics language the construction of order-disorder operators yields a system of a self-interacting static potential coupled to (unquantized) electric sources and (quantized) "magnetic vortices" [9]. The renormalization theory of the fields can be developed from the observation that the self-energy of the sources diverges logarithmically. The Dirac crossing rule leads to a euclidean ramified covering space for the mixed correlation functions. The Mandelstam spinor ψ is a peculiar "dyon" fulfilling a simple equation of motion. The picture obtained for Lagrangians with multiplicative symmetries leads to the Bohm-Aharonov flux situation which is quite different and mathematically more complicated. Here the multivaluedness of the mixed euclidean function is simply the result of a path shift by a gauge transformation. Under such a transformation, the matter field ϕ picks up a constant phase. Returning to operators in Minkowski space by a Wightman boundary prescription, one sees that the two reversed orders in (3) are obtained by once continuing from the (by definition) zero'th sheet respectively from the first covering sheet. This argument has been derived in detail for a simple system with an additive symmetry [9], but it holds in general.

Now we come to the explicit construction of μ in the case of simple models. Consider the Lagrangian of a complex free field. The Bohm-Aharonov spectrum for the presence of just one flux of strength α at the origin is given by the eigenvalue equation

$$\left[-\left(\partial_r^2 + \frac{1}{r}\partial_r\right) - \frac{1}{r^2}(\partial_\varphi + i\alpha)^2 + M^2 \right] \phi = \lambda\phi . \tag{22}$$

Hence

$$\phi_{m,k}(\tau) = \frac{1}{\sqrt{2\pi}}\, e^{im\varphi}\, J_{|m+\alpha|}(k\tau) \quad , \quad \lambda = k^2 + M^2 \, .$$

The λ - integration in the Green's function

$$G(x,y) = \frac{1}{2\pi}\int \frac{k\,dk}{k^2+M^2} \sum_m e^{i(m+\alpha)(\varphi_x-\varphi_y)}\, J_{|m+\alpha|}(k\tau_x)\, J_{|m+\alpha|}(k\tau_y) \tag{23}$$

can be carried out:

$$\int \frac{k\,dk}{k^2+M^2}\, J_\nu(k\tau_x)\, J_\nu(k\tau_y) = I_\nu(M\tau_<)\, K_\nu(M\tau_>)$$

$$\tau_< = \min(\tau_x,\tau_y) \quad , \quad \tau_> = \max(\tau_x,\tau_y)$$

Without loss of generality we may assume

$$0 \leq \alpha < 1 \, .$$

Performing a number of contour changes in the Sommerfeld representation of I_ν and K_ν one arrives at a situation for which the m - summation has been explicitly carried out. Without giving the details of the elementary but somewhat tricky calculations we quote the result (ν = number of sheet):

$$G(x,y) = e^{2\pi i\alpha\nu}\Big\{ G_0(x-y) - \frac{\sin\pi\alpha}{(2\pi)^2}\int_{-\infty}^{\infty} dv \int_{-\infty}^{\infty} dw \cdot \frac{e^{\alpha(v-w+i\varphi_x-i\varphi_y)}}{1+e^{v-w+i\varphi_x-i\varphi_y}}\, e^{-M\tau_x\cosh w - M\tau_y\cosh v}\Big\} \tag{24}$$

The field theoretic Green's function has the form

$$< \mu(0)\,\phi(x)\,\phi^+(y) > \equiv < \phi(x)\,\phi^+(y) >_{BA} = C\; G(x,y) \quad , \tag{25}$$

where C is an x,y-independent factor which collects the contribution of the Bohm-Aharonov functional determinant of the matter field. The euclidean correlation functions involving more ϕ's but the same Bohm-Aharonov flux yield the same constant C and contain sums over products of the above classical Green's functions with the Wick contraction combinatorics. From this it follows that in Fock space, the operator μ has the exponential form

$$\mu(x) = C : \exp K_\alpha(x) : \tag{26 a}$$

Now comparing the various one-particle formfactors of $\mu(0)$ with the Green's function, analytically continued to the time-like or space-like Minkowski region, one determines the kernel K_α. Using the rapidity θ, $p = M(\cosh\theta, \sinh\theta)$, one obtains:

$$K_\alpha(p,q) = \frac{\sin\alpha\pi}{2\pi}\int_{-\infty}^{\infty} d\theta_p \int_{-\infty}^{\infty} d\theta_q\; e^{\alpha(\theta_p-\theta_q)-(\theta_p-\theta_q)/2} \tag{26 b}$$

$$\cdot\left\{\left(\sinh\frac{\theta_p-\theta_q+i\varepsilon}{2}\right)^{-1}\left[e^{-i\pi\alpha}a^+(p)a(q)+e^{i\pi\alpha}b^+(p)b(q)\right]\right.$$

$$\left.-\left(\cosh\frac{\theta_p-\theta_q}{2}\right)^{-1}\left[b(q)a(p)+a^+(p)b^+(q)\right]\right\}$$

The calculation may be done analogously for a complex spinor field. In this case one obtains precisely the operator of Lehmann and Stehr [10] obtained previously by fermion ordering the exponential of the free field axial current potential φ:

$$\bar\psi\gamma_\mu\gamma_5\psi = \frac{1}{\sqrt\pi}\varepsilon_{\mu\nu}\partial_\nu\varphi\;,\quad \mu = \frac{\exp\{-2i\sqrt\pi\alpha\varphi\}}{\langle\exp\{-2i\sqrt\pi\alpha\varphi\}\rangle} = :\exp K_\alpha: \tag{27}$$

A quite different construction of kink operators of free fields has been given by Sato et al.[11] The difference to our euclidean functional formalism is that the framework of Sato et al. uses ab initio very special free field properties; there is no attempt to introduce kinks and dual variables in interacting situations. For the fermion model the euclidean formalism allows the explicit construction of an order variable σ by a short distance limit procedure:

$$\sigma = \text{s.d.l.}\;\psi\mu \tag{28}$$

The two components of ψ lead to the same operator (determined up to a numerical factor). The mixed correlation functions with one σ and arbitrarily many ψ and ψ^+ are easily written down. The dual commutation relation of σ and μ is an algebraic consequence of the definition.

A conceptually intricate situation arises if one tries to formulate the Bohm-Aharonov picture for a real field ϕ or a Majorana spinor $\psi = \psi^c$. There is no possibility of minimally coupling an A_μ potential to such a field. Nevertheless one still can have minus boundary conditions in the eigenvalue equation (22). Although the information of twisted boundary conditions cannot be coded into vector potentials A_μ, there exists the mathematical framework of flat but nontrivial Z_2 vector bundles on the topolo-

gically nontrivial base manifold of a pointed R^2.
Consider a μ at the origin. The space $R^2 - \{0\}$ may be covered by two contractible neighbourhoods $U_i = R^2 - S_i$, where the S_i are small angular segments surrounding the positive and negative y - axis. So $U_1 \cap U_2$ consists of two pieces and in one we choose the gauge transition function +1 and in the other -1.
A global Z_2 gauge transformation does not change the spectrum and Green's function. This construction may be generalized to correlation functions involving several μ's ; e.g. for two μ's the minimal patching system has three pieces. We obtain a beautiful illustration of an idea of T.T. Wu and C.N. Yang [12] that there may exist quantum fields whose functional integral construction requires nontrivial fibre bundles. In the Z_2 model at hand there is no possibility to trade the nontrivial patching functions for a singular vector potential. It is well known that the Z_2-kink of a two-dimensional Majorana field describes the disorder variable of the Ising model in the scaling limit $T_c + 0$ [13]. The order variable σ in the same scaling limit is obtained by a short-distance expansion of $\psi\mu$ as previously described. The "doubling" of the Majorana field, i.e. the introduction of a Dirac field

$$\psi = \psi_1 + i\psi_2 \tag{29}$$

yields a trivialization of the Z_2 bundle since we can now use a regular vector potential in the pointed R^2. From the doubling of the eigenstates we obtain a square relation between the Majorana and Dirac fermion determinants:

$$\langle \mu(x_1) \ldots \mu(x_n) \rangle_D = \langle \mu(x_1) \ldots \mu(x_n) \rangle_M^2 \tag{30}$$

The doubling allows to investigate the short distance behaviour of the μ and the σ in an elementary way. With the Lehmann-Stehr formula one obtains a resummation of μ in terms of the axial current potential (27). The massless limit of these operators can be expressed in terms of exponential functions of a free massless scalar field. In fact the taking of the "square root" of the Lehmann-Stehr operators was how Truong and I obtained the Ising field theory as an alternative to the Clifford algebra construction of μ by Sato et al.[11]. Only now after the functional integral approach is there a deeper understanding of our doubling procedure.

Note that the direct application of the euclidean formalism only gives μ's whose Bohm-Aharonov strength is restricted to the unit α - interval. However by "fusing" several μ's with low α's via short distance limits, one may build up μ's with larger α's. It turns out that these operators are the same as the operators μ_α obtained by simply continuing the formula (26) in α beyond the unit interval in α for which it has been derived. It can be shown that all these operators are of the form

$$\mu_\alpha = \text{s.d.l. } P(\phi,\phi^*)\, \mu_{\alpha_r}\,, \tag{31}$$

where α_r is α reduced to the unit interval and P is a local polynomial in the fields.

More interesting applications of the formalism to interacting fields have not yet been carried out. For example one may want to calculate the correlation function

$$< \mu(x) \prod_i \phi(x_i)\, \mu(y) > \;\equiv\; < \prod_i \phi(x_i) >_{BA} \tag{32}$$

in the perturbation theory for ϕ^4 using a Higgs-Goldstone potential. For the vacuum problem one has a perturbation theory obtained by shifting the field by its vacuum expectation value. Analogously one would like to obtain the systematics of (32) via a shift in the action by $<\phi>_{BA}$. In lowest order one seeks a solution for the classical euclidean field equation which has two vortices at x and y with a string in between. Mathematically this is a line of ramification linking the two sheets of a covering space, and the values of the solution at corresponding points of the two sheets differ simply by a - sign. Moreover, one imposes asymptotic approach to the + vacuum in the first sheet (and hence to the - vacuum in the second sheet) as a boundary condition. The determination of such a two-vortex solution is certainly more difficult than that of the stationary solution in the semiclassical approach. However it is only on this level that our formalism makes contact with classical nonlinear equations.

There are other important open problems, such as the reconstruction of order and disorder operators from their euclidean expectation values (the Osterwalder-Schrader problem for mixed correlation functions) and generalizations to higher dimensions.

The results obtained so far are somewhat meager compared with the potentialities of the underlying ideas. This is certainly related to the sudden death of J.A. Swieca at the end of last year with whom I collaborated on these problems. Only very recently Eduardo Marino and I were able to collect and extend the results of last year. I am indebted to the organizers of this Summer Institute for giving me the opportunity to present these ideas.

References:

1. L.D. Faddeev and V.E. Korepin: Phys. Rep. 42, 1 (1978) and literature quoted there.

2. R. Jackiw: Rev. of Mod. Phys. 49, 681 (1977) and literature quoted there.
3. L.P. Kadanoff and H. Ceva: Phys. Rev. B 3, 3918 (1971) and previous work by L.P. Kadanoff quoted there.
4. G. 't Hooft: Nucl. Phys. B 138, 1 (1978)
5. E.C. Marino, B. Schroer and J.A. Swieca: "Euclidean Functional Integral Approach For Disorder Variables and Kinks", PUC Rio de Janeiro preprint /81.
6. J. Fröhlich: Commun. Math. Phys. 47, 269 (1976)
7. D. Bohm and Y. Aharonov: Phys. Rev. 115, 485 (1959)
8. S. Mandelstam, Phys. Rev. D 11, 3026 (1975)
9. E.C. Marino and J.A. Swieca: Nucl. Phys. B 170 (FS), 175 (1980)
E.C. Marino: PhD Thesis, PUC Rio de Janeiro (1980)
J.A. Swieca: PUC Rio de Janeiro preprint 10/80, Lectures presented at the XVII Winter School of Theoretical Physics in Karpacz, Poland, February 1980
10. H. Lehmann and J. Stehr: DESY preprint 29/76, unpublished
B. Schroer and T.T. Truong: Phys.Rev. D 15, 1684 (1977)
11. M. Sato, T. Miwa and M. Jimbo: Publications of the Research Institute for Mathematical Sciences, Vol 16,2, 531 (1980) and previous publications in this series.
12. T.T. Wu and C.N. Yang: Nucl. Phys. B 107, 365 (1976) and previous publications quoted there.

CONTINUUM (SCALING) LIMITS OF LATTICE FIELD THEORIES

(TRIVIALITY OF $\lambda\varphi^4$ IN $d \underset{(=)}{>} 4$ DIMENSIONS)

Jürg Fröhlich

Institut des Hautes Etudes Scientifiques
35, Route de Chartres
F-91440 Bures-sur-Yvette

SUMMARY

I describe some recent techniques for constructing the continuum (= scaling) limit of lattice field theories, including the one- and two- component $\lambda|\vec{\varphi}|^4$ theories and the Ising and rotator models in a space (- imaginary time) of dimension $d \underset{(=)}{>} 4$. These techniques should have applications to other related models, like the self-avoiding random walk in five or more dimensions and bond percolation in seven or more dimensions. Some plausible conjectures concerning the Gaussian nature of the scaling limit of the $d \geq 2$ dimensional rotator model and the $d \geq 4$ dimensional U(1) lattice gauge theory in the low temperature (weak coupling) phase are described.

1. INTRODUCTORY REMARKS, RESULTS AND CONJECTURES

An important topic in Euclidean (quantum) field theory and statistical mechanics is the study of the scaling = continuum limit of lattice field theories. That limit corresponds to the large distance limit of rescaled lattice correlation functions at values of the inverse square coupling constant β (= field strength, or inverse temperature) approaching a critical value, as the distance scale tends to infinity. Given a family of lattice field theories, e.g. lattice $\lambda\varphi^4$ theories or Ising models, indexed by the dimension d of the (space - imaginary time) lattice $\mathbb{Z}^d$, we define the upper critical dimension $\overline{d}_c$ by the property that, for $d > \overline{d}_c$, the scaling limit of the corresponding lattice field theory is trivial (Gaussian, in the case of a scalar field theory), and an appropriate version of mean field theory provides an exact description of the approach to the critical point.

The lower critical dimension $\underline{d}_c$ of those families of models is defined by the property that in dimension $d > \underline{d}_c$ there exists a critical point, $\beta_c < \infty$, of β at which some correlation length diverges.

It is often a subtle problem to determine the behaviour of a field theory when $d = \underline{d}_c$ or $d = \overline{d}_c$. For example, in the N-vector models (O(N) non-linear sigma models on the lattice) $\underline{d}_c = 2$ and $\overline{d}_c = 4$. For the $N = 2$ model, i.e. the rotator or classical XY model, it has recently been proven rigorously, by T. Spencer and the author[1], that there exists a Kosterlitz-Thouless transition, in particular that $\beta_c < \infty$ and that the susceptibility diverges as $\beta \nearrow \beta_c$, in dimension $d = \underline{d}_c = 2$. The proof is, however, fairly complicated. For $N \geq 3$ and $d = \underline{d}_c = 2$, it is conjectured that $\beta_c = \infty$ (asymptotic freedom), but no rigorous proof is known. Moreover the nature of the scaling limit, as $\beta \nearrow \beta_c$, is unknown, except in the two-dimensional Ising model (N=1). For all these models, the analysis of the scaling limit in dimension $d = \overline{d}_c = 4$ is incomplete, although for $N = 1,2$ there are promising partial results. (When $N \geq 3$, not even the fact that $\overline{d}_c = 4$ has been proven rigorously, although that result appears to be within reach of present mathematical methods; $\underline{d}_c = 2$ follows from[2]). The analysis of these models in dimension $d = \underline{d}_c$ and $d = \overline{d}_c$ is important for the study of the scaling limit and the approach to the critical point (critical exponents) in dimension d, with $\underline{d}_c < d < \overline{d}_c$ by means of a $2+\varepsilon$ - or $4-\varepsilon$ expansion[3,4]. At present, not much is known about how to study the approach to the critical point directly when $d < \overline{d}_c$.

Fortunately, this is not always necessary for the construction of the continuum limit of lattice field theories in dimension $d < \overline{d}_c$. We may think of the massive, weakly coupled $\lambda\varphi^4$ theories in two or three dimensions the continuum limit of which is under rigorous mathematical and quantitative control (see e.g.[5-10]), although we are not able to calculate e.g. the critical exponents for these models.

In these notes we discuss the following rigorous <u>results</u> :

1) For one - and two - component $\lambda|\vec{\Phi}|^4$ theories, the Ising - and the rotator model

$\overline{d}_c = 4$, [11,12] .

2) For $d > \overline{d}_c$, the continuum limit in the single phase region of these models is <u>Gaussian</u> (i.e. a free or generalized free field), and the critical exponent γ of the susceptibility takes its mean field value, i.e. $\gamma = 1$, [11,12].

3) For $d = \overline{d}_c = 4$, the continuum limit in the single phase region is Gaussian if field strength renormalization is infinite, i.e. if the ultraviolet dimension of the fields is not canonical, [12].

4) If hyperscaling holds for $\underline{d}_c \leq d \leq \overline{d}_c$ then the critical exponent η of the two-point function at $\beta = \beta_c$ satisfies

$0 \leq \eta \leq 2 - \frac{d}{2}$, [12] .

For the one-component models most of these results were first obtained by Aizenman[11] who invented some very beautiful and clever inequalities. He also discovered a very simple proof of hyperscaling in two-dimensional Ising models[11]. Subsequently, the author found new proofs and extensions of some of Aizenman's results, in particular proofs of 1) - 4), [12], by adapting a technique developed with D. Brydges and T. Spencer which was inspired by ideas due to Symanzik[13]. See[14]; and[15,16] for some related results.

There are also partial results towards proving the following <u>conjectures</u> :

5) For the models introduced in 1) - 4)

$\eta = 0$, for $d > \bar{d}_c$.

(see[12] for a discussion).

6) When $d = \bar{d}_c = 4$, the continuum limit is Gaussian,[12,17].

7) For the self-avoiding random walk $\bar{d}_c = 4$; for bond percolation $\bar{d}_c = 6$,[18].

8) The scaling limit of the rotator model in the low temperature (multiple phase) region, i.e. $\beta > \beta_c$, is Gaussian in dimension $d \geq 2$ [19]. An analogous result is expected for the weakly coupled U(1) lattice gauge theory in dimension $d \geq 4$.

An interesting open problem is to analyze the scaling limits of the $d = 3$ Ising - and the $d = 2,3$ rotator models and of the U(1) lattice gauge theory in dimension $d = 4$, as $\beta \nearrow \beta_c$. We conjecture that these scaling limits are non-trivial (hyperscaling).

It might be mentioned that, in addition to 1) - 4) , there are rigorous results on the scaling limits of lattice field theories with long range ferromagnetic two-body interactions[12,18]. As an example we mention that the scaling limit of the one-dimensional Ising model with ferromagnetic $1/r^2$ interaction energy is Gaussian for $\beta < \beta_c$. (The existence of a phase transition, i.e. $\beta_c < \infty$, and spontaneous magentization at large values of β has recently been proven in[20]).

ACKNOWLEDGEMENTS

I am very much indebted to T. Spencer for having provided most of the insight underlieing the results and techniques discussed here. I thank him, D. Brydges and A. Sokal for stimulating and enjoyable collaborations.

2. GENERAL REMARKS ON SCALING LIMITS

To be specific, we consider a real, scalar field φ ,

$$\varphi : j \in \mathbb{Z}^d \longrightarrow \varphi(j) \in I , \tag{1}$$

with $I \subseteq \mathbb{R}$. The Ising model corresponds to $I = \{-1,1\}$, a lattice $\lambda\varphi^4$ theory to $I = \mathbb{R}$. We imagine the field $\varphi = \{\varphi(j)\}_{j \in \mathbb{Z}^d}$ is distributed according to some probability measure $d\mu_\beta(\varphi)$ (typically a Gibbs measure) depending on a real parameter β interpreted as the inverse temperature or field strength. We define $\varphi_x(j) = \varphi(j+x)$, $x \in \mathbb{Z}^d$, and assume that $d\mu_\beta(\varphi)$ is translation - invariant, i.e.

$$d\mu_\beta(\varphi_x) = d\mu_\beta(\varphi) \text{ , for all } x . \tag{2}$$

The correlation functions of this lattice theory are defined as the moments of $d\mu_\beta$, i.e.

$$< \varphi(x_1) \dots \varphi(x_n) >_\beta = \int \prod_{k=1}^{n} \varphi(x_k) \, d\mu_\beta(\varphi) . \tag{3}$$

We may assume that $< \varphi(x) >_\beta = 0$. Of particular importance are the two-point function

$$< \varphi(x) \varphi(y) >_\beta = \int \varphi(x) \varphi(y) \, d\mu_\beta(\varphi) \tag{4}$$

and the susceptibility

$$\chi(\beta) = \sum_{x \in \mathbb{Z}^d} < \varphi(0) \varphi(x) >_\beta . \tag{5}$$

In the following we are interested in analyzing the long distance limit of the correlations defined in (3). We suppose there exists some value β_c of β such that for $\beta < \beta_c$ there exists a constant $m(\beta) > 0$, the inverse correlation length or mass, with the property that

$$< \varphi(x) \varphi(y) >_\beta \leq \text{const.} \, e^{-m(\beta)|x-y|} , \tag{6}$$

as $|x-y| \to \infty$, and

$$m(\beta) \searrow 0 \text{ , as } \beta \nearrow \beta_c . \tag{7}$$

These assumptions are known to hold for Ising models and lattice $\lambda\varphi^4$ fields[21]. We now define the scaled correlations

$$G_\theta(x_1,\dots,x_n) = \alpha(\theta)^n < \varphi(\theta x_1) \dots \varphi(\theta x_n) >_{\beta(\theta)} , \tag{8}$$

where $1 < \theta < \infty$, $x_j \in \mathbb{Z}^d_{\theta^{-1}} \equiv \{y : \theta y \in \mathbb{Z}^d\}$,

$j = 1,\dots,n$, and $\beta(\theta) < \beta_c$, $\alpha(\theta)$ are functions of θ determined by the requirements that for $0 < |x-y| < \infty$

$$0 < \lim_{\theta\to\infty} G_\theta(x,y) \equiv G^*(x-y) < \infty . \tag{9}$$

It follows from (6), (7) and (9) that

$$\beta(\theta) \nearrow \beta_c , \quad \text{as} \quad \theta \to \infty \tag{10}$$

If we want to construct a massive continuum $(\theta\to\infty)$ limit we choose $\beta(\theta)$ such that

$$\theta m(\beta(\theta)) \longrightarrow m^* > 0 , \text{ as } \theta \longrightarrow \infty ; \tag{11}$$

if we try to construct a scale-invariant continuum limit we set $\beta(\theta) = \beta_c$ for all θ. (There are other possible conditions fixing the choice of $\beta(\theta)$. See e.g.[22]). Once $\beta(\theta)$ is chosen, $\alpha(\theta)$ is essentially determined by (9). The correlation functions in the continuum limit are then given by

$$G^*(x_1,\ldots,x_n) = \lim_{\theta\to\infty} G_\theta(x_1,\ldots,x_n) , \; n = 2,3,4,\ldots . \tag{12}$$

Thus, in order to construct the continuum limit, it is crucial to know the behaviour of the two-point function. In many models (e.g. the self-avoiding random walk or bond percolation) this turns out to be very hard. We now elaborate on this point. Suppose one can prove a power law a priori bound on $< \varphi(x)\varphi(y) >_\beta$, for $\beta < \beta_c$, e.g.

$$< \varphi(x)\varphi(y) >_\beta \leq c(\beta) \; |x-y|^{-(d-2+\eta)} , \tag{13}$$

with $\sup_{\beta<\beta_c} c(\beta) < \infty$. Then condition (9) imposes the following lower bound on $\alpha(\theta)$:

$$\alpha(\theta) \geq \text{const. } \theta^{(d-2+\eta)/2} . \tag{13'}$$

One may attempt to analyze the two-point function by studiing its operator inverse, the two-point vertex function $\Gamma_\beta(x-y)$ which, in perturbation theory, is expressed as a sum of one-particle irreducible diagrams. For this reason one may hope to estimate it by means of a convergent, infrared-finite expansion if the dimension d is large enough. This is a difficult analytical problem. Fortunately, for the models studied in the following and in[11,12,14], an a priori bound of the form (12), with $\eta \geq 0$, has been established in[2,22] (the infrared - or spin - wave bound), and this turns out to be sufficient to show that $\bar{d}_c = 4$ and that the continuum limit is Gaussian when $d > 4$, thanks to new correlation inequalities[11,12] discussed in the next section.

A more systematic procedure to determine $\beta(\theta)$ and $\alpha(\theta)$ relies on the <u>renormalization group</u> e.g. in the form of <u>Kadanoff Block spin transformations</u>, (calculation of large scale effective Hamiltonian or action) :

Let κ be a function on $\mathbb{R}^d$ defined by

$$\kappa(x) = \begin{cases} \varepsilon^{-d}, & -\frac{\varepsilon}{2} \leq x^\alpha \leq \frac{\varepsilon}{2}, \ \alpha = 1,\ldots,d \\ 0, & \text{otherwise,} \end{cases}$$

where x^α is the αth component of $x \in \mathbb{R}^d$. Let $\kappa_x(y) = \kappa(y-x)$, $x \in \mathbb{Z}_\varepsilon^d$. We consider

$$G_\theta(\kappa_{x_1},\ldots,\kappa_{x_n}) = \sum_{y_1,\ldots,y_n \in \mathbb{Z}^d_{\theta^{-1}}} G_\theta(y_1,\ldots,y_n) \prod_{k=1}^{n} \theta^{-d} \kappa_{x_k}(y_k), \tag{14}$$

$x_j \in \mathbb{Z}_\varepsilon^d$, $j = 1,\ldots,n$. Now, note that $G_\theta(\kappa_{x_1},\ldots,\kappa_{x_n})$ depends only on the variables

$$\{\varphi_\theta(x_j) : x_j \in \mathbb{Z}_\varepsilon^d\}$$

where

$$\varphi_\theta(x) = \frac{\alpha(\theta)}{\theta^d \varepsilon^d} \sum_{\substack{y \in \mathbb{Z}^d \\ -\frac{\varepsilon}{2} \leq \theta^{-1} y^\alpha - x^\alpha \leq \frac{\varepsilon}{2}}} \varphi(y), \quad x \in \mathbb{Z}_\varepsilon^d, \tag{15}$$

and $\theta = \varepsilon^{-1} L^m$, L is some positive integer and $m = 1,2,3,\ldots$. Given $d\mu_{\beta(\theta)}(\varphi)$, let $d\mu_\theta$ be the unique measure on the configurations of the "<u>block field</u>" $\varphi_\theta = \{\varphi_\theta(x)\}_{x \in \mathbb{Z}_\varepsilon}$ with the property that

$$\int \prod_{k=1}^{n} \varphi_\theta(x_k) d\mu_\theta(\varphi_\theta) = \int \prod_{k=1}^{n} \varphi_\theta(x_k) d\mu_{\beta(\theta)}(\varphi), \tag{16}$$

for all $x_1,\ldots,x_n$ in $\mathbb{Z}_\varepsilon^d$ and all n . If $d\mu_\beta(\varphi)$ is a Gibbs measure for all β one expects that $d\mu_\theta(\varphi_\theta)$ is again a Gibbs measure, i.e. $d\mu_\theta$ is given in terms of an effective Hamilton function, or action, on a scale of $\varepsilon\theta = L^m$. The calculation of the effective Hamilton function proceeds by a <u>succession of Block spin</u> (or - <u>field</u>) <u>transformations</u>, and each such transformation increases the scale by a factor of L. A mathematical description of the general features of that technique may be found in[23] , explicit examples have been studied in[24].

In this approach the functions $\beta(\theta)$ and $\alpha(\theta)$ are determined by the requirements that

$$\theta m(\beta(\theta)) \equiv m^* = \text{const.} \geq 0 \text{ , and that} \qquad (17)$$

$$\lim_{\theta\to\infty} d\mu_\theta(\varphi_\theta) \equiv d\mu^*(\varphi^*)$$

is a well-defined probability measure with moments $\neq 0,\infty$. It is hoped that $d\mu^*$ is again a Gibbs measure, but this does not always seem to be the case. Condition (17) and the functions $\beta(\theta)$ and $\alpha(\theta)$ determine the exponents ν and γ of the mass and the susceptibility, respectively. If $m^* = 0$ and $d\mu^*$ is scale-invariant then $d\mu^*$ is a fixed point of the Block spin transformations, and the fall-off of the two-point function is determined by $\alpha(\theta)$.

The smeared continuum correlation functions are given by

$$G^*(\kappa_{x_1},\ldots,\kappa_{x_n}) = \int \prod_{k=1}^n \varphi^*(x_k)d\mu^*(\varphi^*) \ .$$

In practice, it turns out that it is usually very hard to construct the limiting measures $d\mu^*$ and to show that they are Gibbs measures, but there are now some examples where the Block spin transformations can be made to work in a rigorous fashion[24]. However, in these examples $d\mu^*$ is a massless Gaussian.

In the following, we investigate the scaling limits of the Ising model and lattice $\lambda\varphi^4$ models in $d \geq 4$ dimensions using merely an a priori bound of the form (13) with $\eta \geq 0$ and new inequalities on the four-point Ursell function which permit us to avoid applying Block spin transformations.

3. THE CONTINUUM (= SCALING) LIMIT OF THE $\lambda\varphi_d^4$ LATTICE FIELD THEORY AND THE ISING MODEL IN $d \underset{(=)}{>} 4$ DIMENSIONS

We now explain some basic ideas in the proofs of results 1) - 4) described in Sect. 1. Let φ be a real scalar lattice field (or a classical spin) with action (Hamilton function)

$$H(\varphi) = - \sum_{(jj')} \varphi(j)\, \varphi(j') \,, \tag{18}$$

where (jj') are nearest neighbor pairs in $\mathbb{Z}^d$. The measure $d\mu_\beta$, i.e. the Euclidean vacuum functional (or Gibbs state) is given by

$$d\mu_\beta(\varphi) = Z_\beta^{-1}\, e^{-\beta H(\varphi)} \prod_j d\lambda(\varphi(j)) \,, \tag{19}$$

where

$$d\lambda(\varphi) = \exp[-\tfrac{\lambda}{4}\, \varphi^4 + \tfrac{\mu}{2}\, \varphi^2 - \varepsilon]d\varphi \,, \tag{20}$$

$0 < \beta < \infty$, $\lambda > 0$, μ and ε real, and Z_β is the partition function. Equation (19) is to be understood as the thermodynamic limit of measures associated with finite sublattices. The limit exists for a large class of boundary conditions by correlation inequalities,[6,7]. If $\mu = \lambda$, $\varepsilon = \frac{\lambda}{4}$ and $\lambda \to \infty$ we obtain the Ising model. For all such models it is shown in[22], by using the infrared or spin wave bound of[2], that

$$< \varphi(x)\varphi(y) >_\beta \le \frac{c_d}{\beta}\, |x-y|^{2-d} \,, \; d \ge 3 \,, \tag{21}$$

for some geometric constant c_d independent of β, λ and μ, as long as $\beta < \beta_c$. (It is shown e.g. in[2] that the critical inverse temperature, β_c, is finite, and properties (6) and (7) are proven in[21,25]). As shown in Sect. 2, $\alpha(\theta)$ must therefore satisfy the lower bound

$$\alpha(\theta) \ge \text{const.}\ (\beta(\theta)\theta^{d-2})^{1/2} \,, \tag{22}$$

or if $\beta_c < \infty$

$$\alpha(\theta) \geq (\beta_c \theta^{d-2})^{1/2} \quad . \tag{23}$$

The four-point Ursell function, u_4, is defined by

$$u_{4,\beta}(x_1 \dots, x_4) = < \varphi(x_1) \dots \varphi(x_4) >_\beta - \sum_p < \varphi(x_{p(1)}) \varphi(x_{p(2)}) >_\beta \cdot \\ \cdot < \varphi(x_{p(3)}) \varphi(x_{p(4)}) >_\beta \tag{24}$$

where $\sum_p$ ranges over all three pairings of $\{1,2,3,4\}$. It satisfies the following inequalities

$$0 \geq u_{4,\beta}(x_1, \dots, x_4) \geq -3\beta^2 \sum_{z_1 \dots z_4}' \prod_{k=1}^{4} < \varphi(x_k)\varphi(z) >_\beta , \tag{25}$$

where z_1 ranges over $\mathbb{Z}^d$, and $|z_\ell - z_1| \leq 1$, $\ell = 2,3,4$. The upper bound is the well-known Lebowitz inequality[26], the lower bound is the new inequality proven in[12] which is closely related to Aizenman's inequality[11]. We define

$$u_{4,\theta}(x_1, \dots, x_4) = \alpha(\theta)^4 \, u_{4,\beta(\theta)}(\theta x_1, \dots, \theta x_4) \; . \tag{26}$$

In order to satisfy as general a class of renormalization conditions as possible we permit λ and α to depend on θ, as well : A minimal condition on $\lambda = \lambda(\theta)$, $\mu = \mu(\theta)$ and on $\beta(\theta)$ and $\alpha(\theta)$ is that inequalities (9), Sect. 2, be satisfied. We now obtain from (25), (26)

$$0 \geq u_{4,\theta}(x_1, \dots, x_4) \geq \alpha(\theta)^{-4} \theta^d 3\beta(\theta)^2 \, \cdot \\ \cdot \sum_{z_1, \dots, z_4}' \theta^{-d} \prod_{k=1}^{4} G_\theta(x_k, \theta^{-1} z_k) \; . \tag{27}$$

The nice feature of (27) is that the upper and lower bounds are <u>independent</u> of $\lambda(\theta)$, $\mu(\theta)$. (In a sense, (27) says that the prediction of the linearized renormalization group provides a rigorous bound). Now, by (21) and (22)

$$\alpha(\theta)^{-4} \theta^d \beta(\theta)^2 \leq \text{const.} \; \theta^{4-d} \tag{28}$$

which tends to 0, as $\theta \to \infty$, in dimension $d > 4$. One can use the infrared bound (21) to show that

$$\sum_{z_1 \ldots z_4}{}' \quad \theta^{-d} \prod_{k=1}^{4} G_\theta(x_k, \theta^{-1} z_k) \leq \text{const.} , \tag{29}$$

uniformly in θ , provided $x_i \neq x_j$, for $i \neq j$, and $d > 4$. See[12,17]. By (27) - (29)

$$u_4^*(x_1, \ldots, x_4) = \lim_{\theta \to \infty} u_{4,\theta}(x_1, \ldots, x_4) = 0 , \tag{30}$$

at non-coinciding arguments, in $d > 4$ dimensions. Thus $\bar{d}_c = 4$, and the continuum limit is a free or generalized free field, provided $d > \bar{d}_c = 4$. For $d = 4$, the same result follows if either

$$\text{(i)} \quad \alpha(\theta)^2 / \beta(\theta)\theta^{d-2} \longrightarrow \infty \quad , \text{ or} \tag{31}$$

(ii) the limiting theory is scale-invariant. (Some uniformity in the limit of the two-point function is assumed; see[12]). In case (i), (27) and (31) yield (30), and the limiting theory has non-canonical short distance behaviour. In case (ii), triviality follows from a theorem of Pohlmeyer[27]. (Thus, in the language of the Callan-Symanzik equation, $\lambda\varphi_4^4$ is trivial, unless the β- and γ functions have a non-trivial common zero, and the corresponding theory is not scale-invariant). In[12] we have also proven a sharper form of (25) which suggests that the continuum limit in $d = 4$ is trivial, without any additional hypotheses, but we do not have a complete argument. Inequality (25) can also be used to calculate critical exponents. Combining it with an argument due to Glimm and Jaffe[21] one shows that the critical exponent γ of the susceptibility χ has its mean field value, $\gamma = 1$, in $d \geq 5$ dimensions[17,12]. It follows directly from (25) that if hyper scaling holds (i.e. the critical theory is non-trivial) then $\eta \leq 2 - \frac{d}{2}$, $d = 2,3,4$. See[12].

We conclude with some brief comments on the proof of the basic inequality (25). The proof in[12] relies on Symanzik's random walk -, or polymer representation of scalar Euclidean field theories[13,14] : One reexpresses a lattice field theory as a gas of random walks interacting through some soft core repulsion determined by $d\lambda(\varphi)$. Let $\omega_1, \ldots, \omega_n$ be n arbitrary random walks immersed in that gas and interacting with each other and with the (closed) random walks in the gas through the same soft core repulsion. Let $z(\omega_1,\dots,\omega_n)$ be their joint correlation. Then

$$< \varphi(x)\varphi(y) >_\beta = \sum_{\omega:x\to y} z(\omega) \ , \text{ and}$$

$$u_{4,\beta}(x_1,\dots,x_4) = \sum_p \sum_{\substack{\omega_1:x_{p(1)}\to x_{p(2)} \\ \omega_2:x_{p(3)}\to x_{p(4)}}} [z(\omega_1,\omega_2) - z(\omega_1)z(\omega_2)] \ . \tag{32}$$

By a correlation inequality[12,14]

$$z(\omega_1,\omega_2) \geq z(\omega_1)z(\omega_2) \ , \tag{33}$$

<u>unless</u> ω_1 and ω_2 intersect. Thus

$$0 \geq u_{4,\beta}(x_1,\dots,x_4) \geq - \sum_{p;\omega_1,\omega_2} z(\omega_1)z(\omega_2)\chi(\{\omega_1,\omega_2:\omega_1 \cap \omega_2 \neq \emptyset\}), \tag{34}$$

where $\sum_{p;\omega_1,\omega_2}$ is a short hand for the sum on the r.s. of (32). By requiring that $\omega_1 \cap \omega_2$ contains some lattice point z and then summing over all possible points z one can, after splitting ω_i into two walks, $i = 1,2$, and applying a Simon-type inequality[25,14], resum the r.s. of (34) to obtain (25). See[12], and[17] for an alternate, prior proof of a related inequality. Rather than discussing these technical aspects we emphasize that the r.s. of (34) should really vanish in the continuum limit, in dimension $d \geq 4$, because the random paths in the continuum are expected to have a Hausdorff dimension $D_H \leq 2$, so that, for $d \underset{(=)}{>} 4 \geq 2D_H$, two random paths do not intersect with probability 1. This is, in fact, a known theorem for Brownian paths in four or more dimensions[28]. Because of the repulsive character of the self-interaction, the field theoretic paths appear to have rather less tendency to intersect each other than Brownian paths and thus the r.s. of (34) is expected to vanish in the continuum limit in four dimensions. (I am indebted to T. Spencer for explaining such arguments to me).

REFERENCES

1. J. Fröhlich and T. Spencer, Commun. Math. Phys. 81, 527 (1981).

2. J. Fröhlich, B. Simon and T. Spencer, Commun. Math. Phys. 50, 79 (1976).

3. K. Wilson and J. Kogut, Physics Reports 12C, No. 2, 76 (1974).

4. E. Brézin, J.C. Le Guillou and J. Zinn-Justin, Phys. Rev. D14, 2615 (1976).

5. J. Glimm and A. Jaffe, Quantum Physics, Berlin-Heidelberg-New York : Springer-Verlag 1981.

6. B. Simon, The $P(\phi)_2$ Euclidean (Quantum) Field Theory, Princeton : Princeton University Press 1974.

7. Constructive Quantum Field Theory, G. Velo and A.S. Wightman, eds., Berlin-Heidelberg-New York : Springer Lecture Notes in Physics 25, 1973.

8. J. Glimm and A. Jaffe, Fortschr. Phys. 21, 327 (1973).

9. G. Benfatto, M. Cassandro, G. Gallavotti,..., Commun. math. Phys. 71, 95 (1980).
G. Benfatto, G. Gallavotti and F. Nicoló, J. Funct. Anal. 36, 343 (1980).
J. Feldman and K. Osterwalder, Ann. Phys. (NY) 97, 80 (1976).
J. Magnen and R. Sénéor, Ann. Inst. H. Poincaré 24, 95 (1976), Commun. Math. Phys. 56, 237 (1977).

10. T. Spencer, Commun. Math. Phys. 39, 63 (1974), 44, 143 (1975); T. Spencer and F. Zirilli, Commun. Math. Phys. 49, 1 (1976).

11. M. Aizenman, Phys. Rev. Lett. 47, 1 (1981).

12. J. Fröhlich, Nucl. Phys. B, in press; and in preparation.

13. K. Symanzik, in : Local Quantum Theory, R. Jost, ed., New York : Academic Press, 1969.

14. D. Brydges, J. Fröhlich and T. Spencer, Commun. Math. Phys., in press.

15. D. Brydges and P. Federbush, Commun. Math. Phys. 62, 79 (1978).

16. D. Brydges, J. Fröhlich and A. Sokal, in preparation.

17. M. Aizenman, in preparation, and private communication.

18. Some preliminary results have been obtained by A. Sokal, T. Spencer and the author.

19. The conjectures are based on results contained in refs. 1,24, 27 and in J. Fröhlich and T. Spencer, J. Stat. Phys. 24, 617 (1981), Commun. Math. Phys., in press.

20. J. Fröhlich and T. Spencer, to appear in Commun. Math. Phys..

21. O. McBryan and J. Rosen, Commun. Math. Phys. 51, 97 (1976).
J. Glimm and A. Jaffe, Commun. Math. Phys. 52,203 (1977).

22. A. Sokal, Ph.D. thesis, Princeton University, Jan. 1981.

23. Ya. G. Sinai, in : Mathematical Problems in Theoretical Physics, G. Dell'Antonio, S. Doplicher and G. Jona-Lasinio, eds. Berlin-Heidelberg-New York: Springer Lecture Notes in Physics 80, 1978; and references therein.
G. Jona-Lasinio, Nuovo Cimento 26B, 99 (1975).

24. K. Gawedzki and A. Kupiainen, Commun. Math. Phys. 77, 31 (1980), Renormalization Group Study of a Critical Lattice Model I, II, Commun. Math. Phys. to appear.

25. B. Simon, Commun. Math. Phys. 77, 111 (1980).

26. J. Lebowitz, Commun. Math. Phys. 35, 87 (1974).

27. K. Pohlmeyer, Commun. Math. Phys. 12, 204 (1969).

28. A Dvoretzky, P. Erdös and S. Kakutani, Acta Sci. Math. (Szeged) 12B, 75 (1950).

SCHRÖDINGER REPRESENTATION IN RENORMALIZABLE QUANTUM FIELD THEORY

K. Symanzik

Deutsches Elektronen-Synchrotron DESY
Hamburg, Germany

This seminar is based on ref. [1] and discusses some of its points more broadly. That reference should be consulted for all details.

1. Introduction

L. Fadde'ev reported on A.M. Polyakov's recent work [2] on integration over surfaces. The problem of the Schrödinger representation arose from work [3] on the older Nambu-Gōto Ansatz for such integration: Let the (Euclidean) Wilson loop in D dimensions be

$$C: s \to x_\mu(s), \quad 0 \leq s \leq 2\pi, \quad \mu = 1,\ldots,D, \quad x_\mu(0) = x_\mu(2\pi).$$

With a D-component function $\phi_\mu(z)$ defined on a parametrization region $\Gamma \in R^2$ with boundary $\partial\Gamma: s \to z(s)$, set*

(1.1) $$\Psi(C) := \left\langle Tr\left(\exp\left[\int_0^{2\pi} ds\, \dot{x}_\mu(s) A_\mu(x(s))\right]\right)_+ \right\rangle =$$

$$= \int_{\phi_\mu(z(s)) = x_\mu(s)} \mathcal{D}\phi \exp\left[-M^2 \int_\Gamma d^2z \sqrt{g_\phi(z)}\right].$$

Here the + subscript means s-ordering w.r.t. matrix indices,

$$g_\phi(z) := \det g_{..}(z), \quad g_{ij}(z) := \sum_\mu \frac{\partial \phi_\mu}{\partial z_i} \frac{\partial \phi_\mu}{\partial z_j},$$

*We disregard here a subtlety [3] concerning the boundary condition.

and apart from the boundary condition, the measure $\mathcal{D}\phi$ should be translation invariant in surface space.

The exponent in the ϕ-integral in (1.1) is largest for $\phi_\mu(z) = \varphi_\mu(z)$, the minimal surface spanning C in some parametrization with area A(C). One sets $\phi_\mu(z) = \varphi_\mu(z) + \eta_\mu(z)$. Then

$$(1.2) \quad \Psi(C) = \exp[-M^2 A(C)] \cdot$$

$$\cdot \int\limits_{\eta_\mu(z(s))=0} \mathcal{D}\eta \exp\left[-M^2 \int_\Gamma d^2z \left\{\sqrt{g_{\varphi+\eta}(z)} - \sqrt{g_\varphi(z)}\right\}\right].$$

Expanding the curly bracket in powers of η yields the φ-dependent action density of a two-dimensional nonrenormalizable local theory of a D-component field (of which, due to reparametrization invariance, two components carry only "gauge" degrees of freedom), with homogeneous Dirichlet conditions on $\partial\Gamma$. The "string potential"

$$(1.3) \quad V(R) := -\lim_{T\to\infty} [T^{-1} \ln \Psi(C = \partial(T \otimes R))],$$

where C is the boundary of a rectangle with sides T and R, is then the Casimir potential of a pair of points in the one-(space-)dimensional world.

In ref. [3], the semiclassical approximation (SA) to Ψ(C) from (1.2) is constructed and the counter terms needed to render it finite are determined. (Actually, in [3] instead of (1.2) the alternative Ansatz for Ψ(C) due to T. Eguchi [4] was used; in the SA the result is the same.) Trying to go beyond the SA leads to the problems a) of nonrenormalizability, b) of whether Dirichlet boundary conditions can be imposed in a (Euclidean) quantum field theory.

Concerning a): If a nonrenormalizable theory exists at all, its true small-coupling expansion (here, the coupling is prop. M^{-2}) will involve, besides powers of the coupling constant, also its logarithm [5]. Such a double series expansion can so far only be constructed if there is available an alternative renormalizable expansion in a dimensionless parameter. (An important example hereof is the 1/N expansion for the O(N) vector model in three dimensions.) The candidate for such parameter in the present case is $(D-2)^{-1}$, renormalizability has, however, not been shown so far (cp. ref. [6]). - Concerning b): If problem a) is solved in terms of a renormalizable expansion (the only known way to solve it), the problem of imposing Dirichlet conditions at a boundary is not expected to differ significantly from the same problem for any renormalizable theory, in perturbation theory at least. This led the author to consider the simplest renormalizable model, ϕ_4^4. This theory is conjectured [7] not to exist with $g_{ren} \neq 0$, however, since the analysis is done in perturbation theory only, this plays no role. The principles abstracted apply to any renormalizable theory in its perturbation expansion, and should apply also outside perturbation theory if the unlimited-space theory itself does exist outside the perturbation expansion.

2. Schrödinger representation in QFT

The Schrödinger wave function $\psi(\underline{x},t)$ of nonrelativistic QM (of, e.g., a particle of mass m in a time-independent potential $V(\underline{x})$) is the spectral density of the decomposition of the (Heisenberg) state in eigenstates of the position operator $\underline{q}(t)$ at time t. (One says that the position operator at time t is "diagonalized".) The canonically conjugate operator $\underline{p}(t)$ (here $= m\dot{\underline{q}}(t)$) is then represented (setting $\hbar = 1$) by $-i\partial/\partial\underline{x}$. $i\dot{\psi}(\underline{x},t)$ is obtained by acting with the Schrödinger operator $H(\underline{x},-i\partial/\partial\underline{x})$ (here, $= -(2m)^{-1}\Delta + V(\underline{x})$) on $\psi(\underline{x},t)$.

In QFT, instead of the position operator, the itself $\underline{x}$-dependent field operator at time t (e.g., in QED the electromagnetic field $\underline{A}(\underline{x},t)$ and four components of the eight-component hermitean Dirac field) is diagonalized. The spectral density is now a functional of the eigenvalue-function of the field operator at time t. The canonically conjugate field (in QED, the time derivatives $\dot{\underline{A}}(\underline{x},t)$ of the vector potential and the remaining four components of the Dirac field) is represented by functional differentiation. The time derivative (times i) of the Schrödinger functional is obtained by acting on it with the Schrödinger (functional differential) operator.

The attempt to carry out this construction in renormalizable QFT (e.g. QED, QCD, ϕ_4^4) leads to UV divergences that are not removed by the Poincaré invariant counter terms that effect renormalization in unlimited space-time (e.g., ref. [8]). The correct additional renormalization is found by constructing the Schrödinger representation in a way where the well-known principles of general renormalization theory (e.g., refs. [9], [10]) can be referred to. As indicated before, we shall discuss mainly ϕ_4^4 theory but will also give results for theories involving a Majorana field, from which the ones for a Dirac field follow simply e.g. by doubling the number of components.

3. Renormalization

The original Lagrangian density for (Euclidean) ϕ^4_4 theory with external source is

$$(3.1)\quad L = -\tfrac{1}{2}\partial_\mu\phi\partial_\mu\phi - \tfrac{1}{2}m^2\phi^2 - \tfrac{1}{4!}g\phi^4 + J\phi$$

The quadratic terms lead to the (Euclidean) Feynman propagator $G(x_1x_2) = (4\pi^2)^{-1}|x_1-x_2|^{-2}$ + less singular m-dependent terms, and the quartic term yields the bare vertex $-g\delta(x_1-x_2)\delta(x_1-x_3)\delta(x_1-x_4)$. Due to the singularity of the propagator at coinciding arguments, divergences arise, which are cancelled by adding to L the counter-terms

$$(3.2)\quad \Delta L = -\tfrac{1}{2}(Z_3-1)\partial_\mu\phi\partial_\mu\phi - \tfrac{1}{2}(Z_2-1)m^2\phi^2 - \tfrac{1}{2}m^2_{Bo}Z_3\phi^2 - \tfrac{1}{4!}(Z_1-1)g\phi^4.$$

To render these terms meaningful, one must regularize. The most convenient regularization is the dimensional one [11], which here, for the purpose of the Euclidean Schrödinger representation, we understand as computing in $3-\varepsilon$ space- and 1 "time" dimension. (To keep g dimensionless, one must in (3.1) and (3.2) write $g\mu^\varepsilon$ instead, with μ the normalization mass.) Then m^2_{Bo}, the generally quadratically divergent bare mass of the massless theory, is zero, and for the "logarithmically divergent" constants $Z_{1,3,2}$ the "minimal" form

$$(3.3)\quad Z_i(g,\varepsilon) = 1 + \varepsilon^{-1}f_{i1}(g) + \varepsilon^{-2}f_{i2}(g) + \ldots$$

can be chosen. The limit $\varepsilon \searrow 0$ then exists for the Green's functions.

The insertion into all Green's functions of any local monomial in the field and its (possibly, higher) derivatives can be made $\varepsilon \searrow 0$ -finite by adding to that monomial local counter terms of at most the same operator degree. These terms effect "final subtraction" of the divergences of all graphs that involve the original monomial as vertex and have all other vertices (from ϕ^4 in (3.1) and (3.2)) coalesce at the location of the original monomial. Divergences from coalescence of a subset of the vertices of the graph are removed already by counter terms of lower order.

The insertion of a monomial into the Lagrange function lets that monomial act as insertion an arbitrary number of times. If it has a dimensionless coefficient, counter terms must involve that coefficient to arbitrarily high order, while only polynomially if the coefficient has positive mass dimension [12]. (Thus, in (3.2)

m^2 appears only to first order since the c-number term prop. m^4 is omitted.) A monomial with coefficient of inverse mass dimension is excluded as insertion into the Lagrange function since it would lead to nonrenormalizability.

4. Schrödinger functional from surface terms

The scheme of sect. 3 applies to the renormalization of the Schrödinger representation, since the Schrödinger wave functional can be obtained from insertion into the Lagrange function. We label the space coordinates $\underline{x} = (x_1, .. x_{3-\varepsilon})$ and the "time" coordinate y. For the free theory, (3.1) with g = 0, the appropriate addition to the Lagrange density is

$$(4.1) \quad L_{\partial\Gamma} - L = - \lim_{\eta \downarrow 0} \delta(y)\, \phi(\underline{x}\,(-\eta))\, \partial_y \phi(\underline{x}\, y).$$

Gaussian integration shows that the propagator is now for positive y,y' the Dirichlet one and for negative y,y' the Neumann one:

$$(4.2) \quad G_{\substack{Dir \\ Neu}}(x\, x') = \frac{1}{4}\pi^{-2+\varepsilon/2} \cdot$$

$$\cdot \left\{ \left[|\underline{x}-\underline{x}'|^2 + (y-y')^2 \right]^{-1+\frac{\varepsilon}{2}} \mp \left[|\underline{x}-\underline{x}'|^2 + (y+y')^2 \right]^{-1+\frac{\varepsilon}{2}} \right\}$$

plus less singular m^2-dependent terms irrelevant for the following. The propagator between points with $yy' < 0$ is zero, i.e. the two half spaces are decoupled by the surface interaction (4.1). From the propagators follows

$$(4.3a) \quad \lim_{y \downarrow 0} \phi(\underline{x}\, y) = 0, \qquad (4.3b) \quad \lim_{y \uparrow 0} \partial_y \phi(\underline{x}\, y) = 0$$

in the sense of functional averages, i.e. of arguments in Green's functions (if at $(\underline{x}\ 0)$ there is not already another argument of the function). Inhomogeneous Dirichlet conditions at $y = +0$ are implemented by the additional surface term

$$(4.4) \quad L_{\partial\Gamma A} - L_{\partial\Gamma} = \lim_{y \downarrow 0} \delta(y)\, A(\underline{x})\, \partial_y \phi(\underline{x}\, y),$$

yielding instead of (4.3a)

$$(4.5) \quad \lim_{y \downarrow 0} \phi(\underline{x}\, y) = A(\underline{x})$$

in the sense explained there. The decoupling of the two half spaces persists if also the interaction term of (3.1) is present: unrenormalized Feynman graphs are formed, in the upper ($y > 0$) and lower ($y < 0$) half spaces, from Dirichlet and Neumann propagators, respectively, with no propagator crossing the $y = 0$ plane. We need, however, determine divergences and counter terms.

To this end, we observe that (4.1) and (4.4) are <u>insertions restricted to $y = 0$,</u> with dimensionless resp. mass-dimension-one coefficients. Thus, away from the $y = 0$ plane, the counter terms (3.2) suffice. On that plane, in view of the symmetry under $\phi \to -\phi$, $J \to -J$, $A \to -A$, only the new counter terms <u>local also in the space argument</u>

(4.6) $$\Delta L_{\partial TA} = -(Z_4 - 1)\delta(y)\,\phi\partial_y\phi + (Z_5 - 1)\delta(y)\,A\partial_y\phi + c_1\Lambda\delta(y)\phi^2 + c_3\Lambda\delta(y)A\phi + c_5\Lambda\delta(y)A^2$$

can be needed. Here, Λ is the cutoff such that the c-terms cannot be written in dimensional regularization. The "logarithmically divergent" constants $Z_{4,5}$, however, can again be chosen in the form (3.3), and one finds

(4.7a) $$Z_4 - 1 = (16\pi^2\varepsilon)^{-1}g + O(g^2),$$

(4.7b) $$Z_5 - 1 = -(32\pi^2\varepsilon)^{-1}g + O(g^2).$$

Observe that the original insertion (4.1) has the special coefficient unity, such that in Z_4, Z_5 that value, which otherwise would have entered there as "(surface) coupling constant", can be suppressed.

The functional integral with $L_{\partial TA} + \Delta L_{\partial TA}$, divided by the same integral with $A = 0$, $J = 0$, factorizes

(4.8) $$\Psi(A|J) = \Psi_{Dir}(A|J)\cdot\Psi_{Neu}(|J)$$

whereby the first factor depends only on J restricted to $y > 0$ and the second only on J restricted to $y < 0$, assuming J to be smooth across $y = 0$. $\Psi_{Dir}(A|J)$ is the <u>Schrödinger wave functional</u> we are interested in, and can be understood as the scalar product of the state $\langle A|$, eigenstate of $Z_5^{-1}Z_3\phi(\underline{x}y)|_{y=+0}$ to eigenvalue $A(\underline{x})$ (see (5.1) below), with the state $\left(\exp\int_0^\infty dy\int d\underline{x}\,J(\underline{x}y)\,\phi(\underline{x}y)\right)_+\rangle$

where $\phi(\underline{x}y) = e^{-Hy}\phi(\underline{x}0)e^{+Hy}$, with $H: = P_o$ the Hamilton operator and $\rangle$ the vacuum state. The subscript + signifies ordering with larger y to the right.

A more complete discussion [1] reveals that the c-terms in (4.6) cannot be set zero since dimensional regularization does not regularize entirely "in y-direction" since this direction remains one-dimensional. A consequence is that the c_1-term in (4.6), which is incompatible with (4.3b), destroys the Neumann property of the functional $\Psi_{Neu}(/J)$ already in first order in g. We shall not be concerned with this functional, however. (See ref. [13] for a recent investigation related to this functional.)

The renormalization of $\Psi(A/J)$ in (4.8) by the counter terms (4.6) can be verified by examining graphs [1]. Concerning the factor $\Psi_{Dir}(A/J)$, which can be computed directly by integrating in the $y > 0$ half space with Dirichlet propagators, the physical reason for finiteness (even without the Z_4-1-term in (4.6))is: the (homogeneous or inhomogeneous) Dirichlet condition constrains the field at the boundary such that fluctuations must go to zero there. This is not so when setting up the interaction representation: there, merely the coupling constant is abruptly set to zero. For only renormalizable couplings, this leads to divergences [14][9] not removable without loss of the meaning of that representation.

5. Behaviour at the boundary

The presence of the factor Z_5 in the sum of (4.4) and (4.6) implies that, if one lets $\phi(\underline{x}y)$ approach the surface $y = 0$ from $y > 0$, it takes the value $Z_5 Z_3^{-1} A(\underline{x})$, due to the canonical commutation relations for $\phi_B(\underline{x}y) = Z_3^{1/2}\phi(\underline{x}y)$, rather than the value $A(\underline{x})$ as it would in the absence of interaction, see (4.5). Thus, since $A(\underline{x})$ is finite by construction, $\phi(\underline{x}y)$ "diverges logarithmically" as $y \searrow 0$. To compensate this divergence at $\varepsilon = 0$, one must introduce a factor

(5.1)

$$\lim_{y \searrow 0} \{a(g, \mu y, my)\phi(\underline{x}y)|A\rangle\} = |A\rangle A(\underline{x})$$

where

$$a(g, \mu y, my) = 1 - \frac{g}{32\pi^2} \ln(\mu y) + O(g^2) \tag{5.2}$$

can actually be chosen m-independent. - Likewise, if one lets $\partial_y \phi(\underline{x}y)$ approach the boundary, the sum of (4.4) and (4.6) shows that one may replace this by $Z_5^{-1}\,\delta/\delta A(\underline{x})$ rather than $\delta/\delta A(\underline{x})$ as in the absence of interaction. Thus, at $\varepsilon = 0$ again a "logarithmically divergent" compensating factor is required:

$$\lim_{y\downarrow 0}\left\{c(g,\mu y,my)\,\partial_y\phi(\underline{x}y)\,|A\rangle\right\} = \frac{\delta}{\delta A(\underline{x})}\,|A\rangle \tag{5.3}$$

in the obvious sense, where

$$c(g,\mu y,my) = 1 + \frac{g}{32\pi^2}\left[\ln(\mu y)+1\right] + O(g^2) \tag{5.4}$$

can also be chosen m-independent.

The meaning of (5.1) and (5.3) is made precise by introducing the Green's functions with also "surface arguments"

$$\ln\Psi_{Dir}(A|J) = \tag{5.5}$$

$$= \sum_{l=0}^{\infty}\frac{1}{l!}\sum_{n=0}^{\infty}\frac{1}{n!}\prod_{j=1}^{l}\int d\underline{z}_j\,A(\underline{z}_j)\prod_{i=1}^{n}\int d\underline{x}_i\int_0^{\infty}dy_i\,J(\underline{x}_i y_i)$$

$$\cdot\, G(\underline{z}_1\dots\underline{z}_l\,|\,\underline{x}_1y_1\dots\underline{x}_ny_n;\mu,g,m).$$

Then, taking into consideration also the c_5 term in (4.6), (5.1) and (5.3) become (in abbreviated notation)

$$\lim_{y\downarrow 0}\left\{a(y)\,G(\underline{z}_1\dots\underline{z}_l\,|\,\underline{x}y\,\underline{x}_1y_1\dots\underline{x}_ny_n)\right\} = \tag{5.6}$$

$$= \delta_{l1}\,\delta_{n0}\,\delta(\underline{x}-\underline{z}_1)$$

and

$$\lim_{y\downarrow 0}\left\{c(y)\left[\partial_y G(\underline{z}_1\dots\underline{z}_l\,|\,\underline{x}y\,\underline{x}_1y_1\dots\underline{x}_ny_n) - \right.\right. \tag{5.7}$$

$$\left.\left. - \delta_{n0}\,\delta_{l1}\,\delta(\underline{x}-\underline{z}_1)\,\partial_y\tilde{G}(\underline{0}\,|\,\underline{0}\,y)\right]\right\} =$$

$$= G(\underline{z}_1\dots\underline{z}_l\,\underline{x}\,|\,\underline{x}_1y_1\dots\underline{x}_ny_n),$$

respectively, where $\tilde{G}(\underline{0}/\underline{0}y) := \int d\underline{x}\, G(\underline{0}/\underline{x}y)$. The validity of (5.6) and (5.7) is best discussed in graphical terms [1]. The factors a(y) and c(y) can be defined in terms of two-point functions with one or both arguments on the boundary. They play a role analogous to (the inverse of) Wilson short-distance expansion coefficients [15]: the "short distance" is here the one from the boundary.

The Green's functions obey renormalization group equations

$$(5.8)\quad \left[\mu\frac{\partial}{\partial\mu} + \beta(g)\frac{\partial}{\partial g} + n\gamma(g) - l\sigma(g) + \eta(g)m^2\frac{\partial}{\partial m^2}\right] G(\underline{z}_1\cdots\underline{z}_l/\underline{x}_1 y_1\cdots\underline{x}_n y_n) = 0$$

involving the new parametric function

$$(5.9)\quad \sigma(g) = \lim_{\varepsilon\searrow 0}\left\{\beta(g,\varepsilon)\frac{\partial}{\partial g}\ln(Z_5 Z_3^{-1})\right\} = (32\pi^2)^{-1}g + O(g^2)$$

The forms (5.2) and (5.4) follow (apart from finite renormalizations and expendable m-dependence) from the equations consistent with (5.6) and (5.7)

$$(5.10)\quad \left[\mu\frac{\partial}{\partial\mu} + \beta(g)\frac{\partial}{\partial g} - \gamma(g) + \sigma(g) + \eta(g)m^2\frac{\partial}{\partial m^2}\right] a(g,\mu y, m y) = 0$$

and

$$(5.11)\quad \left[\mu\frac{\partial}{\partial\mu} + \beta(g)\frac{\partial}{\partial g} - \gamma(g) - \sigma(g) + \eta(g)m^2\frac{\partial}{\partial m^2}\right] c(g,\mu y, m y) = 0.$$

6. Schrödinger equation

Infinitesimal shift of the sources in $\Psi_{Dir}(A\,J)$ is equivalent to infinitesimal shift of the surface on which the boundary value A (in the sense of (5.1)) is specified, in the opposite direction. The infinitesimal surface shift is the "Euclidean time" derivative of the Schrödinger wave functional and expressible by applying the Schrödinger operator, a functional differential operator on A dependence, on $\Psi_{Dir}(A/J)$:

$$(6.1)\quad \int d\underline{x}\int_0^\infty dy\, \partial_y\left[\delta/\delta J(\underline{x}y)\right]\Psi_{Dir}(A/J) = H(A,\delta/\delta A, J)\,\Psi_{Dir}(A/J)$$

where in H only the value of J on the boundary enters.

To obtain the form of H, one first establishes (6.1) under regularization, which yields formally, taking findings in sect. 5 into account,

$$H(A,\delta/\delta A,J) = \int d\underline{z}\,[-\tfrac{1}{2} Z_3 Z_5^{-2} \frac{\delta^2}{\delta A^2} + \tfrac{1}{2} Z_3^{-1} Z_5^2 \partial_i A \partial_i A + \tfrac{1}{2} m^2 Z_3^{-2} Z_2 Z_5^2 A^2 + \tfrac{1}{4!} g\mu^\varepsilon Z_1 Z_3^{-4} Z_5^4 A^4 + Z_5 Z_3^{-1} JA + \text{const}]. \tag{6.2}$$

However, already in the free-field case (and the dimensionally regularized theory has as short-distance behaviour the free one) $\delta^2/\delta A^2$ cannot be applied on the Schrödinger wave functional but requires point splitting

$$\frac{\delta^2}{\delta A(\underline{z})^2} \to \lim_{\underline{\Delta}\to 0} \frac{\delta^2}{\delta A(\underline{z}+\underline{\Delta})\,\delta A(\underline{z})} \tag{6.3}$$

and letting the const in (6.2) diverge quartically as $|\underline{\Delta}| \to 0$. Thus, even in the regularized theory (6.2) has to be modified in the sense (6.3). This point splitting, however, allows also to absorb the Z-factors in (6.2): at their place, similar to (5.1) and (5.3), factors that diverge logarithmically as $|\underline{\Delta}| \to 0$ must be inserted. Their precise behaviour is again governed by renormalization group equations. The final limit form of the Schrödinger operator is too long and uninteresting to reproduce it here. We emphasize, however, that in principle it is not more complicated than in the free-field case where point splitting is also needed.

7. Completeness, unitarity, and computation of expectation values

In elementary QM, scalar products are formed by integrating the product of Schrödinger wave functions over the space of eigenvalues of the diagonalized operator, which here gives an integral over ordinary space. In QFT, the scalar product becomes a functional integral over all functions $A(\underline{x})$. To all orders of perturbation theory, this turns out to be a Gaussian integral, whereby the Gaussian factor is the square of the free-field vacuum functional. The completeness of the Schrödinger states inherent here is, in the Minkowskian frame, related to the unitarity of the time ordered operator $\left(\exp[i\int dx^0 \int d\underline{x}\, J(\underline{x}x^0)\phi(\underline{x}x^0)]\right)_+$.

In QM, the computation of expectation values of operators at the time at which the Schrödinger representation is set up is immediate. In the renormalizable QFT case, however, one can obtain UV-finite results only for renormalized operators, e.g. the vacuum expectation value $\langle \phi(\underline{z}_1)\ldots\phi(\underline{z}_{2n})\rangle$ is finite, and ϕ differs from the diagonalized operator. The interrelation is described by (5.1). It implies that in the computation of the indicated vacuum expectation value by functional integration over A-space, the arguments of the ϕ have first to be taken off the y = 0 plane. Then the state (or states) must be written as A-functionals and the A-integration be carried out. Only then the limit all $y \searrow 0$ can be taken. Hereby, logarithmic divergences will be cancelled by the divergences of explicitly occurring factors such as a(y).

8. Other models

The extension of these methods to other Bose field theories is straightforward and follows unambiguously from their canonical quantization. For models involving Fermi fields one must note that the usual field equations for these are first order in time derivatives such that only half of the field components can be diagonalized simultaneously. It is simplest to write the Dirac field as the direct sum of two canonically independent Majorana fields, and to discuss the Minkowskian case since only there the Majorana field components can be chosen hermitean [16]. The Lagrangian density with surface terms appropriate for the free Majorana field is

$$
\begin{aligned}
\mathcal{L} = {} & \tfrac{1}{2} i \bar{\psi} \gamma^\mu \partial_\mu \psi - \tfrac{1}{2} m \bar{\psi}\psi + \\
& + \tfrac{1}{2} i \delta(x^0)\, \bar{\psi}(\underline{x}(+0))\,(\gamma^0 + i\gamma_5)\psi(\underline{x}0) + \\
& - \tfrac{1}{2} i \delta(x^0)\, \bar{\psi}(\underline{x}(+0))\,(\gamma^0 + i\gamma_5)\,\eta(\underline{x}).
\end{aligned} \tag{8.1}
$$

The first surface term effects decoupling of positive from negative times, and the second one brings about the boundary condition

$$(1 + i\gamma^0\gamma_5)\,\psi(\underline{x}(+0)) = (1 + i\gamma^0\gamma_5)\,\eta(\underline{x}), \tag{8.2a}$$

$$\bar{\psi}(\underline{x}(+0))\,(1 - i\gamma^0\gamma_5) = \bar{\eta}(\underline{x})\,(1 - i\gamma^0\gamma_5). \tag{8.2b}$$

Here, $\eta(\underline{x}) = \bar{\eta}(\underline{x})\gamma^0$ is an effectively two-component Grassmann (anticommuting c-number) function since $1/2(1 + i\gamma^0\gamma_5)$ is a rank-two projector. ((8.2) is straightforward when read in terms of, here antisymmetric, Green's function as in (5.6,7).) One can show that, if a renormalizable interacting is added, the surface terms in (8.1) require only multiplicative renormalization provided the theory is (neglecting mass terms) invariant under discrete γ_5 transformation and under space reflection. Then also relations (8.2) are modified only as in (5.1).

The $(\vec{\phi}^2)^2_{4-\varepsilon}$ theory, $\vec{\phi} = (\phi_1 \ldots \phi_N)$, constructed in the half space $y > 0$ and taken at the Wilson-Fisher fixed point $g^*_W(\varepsilon)$ [17] with $\varepsilon = 1$, presents a semiinfinite classical ferromagnet (with N-component spin) at the critical point. In this context, the parametric function $\sigma(g^*_W(1),1,N)$ to various bd.conds.is a critical exponent governing the behaviour of spins in or close to the surface, which is different from the one of spins "in the bulk". These exponents have recently been computed in ε expansion to second order (see refs. [18], [13] and refs. given there).

9. Final remarks

The Schrödinger representation is, for renormalizable (unregularized) theories, uninviting to do computations in if the theory is originally defined in unlimited space-time, as in QED or QCD (without "bags" at least). Its existence, however, is of some conceptual value if one wants to consider such theories as extensions (to an infinite number of degrees of freedom) of ordinary QM where the Schrödinger representation is the most advantageous description. The problem of the Schrödinger representation, more generally, of allowed boundary conditions poses itself directly, however, in "quantum field theories" that are ab initio defined only with boundaries as explained in the introduction. In particular, a Casimir effect - the dependence on geometry and boundary conditions of the free energy in the Euclidean case, of the ground state energy in the Minkowskian case - can only be computed if such geometries and boundary conditions are mathematically meaningful.

References

1 K. Symanzik, Nucl. Phys. B190 [FS3], 1 (1981)
2 A.M. Polyakov, Phys. Lett. 103B, 207, 211 (1981)
3 M. Lüscher, K. Symanzik, P. Weisz, Nucl. Phys. B173, 365 (1980)
4 T. Eguchi, Phys. Rev. Lett. 44, 126 (1980)
5 K. Symanzik, Comm. Math. Phys. 45, 79 (1975)
6 O. Alvarez, Phys. Rev. D24, 440 (1981)
7 J. Fröhlich, these Proceedings
8 C. Itzykson, J.B. Zuber, "Quantum Field Theory", McGraw-Hill, New York 1980
9 N.N. Bogoliubov, D.V. Shirkov, "Introduction to the Theory of Quantized Fields", Wiley-Interscience, New York 1979
10 W. Zimmermann, in "Lectures on Elementary Particles and Quantum Field Theory", Eds. S. Deser et al., MIT Press, Cambridge, Mass. 1971, Chpt. 4

11 G. 't Hooft, M. Veltman, in "Particle Interactions at Very High Energies", Eds. D. Speiser et al., Plenum Press, New York 1974, part B, p. 177
12 J.C. Collins, Nucl. Phys. B80, 341 (1974)
13 H.W. Diehl, S. Dietrich, Phys. Rev. B24,
14 E.C.G. Stückelberg, Phys. Rev. 81, 130 (1951)
15 K.G. Wilson, Phys. Rev. 179, 1499 (1969)
16 J. Schwinger, Proc. Nat. Acad. Sci. U.S. 44, 956 (1958)
17 K.G. Wilson, J. Kogut, Phys. Rep. 12C, 75 (1974)
18 H.W.Diehl, S. Dietrich, Z. Phys. B42, 65 (1981)

ALL SELF-DUAL MULTIMONOPOLES FOR ARBITRARY GAUGE GROUPS

W. Nahm

CERN, Geneva, Switzerland

ABSTRACT

The ADHM formalism is adapted to self-dual multimonopoles for arbitrary charge and arbitrary gauge group. Each configuration is characterized by a solution of a certain ordinary non-linear differential equation, which has chances to be completely integrable. For axially symmetric configurations it reduces to the integrable Toda lattice equations. The construction of the potential requires the solution of a further ordinary linear differential equation.

INTRODUCTION

The SU(4) supersymmetry (with arbitrary gauge group G) is the only known non-trivial quantum field theory in 3+1 dimensions which has good chances to be explicitly solvable[1]. In particular, there are indications that the conformal invariance of this theory is not broken by any anomalies. However, it may be broken spontaneously by a non-zero expectation value of the scalar field, as supersymmetry guarantees in certain directions of the field space a flat effective potential. Simultaneously, the gauge group G is broken down to a subgroup K, which always contains an Abelian factor. With respect to the latter, the theory contains magnetic monopoles.

These monopoles may be considered as excitations of the field of a dual quantum field theory. Dynamically, they are expected to behave exactly as a supersymmetry multiplet containing massive gauge bosons[2]. More precisely, the dual theory should have the same form as the original one, but with the dual gauge group G^V and a corresponding subgroup K^V [3].

To check these ideas, monopole solutions of the classical theory should be helpful. For simplicity let us consider the case where only one component ϕ of the six SU(4) components of the scalar field has a non-zero value at infinity, which corresponds to a non-zero vacuum expectation value in the quantum field theory. To obtain static solutions we want to find the most general finite energy solution of the equation

$$B_i = D_i\phi \ , \tag{1}$$

where the D_i are covariant derivatives, and B_i is the corresponding magnetic field. Of course, ϕ has to be in the adjoint representation of G.

If one introduces a dummy space direction with coordinate x^0, the scalar field may be considered as the corresponding component of the gauge potential. Then Eq. (1) takes the same form as the self-duality equation for instantons

$$B_i = E_i \ . \tag{2}$$

The ADHM construction yields the most general finite action solution of this equation[4]. In a suitable gauge, the potentials turn out to be algebraic.

THE ADHM CONSTRUCTION FOR MULTIMONOPOLES

The condition of finite energy and x^0-independence is somewhat more difficult to handle than the condition of finite action. Nevertheless we shall see that the ADHM construction can be adapted to our case, though its algebraic character becomes somewhat less obvious.

If one complexifies the R^4 to C^4, self-dual fields yield an analytic vector bundle over the space CP^3-CP^1 of anti-self-dual planes in C^4. The points in a fibre are functions χ from the corresponding plane to the fundamental representation of G which are constant with respect to covariant differentiations in the plane. CP^3-CP^1 can be covered by two charts, and the corresponding transition function of the bundle essentially yields back the potential.

Instead of using a transition function, the ADHM construction embeds the bundle in the direct product of the base with a larger linear space. In this larger space, χ is now defined globally and fulfils the equation[5]

$$\left(D_\mu e^\mu\right)^{A'A}\chi = \tilde{\Omega}^{A'A} + \Omega^{A'}_{B'}x^{B'A} \tag{3}$$

where for fixed B', B both Ω and $\tilde{\Omega}$ have to fulfil the Dirac equation

$$\left(D_\mu e^{+\mu}\right)_{AA'}\tilde{\Omega}^{A'B} = \left(D_\mu e^{+\mu}\right)_{AA'}\Omega^{A'}_{B'} = 0 . \quad (4)$$

Here the e^μ form the usual basis of the quaternions, represented by 2×2 matrices, with $e^0 = 1_2$. Moreover,

$$x = x_\mu e^\mu \quad (5)$$

represents the coordinates in C^4.

In order to avoid confusion, we now shall denote base space coordinates by X^μ. The points in the fibre over an anti-self-dual plane

$$X^{A'A}\pi_A = \omega^{A'} \quad (6)$$

fulfil the additional condition

$$\tilde{\Omega}^{A'A}\pi_A + \Omega^{A'}_{B'}\omega^{B'} = 0 . \quad (7)$$

Still this leaves too much freedom, as there are solutions χ which vanish all over some planes. One may either divide them out or replace the basis directly by R^4, with points in the fibre given by

$$\tilde{\Omega}^{A'A} + \Omega^{A'}_{B'}X^{B'A} = 0 . \quad (8)$$

For $G = SU(N)$, Eq. (8) yields for each X an N-dimensional complex hyperplane in the vector space of normalizable solutions of Eqs. (3) and (4). This vector space has a scalar product, which induces a natural connection on the C^N-bundle over R^4 described by Eq. (8) and in this way gives back the potential on which the construction was based.

Equation (8) may be written in the form

$$\Delta^+(X)v_i(X) = 0 \quad \text{for} \quad i = 1, \ldots, N , \quad (9)$$

where the $v_i(X)$ are orthonormal solutions of Eqs. (3) and (4) and Δ is of the form

$$\Delta = a + bX \quad (10)$$

with X-independent linear operators a,b. The potential is given by

$$A_{\mu}^{ij} = v_i^+ \partial_{\mu} v_j \ . \tag{11}$$

Different choices of the v_i yield gauge equivalent potentials.

All this applies equally for instantons and self-dual monopoles, with the single change that the condition of normalizability requires a four-dimensional integration in the former case and a three-dimensional one in the latter.

Now Eqs. (9) to (11) can be applied without knowing the details of the domain of Δ^+ [given by Eqs. (3) and (4)] or of its range (given by the normalizable solutions of the Dirac equation). One just needs some general information on the linear operators a,b and can show *a posteriori* that all potentials constructed by Eqs. (9) and (10) using such operators yield self-dual fields with the required properties. The main information taken over from the ADHM construction is that $\Delta^+\Delta$ is invertible and commutes with the quaternions. This already guarantees that the resulting field will be regular and self-dual[6].

In the instanton case the only further information used is the dimension of domain and range of Δ^+. The Dirac equation has k linearly independent normalizable solutions, where k is the instanton number, and the number of linearly independent normalizable solutions of Eqs. (3) and (4) is $2k + N$.

In the monopole case we have to take account of the dummy coordinate x^0. We introduce its conjugate momentum z and take for the Dirac equation a basis of solutions of the form

$$\psi(\vec{x},x^0) = \exp\,(ix^0 z)\psi(\vec{x}) \ . \tag{12}$$

The number $k(z)$ of normalizable solutions of this form is known[7,8]. It is zero, if z lies outside the interval spanned by the extremal eigenvalues of $\phi(\infty)$.

For Eqs. (3) and (4), χ and Ω can also be written in the form of Eq. (12). Moreover one obtains

$$\tilde{\Omega}^{A'A} = -\Omega_{B'}^{A'} x^{B'A} + 2\left(D_{\mu} e^{\mu}\right)^{A'A} (D^2)^{-1} \Omega_{B'}^{B'} \ , \tag{13}$$

except for those values of z for which D^2 is not invertible. These values are the eigenvalues of $\phi(\infty)$ and yield the jumping points of $k(z)$.

Let us label the basis of normalizable solutions of Eqs. (3) and (4) by Ω of the form (12). Then in Eq. (13) the multiplication by x^0 may be expressed as a derivation with respect to z. Everything else commutes with ∂_0. Thus our Δ is of the form

$$\Delta = (i\partial_z + X)1_{k(z)} + iA(z) . \tag{14}$$

Here $1_{k(z)}$ is the k(z) dimensional unit matrix and A(z) also is k(z) dimensional. The matrix elements are quaternions.

Let us write

$$A(z) = e^{\mu}T_{\mu}(z) . \tag{15}$$

To obtain a $\Delta^+\Delta$ which commutes with the quaternions, the $T^{\mu}(z)$ must be anti-hermitian. $T^0(z)$ can always be absorbed by an equivalence transformation

$$\Delta \to U(z)^+\Delta U(z) , \quad U(z) \in U(k(z)) , \tag{16}$$

which represents a different choice of the basis of solutions of Eqs. (3) and (4) for each z. For the $T_i(z)$, i = 1, 2, 3, one obtains the differential equations

$$T_i'(z) = \frac{1}{2} \varepsilon_{ijk} \left[T_j(z)T_k(z)\right] . \tag{17}$$

Together with suitable boundary conditions at the jumping points of k(z) these equations yield the most general self-dual multimonopole configuration.

The $T_i(z)$ are meromorphic and can have at most simple poles. For physical values of z those can only occur at the jumping points of k(z). In a mathematically precise formulation, domain and range of Δ are Sobolev spaces, such that Δ and Δ^+ are bounded[9]. Moreover, $\Delta^+\Delta$ can easily be seen to be bounded below by a positive constant, as

$$\Delta^+\Delta = -\partial_z^2 + (x_i + iT_i)^+(x_i + iT_i) . \tag{18}$$

Thus $\Delta^+\Delta$ is invertible and all configurations constructed this way are regular.

According to Eq. (17) the trace part of the $T_i(z)$ is constant and can be absorbed by a translation of X. This defines a unique centre for arbitrary multimonopole configurations. Under rotations, the $T_i(z)$ transform as a vector. The traceless part of the symmetric tensor $tr(T_i(z)T_j(z))$ is independent of z and may be diagonalized by a rotation. This defines a system of natural axes for the multimonopole.

Apart from the translations, the general solution of Eq. (17) depends on $3(k(z)^2-1)$ parameters, of which $k(z)^2-1$ can be absorbed by an equivalence transformation of type (16), but with a z-independent U. For an axially symmetric multimonopole configuration, rotations around the symmetry axis can be represented as such equivalence transformations. If one decomposes SU(k(z)) with respect to a Cartan subalgebra which contains the generator of these equivalence transformations, Eq. (17) yields the completely integrable Toda lattice equations[10]. Note that these equations occurred in a configuration space analysis of spherically symmetric monopoles, while in z space the same equations describe the more general axially symmetric case. In general, the boundary conditions are different.

Let $t_i e^i$ be the residue of A(z), as z approaches some pole z_s. According to Eq. (17) the t_i must form an SU(2) subalgebra of SU(k(z)). The maximal embedding can only occur at the extremal eigenvalues of $\phi(\infty)$, as otherwise the equations above and below z_s decouple completely and yield independent monopoles in two direct factors of G.

THE CASE G = SU(2)

For G = SU(N) the number N is according to Eq. (9) the dimension of the kernel of Δ^+. According to Eq. (18) the dimension of the cokernel vanishes, such that

$$N = \text{index}(\Delta^+) \ . \tag{19}$$

To evaluate this index we have to find all local non-normalizable solutions of Eq. (9). Let us consider at some pole of A(z) an irreducible representation of the t_i of dimension d. With

$$t = t_i e^i \tag{20}$$

one finds

$$t^2 + t = -\sum_i t_i^2 = (d^2-1)/4 \ . \tag{21}$$

As the trace of t vanishes, all eigenvalues are known. One obtains d + 1 solutions of type

$$v(z) \sim (z-z_s)^{(d-1)/2} \tag{22}$$

and d − 1 non-normalizable solutions of type

$$v(z) \sim (z-z_s)^{(-d-1)/2} \ . \tag{23}$$

Let $\phi(\infty)$ only have two distinct eigenvalues, between which k(z) is a constant k. To obtain N = 2, Eq. (9) must have 2k - 2 non-normalizable local solutions. This is only possible if at both jumping points of k(z) the t_i form SU(2) algebras which are maximally embedded into SU(k). In this case, k is the magnetic charge[7,8]. One can check the behaviour of the scalar field by solving Eq. (9) in the limit $r \to \infty$. Up to normalizations the solutions behave like

$$v_{\pm}(z) \sim (z \mp z_s)^{(k-1)/2} \exp(\pm rz) . \tag{24}$$

Equation (11) yields for the eigenvalues of $\phi(\infty)$

$$\phi_{\pm} = \pm\left(z_s - \frac{k}{2r}\right) , \tag{25}$$

as it should be. The case k = 1 yields the simple BPS monopole, to which the ADHM formalism has been applied before[11].

Now let us count the number of multimonopole parameters. Let $T_i(z) + \delta_i(z)$ be an infinitesimal perturbation of a solution of Eq. (17) which still is a solution. We have to find zero modes of the operator

$$(P\delta)_i = \delta_i' - \varepsilon_{ijk}\left[T_j \delta_k\right] , \tag{26}$$

which are non-singular at the boundaries. On δ_i of the form

$$\delta_i = \left[T_i u\right] \tag{27}$$

this operator acts simply as a differentiation of u. Constant u yields an equivalence transformation and may be neglected. Thus we may consider P to be an operator defined on δ_i modulo δ_i of form (27). Then P^+P is bounded away from zero, such that the dimension of the kernel of P is equal to its index. Thus we only have to consider local zero modes of P close to the jumping points. With

$$(p\delta)_i = \varepsilon_{ijk}\left[t_j \delta_k\right] \tag{28}$$

one finds

$$p^2 + p = -\sum t_i^2 . \tag{29}$$

If one includes δ_i of type (27), the trace of p vanishes, such that one can easily calculate all eigenvalues. Under the adjoint action of the t_i, the algebra of SU(k) decomposes into k - 1 irreducible representations of dimensions d_r, r = 1, ..., k - 1. For each one there are $d_r + 2$ solutions of type

$$\delta \sim (z-z_s)^{(d_r-1)/2} \tag{30}$$

and $d_r - 2$ solutions of type

$$\delta \sim (z-z_s)^{-(d_r+1)/2} . \tag{31}$$

Thus

$$\text{index}(P) = 2(k^2-1) - 2\sum_r (d_r - 2) = 4(k-1) . \tag{32}$$

If one adds the translations, one has the correct number of degrees of freedom.

Finally let us solve Eq. (17) for k = 2. If one rotates the configuration to the natural system of axes, the $T_i(z)$ form up to normalizations a standard set of SU(2) generators, and according to Eq. (17) this set does not depend on z. Thus we may write

$$T_i(z) = -if_i(z)\sigma_i/2 . \tag{33}$$

One obtains the first integrals

$$f_i(z)^2 - f_j(z)^2 = C_{ij} \tag{34}$$

and

$$f_i'(z) = \prod_{j\neq 1} \left(f_i^2(z) + C_{ij}\right)^{\frac{1}{2}} . \tag{35}$$

Thus we obtain Jacobi elliptic functions with a pure real and a pure imaginary period.

According to the choice of the z interval, one obtains SU(2), SU(3), and SU(4) configurations. In the latter case, the $f_i(z)$ must stay regular at the boundary of the interval, in the SU(2) case they must have poles at both ends.

The solutions of Eq. (9) can only be multiplied by a constant matrix, when z is shifted by a period. Thus they are essentially elliptic sigma functions.

Axial symmetry yields $C_{12} = 0$ and up to rescaling

$$f_3 = -\cot z , \tag{36}$$

$$f_1 = f_2 = -\frac{1}{\sin z} \tag{37}$$

or the corresponding hyperbolic functions. For this case the solutions of Eq. (9) can be read off from Ref. 9.

The axially symmetric configurations of higher charge can also be obtained from Eqs. (36) and (37), if one replaces the matrices σ_i in Eq. (33) by a maximal embedding of SU(2) in SU(k), i.e. by a k-dimensional representation of SU(2). This procedure also has more general applications.

Spherical symmetry requires $c_{12} = c_{13} = 0$, and

$$f_i(z) = -\frac{1}{z} . \qquad (38)$$

In this case, only SU(3) and SU(4) configurations are possible. The SU(3) solution is known[12], but the SU(4) solution, which contains a free parameter, appears to be new.

OPEN PROBLEMS

Equations (17) and (9) yield all self-dual multimonopole configurations. If $\phi(\infty)$ has more than two different eigenvalues the boundary conditions for A(z) across the corresponding jumping points of k(z) still have to be studied in detail.

Equation (17) is non-linear, but it might be possible to characterize all solutions in a relatively simple way. More precisely, one might hope to represent the solutions as automorphic functions on some compact Riemann surface. The linear Eq. (9) may not allow such a treatment and in the general case may have to be treated numerically.

It will be interesting to compare the ADHM representation of the multimonopoles with other approaches[13,14].

REFERENCES

1. M. Sohnius and P. West., Phys. Lett. 100B (1981) 245.
2. H. Osborn, Phys. Lett. 83B (1979) 321.
3. P. Goddard and D. Olive, DAMTP 81/11 (1981).
4. M. Atiyah, N. Hitchin, V. Drinfeld and Yu. Manin, Phys. Lett. 65A (1978) 185.
5. H. Osborn, CERN-TH 3210 (1981).
6. E. Corrigan, D. Fairlie, P. Goddard and S. Templeton, Nucl. Phys. B140 (1978) 31.
7. C. Callias, Comm. Math. Phys. 62 (1978) 213.
8. R. Bott and R. Seeley, Comm. Math. Phys. 62 (1978) 235.
9. W. Nahm, Phys. Lett. 93B (1980) 42.
10. D. Olive, ICTP 80-81-1 and 80-81-41.

11. W. Nahm, Phys. Lett. 90B (1980) 412.
12. A. Bais and H. Weldon, Phys. Rev. Lett. 41 (1978) 601.
13. E. Corrigan and P. Goddard, DAMTP 81/9 (1981).
14. P. Forgacs, Z. Horvath and L. Palla, KFKI-1981-21, KFKI-1981-23 with further references.

RENORMALIZATION GROUP ASPECTS OF 3-DIMENSIONAL PURE U(1) LATTICE GAUGE THEORY *

M. Göpfert and G. Mack

II. Institut für Theoretische Physik
der Universität Hamburg
(Lecture presented by G. Mack)

We have studied the 3-dimensional pure U(1) lattice gauge theory model with Villain action [1]. It was expected on the basis of the work of Polyakov [2], Banks, Myerson and Kogut [3] and others [4] that this model would show confinement for all values of the coupling constant. We proved that this is indeed the case. There were also some surprises, however, and it is of interest to discuss these from the point of view of the renormalization group theory [5,6].

The model lives on a 3-dimensional cubic lattice of lattice spacing a. It is a classical mechanical system whose random variables are attached to the links b = (x,y) of the lattice

$$U(b) = e^{i\theta_\mu(x)} \quad \text{for } b = (x, x+e_\mu),\ -\pi \leq \theta_\mu < \pi \tag{1a}$$

(e_μ = lattice vector in μ direction). The action is of the form

$$L(U) = \sum_p \mathcal{L}(U(\partial p)) \tag{1b}$$

with $U(\partial p) = U(b_1) \ldots U(b_4)$ if p is the plaquette whose boundary consists of links $b_1 \ldots b_4$, and

$$\mathcal{L}(e^{i\varphi}) = \ln \sum_{m=0,\pm 1,\pm 2,\ldots} \exp\left[-\frac{1}{2ag^2}(\varphi - 2\pi m)^2\right] \tag{1c}$$

g^2 is the unrenormalized electric charge squared. It has dimension

* Work supported in part by Deutsche Forschungsgemeinschaft

of a mass in 3 dimensions. The Boltzmann factor is exp L(U).

Since the gauge group U(1) of this model is abelian, the model can be subject to a Kramers-Wannier duality transformation [3]. As a result one obtains a ferromagnet with a global symmetry group $\mathbb{Z}$. Its random variables n(x) are attached to the sites of a 3-dimensional cubic lattice Λ (the dual of the original one) and assume values which are integer multiples of 2π. The new action is

$$\hat{L}(n) = -\frac{1}{2\beta} \int_x [\nabla_\mu n(x)]^2 \quad \text{with } \beta = 4\pi^2/g^2.$$

We use the standard notations

$$\int_x = a^3 \sum_x , \quad \nabla_\mu f(x) = a^{-1} [f(x+e_\mu) - f(x)]$$

Expectation values are computed with the help of the Boltzmann factor exp $\hat{L}(n)$. This model is known as the "discrete Gaussian model", for obvious reasons. We call it the "$\mathbb{Z}$-ferromagnet" in order to emphasize its symmetry properties.

The global $\mathbb{Z}$-symmetry of this model is always spontaneously broken (if $\langle n(x) \rangle$ exists at all) since the equation

$$\langle n(x) \rangle = \langle n(x) \rangle + 2\pi I, \quad I \in \mathbb{Z},$$

has no solution if $I \neq 0$. The surface tension α of the model is defined as the cost of free energy per unit area of a domain wall which separates two domains whose spontaneous magnetization $\langle n(x) \rangle$ differs by 2π. The duality transformation shows that α equals the string tension of the U(1) gauge model.

It is convenient to introduce the following quantity with the dimension of a mass squared

$$m_D^2 = (2\beta/a^3)\exp[-\beta v_{Cb}(0)/2]$$

where v_{Cb} is the lattice Coulomb potential. As was shown by Banks et al., the model can also be transformed into a (special) Coulomb system. m_D^{-1} is the prediction of a Debye Hückel approximation for the screening length of that system, for large β/a.

Our main results are as follows.

<u>Theorem 1</u> There is a dimensionless constant $C > 0$ such that

$$\alpha \geq C \cdot m_D \beta^{-1} \quad \text{for sufficiently large } \beta/a.$$

Since αa^2 is a monotone decreasing function of β/a, by Guth's inequality [7], it follows that $\alpha > 0$ for all values of the coupling constant $g^2 > 0$. We believe that the r.h.s. of the inequality of theorem 1 represents the true asymptotic behavior of α. The theorem is proven by computing first an effective action which embodies a Pauli-Villars cutoff M of order m_D (it can also be chosen bigger, if desired, but not smaller). A classical approximation to this effective action with neglect of some correction terms in it produces the approximation $\alpha = 8m_D\beta^{-1}$. It is amusing to compare with the leading term of the high temperature expansion, which is valid when β/a is small. It reads $\alpha = 2\pi^2 a^{-1}\beta^{-1} + \ldots$.

The meaning of m_D as an asymptotic mass is clarified by the second result. The effective action L_{eff} mentioned above depends on a real field $\Phi(x)$. It is obtained by integrating out the high frequency components of $\beta^{-1/2}n(x)$. Symbolically we may write

$$\Phi(x) = \beta^{-1/2} n(x) \text{ with Pauli-Villars cutoff M}$$

<u>Theorem 2</u> Consider the correlation functions $\langle\Phi(x_1)\ \ldots\ \Phi(x_n)\rangle$ for fixed distances $m_D|x_i-x_j|$ in units of m_D. They tend to the correlation functions of a massive free field theory with mass m_D as $\beta/a \rightarrow \infty$ and $M/m_D \rightarrow \infty$ (proportional $(\beta/a)^{1/12}$, for instance).

These results were obtained by a rigorous block spin calculation, and are therefore perfectly consistent with the general renormalization group theory [6]. However, they contradict what would be obtained by making simple but popular approximations.

Suppose that one could set up a renormalization group procedure (block spin calculation) for the U(1) gauge theory (1) in such a way that the effective action at each step of this iterative procedure is still (approximately) of the same form (1), except for the replacement of g^2 by a running coupling constant $g^2_{eff}(a')$ and a new value $a' > a$ of the lattice spacing. Suppose moreover that $a'g^2_{eff}(a')$ reaches values in the domain of validity of high temperature expansions after sufficiently many iteration steps (depending on ag^2), no matter how small the bare coupling constant g^2a is. Then it would follow immediately that the string tension α should be proportional to the physical mass (= mass gap) squared.

In contrast, theorem 1 tells us that

$$\alpha/m_D^2 \geq C\cdot(\beta m_D)^{-1} = C\cdot(2\beta^3/a^3)^{-1/2}\exp \tfrac{1}{4}\beta v_{Cb}(0) \quad \rightarrow \infty$$

as $\beta/a = 4\pi^2/g^2a \rightarrow \infty$ (The numerical value of $v_{Cb}(0)$ is found in the literature to be 0.2527 a^{-1} [8].)

In the language of the Z-ferromagnet the reason for the discrepancy can be described as follows.

Our effective action is not of the form of the original action (2) but involves a real field $\Phi(x)$ rather than an integer valued one. It is of the form

$$L_{eff}(\Phi) = -\frac{1}{2}(\Phi,-\Delta(1-\frac{\Delta}{M^2})\Phi) - m_D^2\beta^{-1}\int_{x\in\Lambda}[1-\cos\beta^{1/2}\Phi(x)] + \ldots$$

The first term is the usual kinetic term with Pauli-Villars cutoff M [9]. The correction terms ... are nonlocal, but they are shown to be small in some precise sense, and they do not involve interactions which decay less fast with distance $r = 3^{-1/2}\max|x_\mu|$ than $e^{-(1-\delta Mr)}$ (δ small for β/a large). The effective potential (negative of the 2nd term) has minima $\Phi(x) = 2\pi\beta^{-1/2}\cdot$integer.

If we wanted to obtain an effective action for an integer valued field $h/2\pi$ we could split

$$\Phi(x) = \beta^{-1/2}[h(x) + \Theta(x)]$$

where $h/2\pi$ is integer and constant on cubes (= cells of a block lattice) of lattice spacing $a' \lesssim M^{-1}$. It selects a minimum of the effective potential, whereas Θ describes fluctuations around it. The variables Θ and h describe spin waves and domain walls on the block lattice, respectively. To obtain the effective action for h, the spin waves have to be integrated out.

It turns out that the spin waves have a mass $\approx m_D$ as soon as the cutoff M has been brought down to values of order β^{-1}. If β/a is small, such M is still very much larger than m_D. Integrating out the spin waves will produce interactions of range m_D^{-1} rather than M^{-1}, and therefore the resulting effective action will be very nonlocal when the cutoff M is lowered to $O(\beta^{-1}) \ggg m_D$. In renormalization group theory this is regarded as a catastrophy. It means that h is not a good block spin, in contract with Φ. And in any case, the effective action for it is not approximately of the form of the original action with a running coupling constant, since the original action had only neighbour interaction.

Lesson. To choose a good block spin in a renormalization group procedure one should know what the low lying excitations of the theory are, in order to avoid integrating out some of them by mischief.

Of course, if one wants to obtain the surface tension α, the spin waves have to be integrated out eventually. In our work [1] this is only done after the cutoff M has been brought down to order m_D, in the course of the analysis of a theory with action $L_{eff}(\Phi)$. Such an analysis had already been carried out by Brydges and Federbush, and we could use their result [10]. It uses the Glimm - Jaffe - Spencer expansion of constructive field theory [11].

REFERENCES

1. M. Göpfert and G. Mack, Commun. Math. Phys. (in press), Commun. Math. Phys. 81:97 (1981)
2. A. M. Polyakov, Nucl. Phys. B120:429 (1977)
3. T. Banks, R. Myerson and J. Kogut, Nucl. Phys. B129:493 (1977)
4. T. A. DeGrand and D. Toussaint, Phys. Rev. D22:2498 (1980)
   S. D. Drell, H. R. Quinn, B. Svetitsky and M. Weinstein, Phys. Rev. D19:619 (1979)
   M. E. Peskin, Ann. Phys. (N.Y.) 113:122 (1978)
5. L. D. Kadanoff, Physis 2:263 (1965)
6. K. Wilson, Phys. Rev. D2:1473 (1970)
7. A. H. Guth, Phys. Rev. D21:2291 (1980)
8. A. Ukawa, P. Windey and A. H. Guth, Phys. Rev. D12:1013 (1980), V. F. Müller and W. Rühl, Z. Physik C9:261 (1981)
9. N. N. Bogoliubov and D. V. Shirkov, "Introduction to the theory of quantized fields", Interscience, New York 1959
10. D. Brydges and P. Federbush, Commun. Math. Phys. 73:197 (1980)
11. J. Glimm, A. Jaffe and T. Spencer, Ann. Phys. 101:610, 631 (1975)

MASS SPLITTING OF THE PSEUDOSCALAR MESONS BY MONTE CARLO TECHNIQUE

P. Di Vecchia

Physics Department
University of Wuppertal, D-5600 Wuppertal 1

Talk given at the VIth International Conference on the Problems of Quantum Field Theory, Alushta (1981) and at the 11th NATO Summer Institute, Freiburg (1981).

In the last couple of years it has been shown that a long standing problem of hadron physics, that goes under the name of the $U_A(1)$ problem, can be explicitly solved in large N QCD[1]. The result of this analysis is summarized by the following effective Lagrangian[2] for the nonet of the pseudoscalar mesons:

$$\mathcal{L} = \frac{1}{2}\,\mathrm{Tr}(\partial_\mu U \partial_\mu U^+) + \frac{1}{aF_\pi^2} Q^2 + \frac{i}{2} Q(x)\ \mathrm{Tr}(\log U - \log U^+)$$

$$+ \frac{F_\pi}{2\sqrt{2}}\,\mathrm{Tr}\,[MU + MU^+] \qquad (1)$$

$\mathcal{L}$ contains a field U describing the pseudoscalar mesons as bound states of a quark-antiquark pair

$$U_{ij} \leftrightarrow \bar{\psi}_i (1+\gamma_5)\psi_j \qquad (2)$$

and a field Q(x) corresponding to the topological charge density of the Yang-Mills theory:

$$Q(x) = \frac{g^2}{32\pi^2} F_{\mu\nu} \tilde{F}_{\mu\nu} \tag{3}$$

Since (1) is to be used only for computing current algebra predictions, a potential term has been omitted and the meson field U has to satisfy the following constraint:

$$U\,U^{+} = \frac{F_\pi^{\,2}}{2} \mathbb{1} \tag{4}$$

In addition to the quark masses included in the mass matrix M, that explicitly breaks chiral invariance, and the pion decay constant F_π, the effective Lagrangian (1) contains a new dimensional constant A, that can be computed in pure Yang-Mills theory and that is given by:

$$\int dx \left. \langle o|\, Q(x) Q(o)\, |o\rangle \right|_{\text{No quarks}} = A = \frac{1}{2}\, aF_\pi^2 \tag{5}$$

A has been estimated from the spectrum of the pseudoscalar mesons and the following relation (for L = 3) has been obtained[3]:

$$\frac{2\,L\,A}{F_\pi^{\,2}} = m_\eta^2 + m_{\eta'}^2 - 2m_k^2 \underset{\text{Exp.}}{=\!=} 0.726\ (\text{GeV})^2 \tag{6}$$
$$A \approx (180\ \text{MeV})^4$$

The resolution of the $U_A(1)$ problem as well as successful description of the dynamics of the pseudoscalar mesons is then reduced to explain why $A \neq o$ in the quarkless theory, although it vanishes in perturbation theory.

The most promising non-perturbative way to compute A is by using the lattice gauge theory. In the following I want to discuss a recent attempt[4] to extract A with the Monte Carlo technique in the case of a SU(2) lattice gauge theory.

One must first of all give a definition of the topological charge on a lattice. A natural definition is a generalization (in order to have the right behaviour under parity) of the one constructed by Peskin[5]) some time ago that it is given by:

$$Q_n^{(2)} = -\sum_{(\mu\nu\rho\sigma)=\pm 1}^{\pm 4} \frac{\tilde{\epsilon}_{\mu\nu\rho\sigma}}{2^4 32\pi^2} \mathrm{Tr}\left(U_{n;\mu\nu}\, U_{n;\rho\sigma}\right) \tag{7}$$

where $U_{n;\mu\nu}$ is the usual plaquette variable constructed out of the four links:

$$U_{n;\mu\nu} = U_{n;\mu}\, U_{n+\mu;\nu}\, U^{+}_{n+\nu;\mu}\, U^{+}_{n;\nu} \tag{8}$$

and $1 = \tilde{\epsilon}_{1234} = -\tilde{\epsilon}_{2134} = -\tilde{\epsilon}_{-1234}$ etc.

It is easy to check that in the naive continuum limit $[a \to o]$ (7) reproduces the density of topological charge given by (3).

Another definition of the density of topological charge on a lattice, that has also the right naive continuum limit, is given by:

$$Q_n^{(1)} = -\frac{1}{2^9\pi^2}\sum_{(\mu\nu\rho\sigma)=\pm 1}^{\pm} \tilde{\epsilon}_{\mu\nu\rho\sigma}\, \mathrm{Tr}\{U_{n\mu}\, U_{n+\mu,\nu}\, U_{n+\mu+\nu,\rho}\cdot$$

$$U_{n+\mu+\nu+\rho,\sigma}\, U^{+}_{n+\nu+\rho+\sigma;\mu}\, U^{+}_{n+\rho+\sigma;\nu}\, U^{+}_{n+\sigma;\rho}\, U^{+}_{n;\sigma}\} \tag{9}$$

Both definitions (7) and (9) can be used to compute the matrix element (5) and we can therefore construct the following three lattice definitions of A:

$$A_{11} = \frac{1}{a^4}\sum_n \langle o|\, Q_n^{(1)}\, Q_o^{(1)}\, |o\rangle \tag{10a}$$

$$A_{22} = \frac{1}{a^4}\sum_n \langle o|\, Q_n^{(2)}\, Q_o^{(2)}\, |o\rangle \tag{10b}$$

$$A_{12} = \frac{1}{a^4}\sum_n \langle o|\, Q_n^{(1)}\, Q_o^{(2)}\, |o\rangle \tag{10c}$$

The three quantities (10) have been computed with the Monte Carlo technique on a 4^4 lattice with periodic boundary conditions using a computer program based on that of Ref. (6), which uses a 120 element subgroup of SU(2).

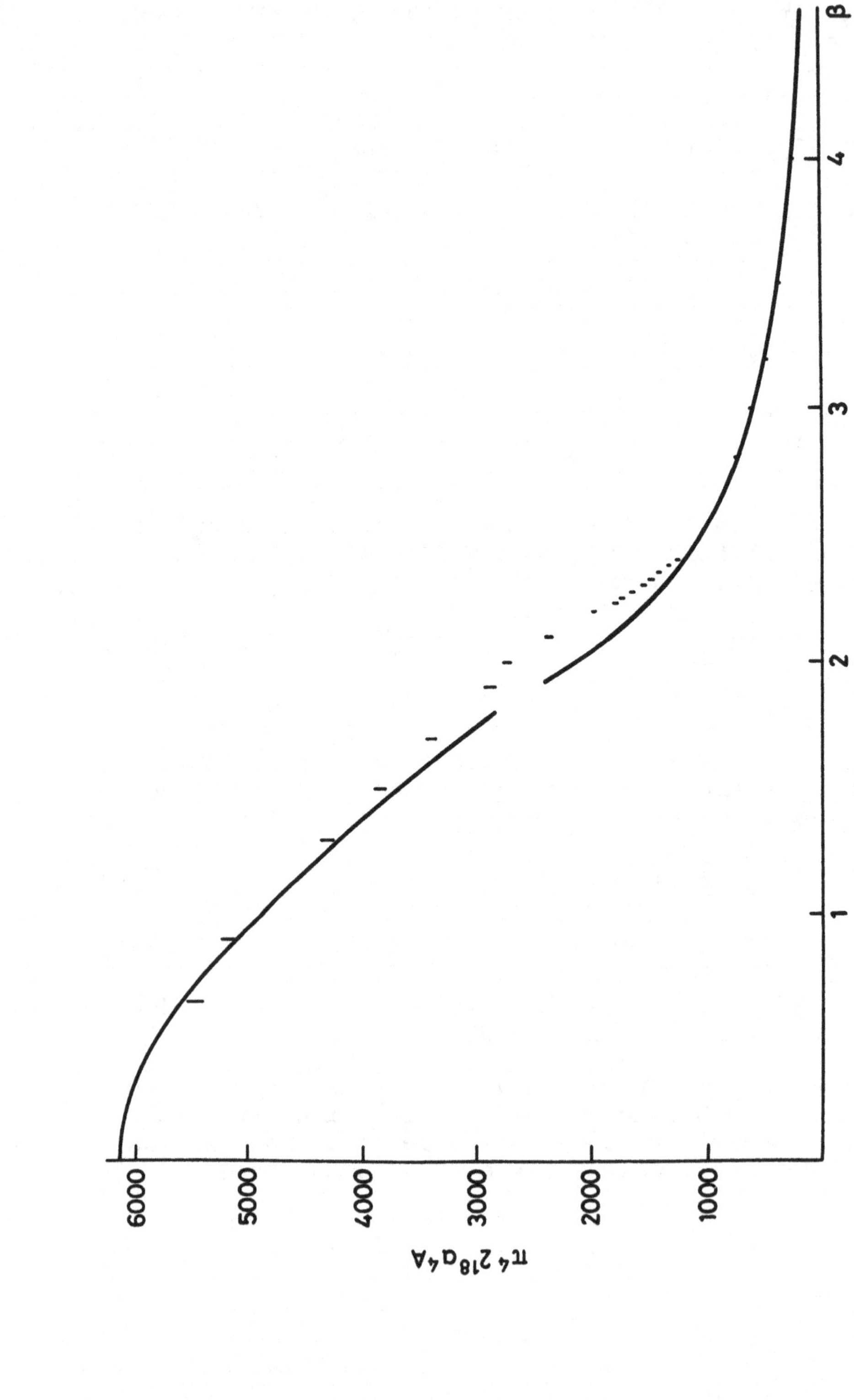

Fig. 1a: Monte Carlo data, strong coupling expansion and perturbative tail for A_{11}.

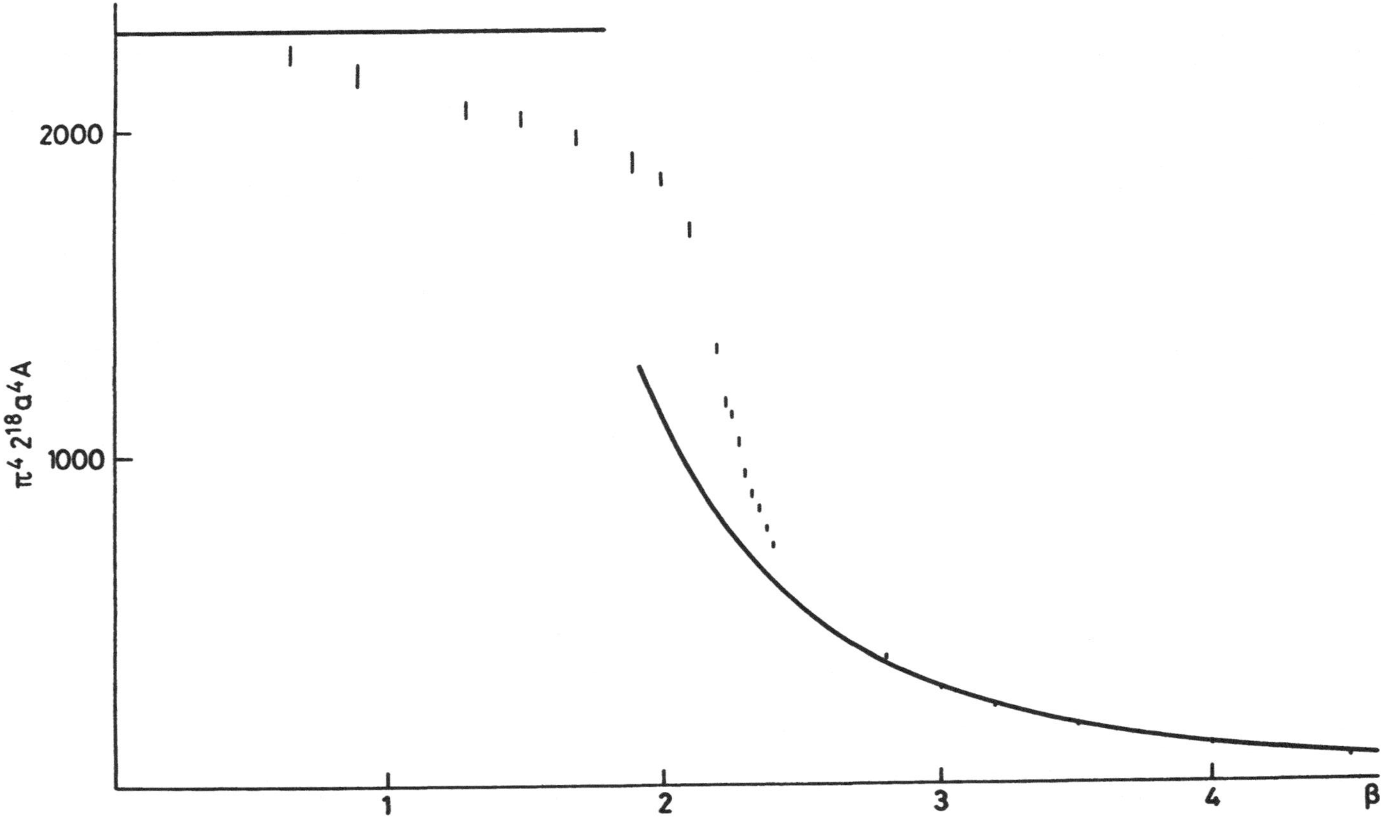

Fig. 1b: Monte Carlo data, strong coupling expansion and perturbative tail for A_{22}.

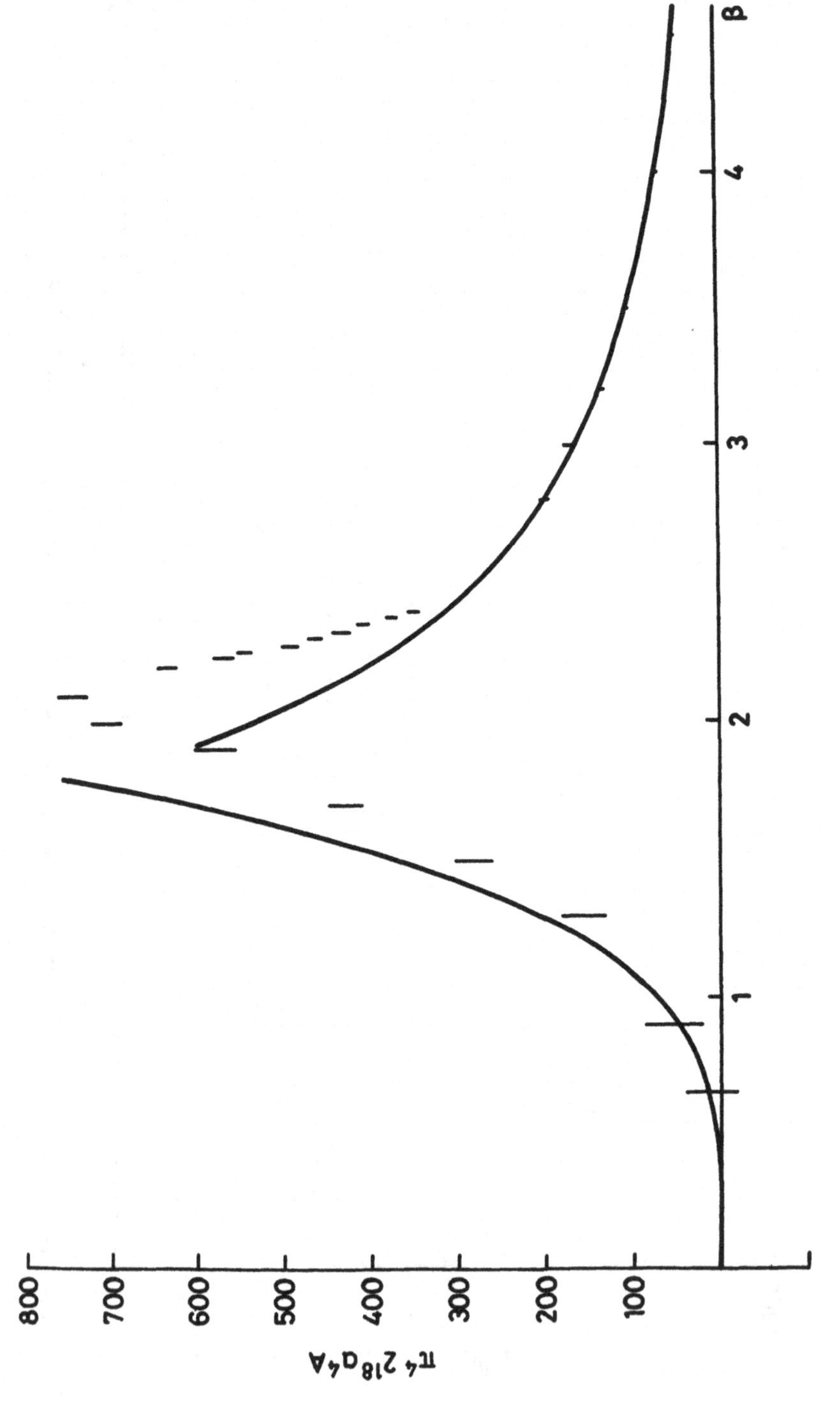

Fig. 1c: Monte Carlo data, strong coupling expansion and perturbative tail for A_{12}.

The data for $\pi^4 2^{18} a^4 A$ are presented in figs. 1a, 1b and 1c for the three quantities defined respectively in eqs. (10a), (10b) and (10c) with their statistical errors as function of $\beta = 4/g^2$. They are in good agreement with the curve of strong coupling for $\beta < 1.5$.

In order to extract the value of a dimensional constant as A from a lattice calculation one must find a region of large β where $A^{1/4}$ shows the dependence on β predicted by the renormalization group. In the case of the quantity A the renormalization group predicts the following behaviour for large β:

$$A^{1/4} a \sim \exp. \left(- \frac{3\pi^2}{11} (\beta - \beta_A) \right) \tag{11}$$

The slope of the exponential is a consequence of the asymptotic freedom behaviour of the theory, while the parameter β_A cannot be predicted with the renormalization group, but can be measured by use of the Monte-Carlo technique.

If we examine however the Monte-Carlo data of figs. 1a, 1b and 1c it is not possible to find a region of β where they follow a behaviour of the type (11). This is a consequence of the fact that both definitions (7) and (9) have a perturbative tail, that covers the exponential non-perturbative behaviour.

In order to extract the non-perturbative exponential (11) the authors of Ref.[4)] have subtracted from the Monte-Carlo data the first two terms of the perturbative tail. The coefficient of the first term behaving as $1/\beta^3$ has been explicitly computed, while the coefficient of the term behaving as $1/\beta^4$ has been fitted from the data in the region $2.8 \leq \beta \leq 4.5$. The weak coupling tails computed in this way are shown in figs. 1a, 1b and 1c. They have been then continued in the region $2.1 \leq \beta \leq 2.4$ and have been subtracted from the data.

The logarithm of the differences between the data and the weak coupling tails are shown in fig. 2 and they seem to follow rather well the straight line predicted by the renormalization group. Comparing these results with those on the string tension[7)], we get:

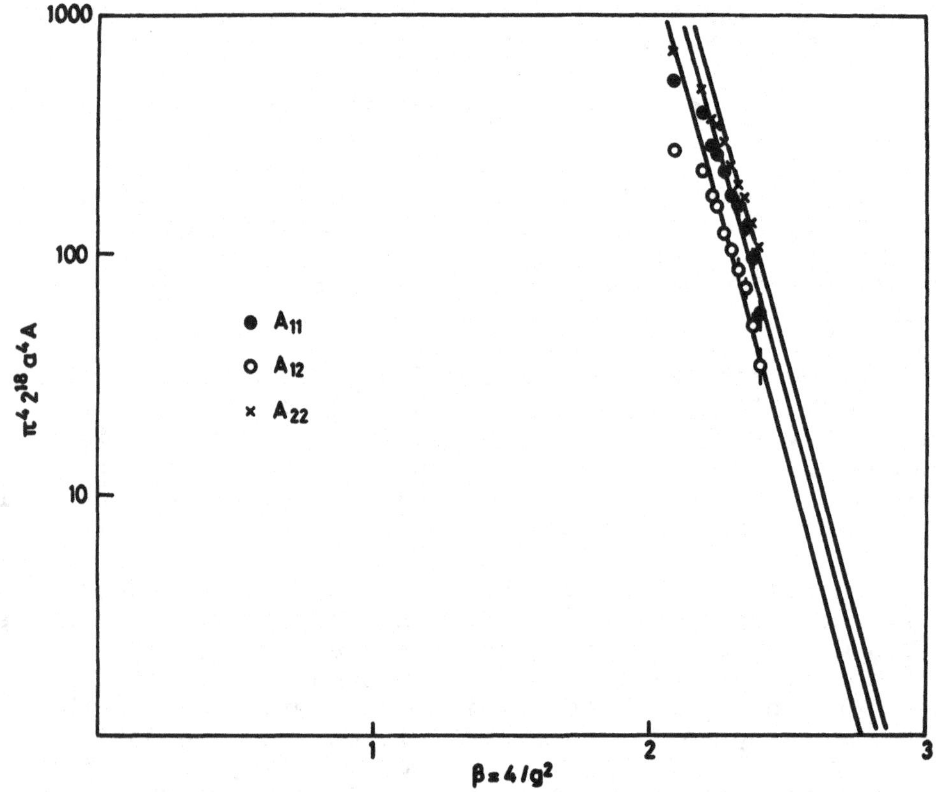

Fig. 2: Subtracted data in the region $2.1 < \beta < 2.4$ for the three definitions A_{11}, A_{22} and A_{12}.

$$A^{1/4} K^{-1/2} = \begin{array}{ll} 0.13 \pm 0.02 & \text{for } A_{11} \\ 0.14 \pm 0.02 & \text{for } A_{22} \\ 0.11 \pm 0.02 & \text{for } A_{12} \end{array} \qquad (12)$$

in reasonable agreement with each other and with the results for A_{11} for the full SU(2) group.

Taking for $\sqrt{k}$ the string model values $\frac{1}{\sqrt{2\pi\alpha'}} = 420$ MeV one obtains the following value for A:

$$A^{1/4} = (55 \pm 10)\ \text{MeV} \qquad (13)$$

The value found with the Monte-Carlo technique is too small with respect to the value (6) found from the spectrum of the pseudoscalar mesons.

This fact can be attributed to several reasons that are discussed in some detail in Ref. (4).

Although the results we have presented here are quite encouraging two weak points should be mentioned.

The first one is the subtraction of the perturbative tail, that is quite delicate due to the uncertainty of the higher order terms. Recently Lüscher[8)] has constructed a definition of the topological change density on a lattice, which is free from a perturbative tail. It would be very interesting to use his definition to compute A.

The other one has to do with the smallness of the lattice that has been used. Hopefully the calculation can be repeated with a bigger lattice.

REFERENCES

1. E. Witten, Nucl. Phys. B156 (1979) 269
   G. Veneziano, Nucl. Phys. B159 (1979) 213
   P. Di Vecchia. Phys. Letters 85B (1979) 213
2. C. Rosenzweig, J. Schechter and G. Trahern, Phys. Rev. D21 (1980) 3388

P. Di Vecchia and G. Veneziano, Nucl. Phys. B171 (1980) 253
E. Witten, Ann. of Phys. (N.Y.) 128 (1980) 363
R. Arnowitt and P. Nath, Northeastern University preprint NUB 2494 (1981).
3. G. Veneziano, Ref. (1)
P. Di Vecchia, F. Nicodemi, R. Pettorino and G. Veneziano, Nucl. Phys. B181 (1981) 318
K. Kawarabayashi and N. Ohta, Nucl. Phys.B175(1980) 477
D. I. Dyakanov and M. J. Eides, Leningrad Nucl. Phys. Inst. preprint 639 (1981)
4. P. Di Vecchia, K. Fabricius, G.C. Rossi and G. Veneziano, Cern preprints TH 3091 and TH 3180 (1981)
5. M. Peskin, Cornell University preprint CLNS 395 (1978) (Thesis)
6. C. Rebbi, Phys. Rev. D21 (1980) 3350
G.B. Bhanot and C. Rebbi, Nucl. Phys. B180 (1981) 469
7. M. Creutz, Phys. Rev. Letters 43 (1979) 553; Phys. Rev. D21 (1980) 2308; Phys. Rev. Letters 45 (1980) 313
8. M. Lüscher, Bern preprint BUTP-4/1981

RADIATIVE SYMMETRY BREAKING

IN GRAND UNIFIED THEORIES

Norbert Dragon and Berthold Stech

Institut für Theoretische Physik
Universität Heidelberg
Philosophenweg 16, D-6900 Heidelberg

ABSTRACT

Motivated by composite models, we start from O(n)-invariant tree potentials for the scalar fields of unified theories and calculate the dynamical symmetry breaking caused by radiative effects. A phenomenologically interesting variety of breaking patterns is obtained. In reducible representations of the scalar fields only the field with the largest Casimir number obtains vacuum expectation values. Right-left symmetric models experience a spontaneous parity breakdown for a range of values of the coupling parameter.

1. INTRODUCTION AND MOTIVATION

Symmetry breaking in unified and grand unified theories poses still a fundamental and so far unsolved problem. The only known mechanism is the Higgs mechanism which involves scalar fields. The representations of these scalar fields must be chosen by hand. Even more annoying, there are several unknown parameters in the Higgs potentials and the Yukawa interactions. These parameters have to be adjusted to obtain the desired breaking pattern and masses and it is not at all obvious why nature should select these special values. In particular, one would like to understand how and why the Glashow-Weinberg-Salam group with its special chiral properties emerges from a grand unified theory. A dynamical explanation could so far not be given but would be very welcome.

It is often assumed that quarks and leptons are composite particles formed (at a scale of 10^{15} GeV or higher) by more fundamental objects (preons...), which are held together by a superstrong "hypercolor" force. Also the scalar fields are then composite fields. In this scenario a unified or grand unified theory is an effective theory which describes the remaining "hyperflavour" interactions between the tightly bound "hypercolor" singlet particles, the quarks, leptons and scalars[1]. We would like to point out that such a picture requires a very simple form of the tree part of the Higgs potential: Because the "hyperflavour" interactions are weak at high energies and can be treated perturbatively, and because the hypercolor force is blind with respect to these interactions, the tree potential for the scalars must be O(n)-invariant where n is the number of real scalar fields. Consequently, the tree potential of a renormalizable interaction consists of only two terms:

$$V_{tree} = \mu^2 \sum_{i=1}^{n} \phi_i\phi_i + r\left(\sum_{i=1}^{n} \phi_i\phi_i\right)^2 \tag{1}$$

For $\mu^2 < 0$ this potential induces a symmetry breaking with $n - 1$ massless Goldstone bosons. The inclusion of the "hyperflavour" interaction, i.e. the forces obtained from the unifying group, will now decisively determine the physics at lower energies: The breaking pattern and the vector boson masses of the unifying group are caused by its own radiative effects. The corresponding expression which has to be added to the tree potential is the famous Coleman-Weinberg potential[2]. The total effective potential is, therefore, in one-loop approximation

$$V_{eff} = V_{tree} + V_{CW}$$

$$V_{CW} = \frac{3}{64\pi^2} \, \mathrm{Tr}\, M^4 \ln \frac{M^2}{\mu_o^2} \tag{2}$$

+ contributions from scalar particle and fermion exchanges

The matrix $M^2(\phi)$ is given by the expression

$$M^{2\,ab}(\phi) = \frac{g^2}{2} \langle\phi \,|\, t^a t^b + t^b t^a \,|\, \phi\rangle$$

where t^a are the generators (acting on the real fields ϕ) of the hyperflavour group which includes ordinary flavour and color as a subgroup. At the minimum of the potential $\phi = \phi_o$, $M(\phi_o)$ is the mass matrix of the vector bosons and $g(\mu_o)$ the gauge coupling constant.

We note that in eq. (2) μ_o is not an arbitrary renormalization point here. It is defined by the requirement that V_{tree} depends on $\vec{\phi}^2$ only (eq. (1)). μ_o parametrizes via the M^4 terms in (2) the ϕ^4 contributions which are not of the form $(\vec{\phi}^2)^2$. Reasonable values of μ_o should lead to small values for the logarithm near the minimum of the potential.

In the following we will neglect the contribution from scalar boson and fermion loops. The former is - as a consequence of (1) - a function of $\vec{\phi}^2$ only and causes therefore no essential modification of V_{eff}. The latter is generally assumed to have very small couplings and, if so, cannot influence the breaking pattern except when ϕ_o turns out to be very close to a singular point (e.g. if one scalar meson becomes very light).

In the present paper we study the breaking pattern which arises from the effective potential (2) with an O(n) invariant tree potential. We investigate the unifying groups[3] SU(5), SO(10), E6 and U(1) x SU(2). In particular, we study large Higgs representations which offer many possibilities for inequivalent symmetry breaking. Such representations are phenomenologically required in SO(10) and E6 models. We also address ourselves to the question of a spontaneous breakdown of parity conservation.

2. PROPERTIES OF THE EFFECTIVE POTENTIAL

Our task is to find the vector $\vec{\phi} = \vec{\phi}_o$ for which the effective potential (2) takes its minimal value. An inspection of $\vec{\phi}_o$ and the multiplet structure of the vector meson masses will then give us the remaining symmetry (the stability group) and the mass spectrum of the gauge bosons. The Higgs masses are obtained from the second derivatives of the potential. We consider separately the two cases:

A) V_{tree} fixes the scale of ϕ^2, $\phi^2 = \phi_o^2 \neq 0$. In this case the effect of the Coleman-Weinberg potential on ϕ_o^2 will be negligible. However, this potential will determine the <u>direction</u> of the vector $\vec{\phi}_o$ and thereby the specific breaking pattern.

B) The tree potential contains no mass term, i.e. $\mu^2 = 0$ in eq. (1). In this case the scale parameter μ_o in eq. (2) sets the scale of the breaking.

To simplify the notation, we introduce the mass matrix m^2 by

$$M^2(\phi) = g^2 \phi^2 m^2(\phi)$$

The potential to be minimized is in case A) proportional to

$$V^A_{eff} = \text{Tr}\,(m^4 \ln m^2) - s\,\text{Tr}\,m^4$$
$$s = \ln\,(\mu_o^2/g^2|\phi_0|^2) \tag{3}$$

All results about the breaking pattern and the relative gauge and Higgs boson masses, apart from the very heavy polar Higgs mass, depend on the single coupling parameter s only.

In case B the variation of the potential with respect to ϕ^2 can be performed. The potential to be minimized with respect to the <u>direction</u> of ϕ is then proportional to

$$V^B_{eff} = -\,(\text{Tr}\,m^4)\,\exp\{\,1 - 2\,\frac{\hat{r} + \text{Tr}\,m^4 \ln m^2}{\text{Tr}\,m^4}\,\} \tag{4}$$

Again the only coupling parameter $\hat{r}$ defined by $r = \frac{3}{64\pi^2} g^4 \hat{r}$ (see eq. (1)) determines the stability group and the boson masses.

It is evident from an inspection of (3) or (4) that a straightforward evaluation of ϕ_o for which V^A_{eff} or V^B_{eff} takes its minimal value is feasible only for small Higgs representations. For large Higgs representations the potential depends on too many components of ϕ such that even a modern computer cannot find the true minimum. For example, in the case of the complex 351 representation of E6, which is certainly of interest[4,5], the effective interaction depends on more than 600 real components of ϕ. The 78 x 78 mass matrix (78 is the number of gauge bosons in E6) must be computed and the potential minimized as a function of all these variables. It is, however, not difficult to see the general trend for the breaking pattern. First we note that for an irreducible Higgs representation we have

$$\text{Tr}\,m^2 = C_2 \tag{5}$$

where C_2 is the Casimir number for the Higgs representation. For large positive s or large $\hat{r}$ the stability group will be large because $\text{Tr}\,m^4$ subject to eq. (5) is larger the more masses are zero. For lower values of s or small values of $\hat{r}$ the term $\text{Tr}(m^4 \ln m^2)$ becomes important. It behaves like a negative entropy and favours a breaking to small subgroups of the unifying group.

If we disregard the fact that the vector masses are variables dependent on ϕ but consider them as independent, constrained only by (5), the expressions (3) and (4) define new functions $\tilde{V}^A$ and $\tilde{V}^B$. The minimum of $\tilde{V}^A$ (or $\tilde{V}^B$) is clearly a lower bound for the potential

V^A_{eff} (or V^B_{eff}). If the vector masses in V^A_{eff} (or V^B_{eff}) depend on many independent components of ϕ - as is the case for symmetry breaking to small subgroups with high dimensional Higgs representations - we expect $\tilde{V}^A$ (or $\tilde{V}^B$) to be a good approximation to the potential V^A_{eff} (or V^B_{eff}).

One obtains the lower bound as the minimum of $\tilde{V}^A$ and $\tilde{V}^B$, if N gauge boson masses are different from zero and equal, while the rest remains zero. A variation with respect to N, where we treat N as a continuous parameter, finally gives

$$\tilde{V}^A_{min} = -\frac{C_2^2}{N_A}, \qquad N_A = C_2\, e^{1-s}$$

$$\tilde{V}^B_{min} = -N_B, \qquad N_B = \frac{C_2^2}{2\hat{r}} \tag{6}$$

In case of Higgs fields in a reducible representation a similar procedure can be applied: For two irreducible representations χ, ψ with Casimir numbers $C_2^{(1)}$ and $C_2^{(2)}$ the replacement

$$C_2 = C_2^{(1)} \cos^2\alpha + C_2^{(2)} \sin^2\alpha$$

has to be made in (5) and (6), where $tg^2\alpha = \psi^2/\chi^2$. If α is varied to reach the true minimum of the potential $\tilde{V}^A$ or $\tilde{V}^B$, it follows that ϕ_o will lie entirely in the direction of the Higgs field with the largest Casimir number:

$$\psi_0 = 0 \qquad \text{for} \quad C_2^{(2)} < C_2^{(1)}$$

The preference of the scalar field with the largest Casimir number C_2 is very important phenomenologically. It is not in conflict with the suggestion that condensates are formed in most attractive channels[6], i.e. in channels with small Casimir numbers. These latter numbers refer to the hypercolor group, the group responsable for the formation of the condensate, not its direction in hyperflavour space.

As a next point we observe that the lower bound of the effective potential as given by (6) can indeed be reached in special circumstances, namely if the following three conditions are fulfilled:

i) a subgroup H of G exists, such that the adjoint representation of G decomposes into the adjoint representation of H plus an irreducible representation of H.

ii) the Higgs field ϕ contains a singlet with respect to H.

iii) s (or $\hat{r}$) is such that N = dim G - dim H

with $N = N^A (N=N^B)$ defined in eq. (6). Obviously these conditions are sufficient to reach the lower bound. As an example consider the breaking of E6 by a ϕ_{351} Higgs field. One finds from branching tables[7] that a number of subgroups fulfil the conditions i) and ii). For instance the decomposition with respect to the group Sp(8) reads

$$78_{adj} = 36_{adj} + 42_{irreducible}$$

Thus, for a value of $\hat{r}$ close to $\frac{1}{2}(C_2^{351})^2/42 = 4.15$, E6 will break to the Sp(8) subgroup. A much smaller stability group is, for instance, a special SU(3) for which the decomposition

$$78_{adj} = 8_{adj} + 70_{irreducible}$$

holds. This breaking occurs for $\hat{r} \simeq \frac{1}{2}(C_2^{351})^2/70 = 2.45$. It is obvious from this result that large Higgs representations lead to a complete breaking of the group if the coupling parameter s or $\hat{r}$ has a sufficiently low value.

3. RESULTS

Let us first discuss the unification group SU(5)[8] and its breaking due to a real Higgs field ϕ which transforms according to the 24 (adjoint) representation of SU(5).

<u>SU(5)</u>

We can always bring ϕ_{24} into a standard form by diagonalizing the matrix ϕ with eigenvalues ϕ_i (i = 1, ...5) constrained by $\sum_i^5 \phi_i = 0$. The Casimir number C_2^{24} is 5 for a conventional normalization of the generators[8]. The 24 gauge boson masses are

$$M_{ij}^2 = \frac{g^2}{2} (\phi_i - \phi_j)^2 \qquad (i, j = 1...5)$$

We minimized the potential V^B given by eq. (4) numerically. It was possible to handle this problem by a computer since only 4 parameters had to be varied (for fixed $\hat{r}$). We found that for $\hat{r} < 1.26$ SU(5) breaks to the Glashow-Weinberg-Salam group $G_{GWS} = U(1) \times SU_L(2) \times SU_C(3)$. For $\hat{r} > 1.26$ the stability group is U(1) x SU(4). In both breaking patterns all the Higgs particles which are not eaten by vector bosons become massive. It is quite satisfactory that the Glashow-Weinberg-Salam group can indeed be obtained for a reasonable range of values of our single coupling parameter. This range even includes $\hat{r} = 0$. A breaking to smaller subgroups does not occur here since the number of independent components of ϕ is small and the vector meson masses are far from being independent variables. A breaking to smaller subgroups would lead to a mass splitting of the single massive vector multiplet and is energetically disfavoured.

SO(10)

For the breaking of SO(10)[9] we use a 126 Higgs field which occurs in the product of two 16 dimensional spinor representations. A large vacuum expectation value of this field is necessary phenomenologically in order to give the right-handed neutrinos large Majorana masses which in turn lead to small masses for the ordinary neutrinos[10, 4].

In the reducible 16 x 16 representation ϕ_{126} has the highest value of C_2 ($C_2^{126} = 25/2$) and will thus be the representation which acquires non-vanishing vacuum expectation values according to the result obtained in the previous section. Although the argument was given only for the lower bound $\tilde{V}^{A,B}$ we found it to be true for the potential $V^{A,B}_{eff}$ in all computer test examples: In case of two Higgs fields, the one with the lowest Casimir number always acquired zero vacuum expectation values.

The number of independent components of ϕ_{126} is of the order $126 - 45 = 81$. Thus we can savely assume that for a low value of s or a small value of $\hat{r}$ the SO(10) symmetry will be broken completely. Let us then go to higher values of these coupling parameters. No breaking to the Glashow-Weinberg-Salam group or the Pati-Salam group can occur since ϕ_{126} (as well as ϕ_{10} and ϕ_{120}) contain no vectors which break to these subgroups. Orbits preferred by larger values of s and $\hat{r}$ are SU(5), SO(7), and SO(5) x SO(5). These subgroups have the highest dimensions, namely 24, 21, 20, respectively. It is interesting to ask whether or not a breaking to SU(5) can occur. Among the three groups, only the breaking to SU(5) implies a spontaneous breaking of the right-left symmetry contained in SO(10). We computed numerically the three orbits and obtained - using the potential V^{A}_{eff} - the following breaking pattern: For $s > 3.1$ the SU(5) orbit lies lowest, for $0.97 < s < 3.1$ the symmetry is broken to SO(7), and for $0.25 < s < 0.97$ the remaining symmetry is SO(5) x SO(5). Although we could neither determine the absolute minimum of the potential algebraically nor numerically with <u>all</u> variables at once, from our experience in many test runs and from the comparison with the lower bounds we are fully convinced that the results are conclusive. Thus, a spontaneous parity breaking SO(10) → SU(5) occurs by radiative effects for a reasonable range of the single coupling parameter s. For the potential V^{B}_{eff} the same breaking pattern was obtained as one might have expected.

E6

E6 models have a number of attractive features for which we refer to the literature[11,4,5]. Here we consider the Higgs field sector. A scalar field in the adjoint (78) representation has an interesting peculiarity: There exists no quartic E6 invariant potential besides $|\phi_{78}|^4$ [12]. Thus our basic assumption of an O(n) symmetry (eq. (1)) is automatically fulfilled (n = 78). As a consequence, the corresponding breaking of E6 is independent of s or $\hat{r}$. As was first shown in ref. 12, one obtains

$$E6 \xrightarrow{\phi_{78}} SO(10) \times U(1)$$

Phenomenology requires, however, the use of scalar fields which transform as the product of two fermion representations

$$27 \times 27 = (27^* + 351)_S + 351'_{AS} \quad .$$

ϕ_{351} has the highest Casimir value $C_2^{351} = \frac{56}{3}$. According to our general argument we expected that only[2] ϕ_{351} gets non-vanishing vacuum expectation values and established this numerically by computer calculations. $\phi_{351} \neq 0$ is very welcome in order to obtain a large mass for the right-handed neutrino[4]. $\phi_{351'} \neq 0$ could also account for this mass but - if used solely - would lead to an unacceptable generation degeneracy. We note that ϕ_{351} (and $\phi_{351'}$) contain singlets with respect to the Glashow-Weinberg-Salam group G_{GWS}, which could principally break E6 down to this group.

For the most general vacuum expectation value compatible with G_{GWS} 4 complex components of ϕ_{351} have to be adjusted such that the potential becomes minimal. It turned out, however, that for all values of s or $\hat{r}$ there exists a different stability group which is energetically preferred. The reason is that the 66 = 78 - 12 vector meson masses, which become massive in the GWS case, fall into several multiplets with mass values which cannot become approximately equal. There are several different groups which do better. They fulfil the conditions i) and ii) mentioned in the previous section. Because the non-vanishing vector meson masses are in one irreducible representation and thus of equal mass, they can reach the lower bound at certain values of s or $\hat{r}$. Unfortunately, these groups are too small to be of interest and break color or charge or both. Thus a direct breaking $E6 \rightarrow G_{GWS}$ cannot be obtained by the mechanism considered by us.

For higher values of s or $\hat{r}$, SO(8), Sp(8), F4, and SO(10) orbits compete with each other. We found that for increasing values of s or $\hat{r}$ first the SO(8) orbit, then Sp(8), then F4, and finally the SO(10) orbit has the lowest energy. Therefore, above a critical value the breaking E6 → SO(10) occurs (Fig. 1). Fixed mixtures of ϕ_{351} with ϕ_{351}, have also been investigated. For ϕ_{351}, at large s an SU(5) x SU(2) orbit which breaks parity shows up.[13] (The field ϕ_{351} is of use in connection with the axion problem.)

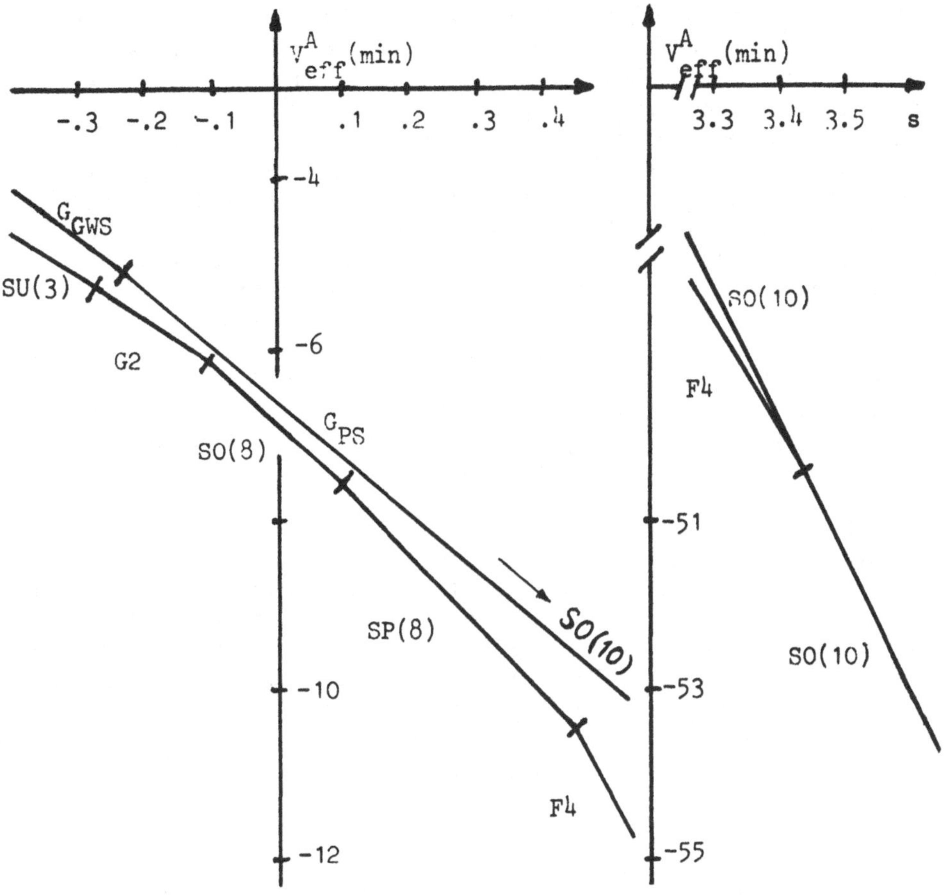

Fig. 1: The minimum of V^A_{eff} for the Higgs field ϕ_{351} of E6.

The lower curve shows the minimum of V^A_{eff} as a function of the coupling parameter s. The resulting stability groups together with the corresponding ranges of s are indicated. The upper curve is obtained by restricting ϕ to components which are singlets with respect to G_{GWS}. With increasing s this orbit turns into the Pati-Salam orbit G_{PS} = $SU_L(2)$ x SU(4) and then bends down. Eventually an G_{PS} = $SU_L(2)$ x $SU_R(2)$ x SU(4) and then bends down. Eventually an SO(10) symmetry appears and the upper and lower curves merge. We note that according to the definition (3) the minimum of V^A_{eff} lies, as a function of s, always on a convex curve.

U(1) x SU(2)

It is conceivable that the mechanism of radiatively induced symmetry breaking discussed here can also be applied to the "low energy" breaking of G_{GWS}. Because of the totally different scale of this breaking compared to the scale of the superstrong breaking, we have to assume that the relevant scalars are quite different from the ones considered above. Perhaps these scalars are composite of composites due to residual hypercolor interactions and additional attractive hyperflavour forces. In any case, it is interesting to apply our method. For a single complex $SU_L(2)$ doublet Higgs field with U(1) charge $Q - I_{3L} = -1/2$, transforming as $(\bar{u}_R u_L)$, $(\bar{u}_R d_L)$, one obviously obtains the desired breaking to the electromagnetic gauge group $U_Q(1)$ independent of s or $\hat{r}$. For the potential V^B, i.e. for $\mu = 0$, one finds, of course, the well-known formula for the Higgs mass[2,14]. If we add another Higgs doublet with $Q - I_{3L} = +1/2$, transforming as $(\bar{d}_R u_L)$ $(\bar{d}_R d_L)$, values for s or $\hat{r}$ above a critical value lead again to the desired electromagnetic $U_Q(1)$ group as the lowest lying orbit. Below the critical value of s or $\hat{r}$ a complete breakdown of the U(1) x $SU_L(2)$ group takes place. In this range of the coupling parameter all Higgs bosons acquire masses.

In the more interesting case of a conserved $U_Q(1)$, one obtains, besides massive Higgs particles of charge ±, 0, two massless pseudo-Goldstone particles, a neutral scalar, and a neutral pseudoscalar (axion type[15]) meson. These particles are massless because of the spontaneous breaking of a global O(3) symmetry of the effective Higgs potential V^A_{eff} or V^B_{eff} at the minimum. The scalar Higgs meson will finally obtain a mass through the fermion loop contributions of eq. (2). The pseudoscalar meson remains massless for Yukawa couplings which respect an axial U(1) symmetry[15].

4. CONCLUSION

The suggested O(n)-invariant tree potential together with radiative effects lead to a surprising variety of breaking patterns of unified groups, although the effective potential depends on a single coupling parameter only. In reducible representations of scalar fields, Higgs fields with large Casimir numbers cause the symmetry breaking. In left-right symmetric groups parity breaking occurs spontaneously for a range of values of the coupling parameter. Scalar fields which transform as the product of fermion fields do not lead to a direct breaking into the Glashow-Weinberg-Salam group foe the unification groups studied here. A step-by-step breaking of E6 down to G_{GWS} is, however, not excluded.

N. D. gratefully acknowledges fruitful discussions during the Summer Institute 1981 in Seattle and financial support from the organizers. B. S. likes to thank the Deutsche Forschungsgemeinschaft for a travel grant and the Institute for Theoretical Physics in Santa Barbara for its hospitality.

REFERENCES

1. For a review on composite particle models and the literature see H. Harari, Proceedings of the Summer Institute on Particle Physics, July 28 - August 8, 1980, SLAC Report No. 239, p. 162 (1981).
   Y. Achiman, Wuppertal preprint WU B81-13 (1981).
2. S. Coleman and E. Weinberg, Phys. Rev. D7, 1888 (1973).
   S. Weinberg, Phys. Rev. D7, 2887 (1973).
3. For a recent review of grand unified theories see J. Ellis, CERN Th 2942 (1980).
4. B. Stech, in "Unification of Fundamental Particle Interactions", Ed.: S. Ferrara, J. Ellis, P. van Nieuvenhuizen, Plenum Press, New York 1980, p. 23.
5. R. Barbieri and D. V. Nanopoulos, Phys. Lett. 91B, 369 (1980).
6. S. Raby, S. Dimopoulos, and L. Susskind, Stanford preprint ITP-653 (1979).
7. W. G. McKay and J. Partera, Lecture notes in pure and applied mathematics V 69, Tables of dimensions, indices, and branching rules for representations of simple Lie algebras, Marcell Dekker Inc., New York 1981.
8. H. Georgi and S. L. Glashow, Phys. Rev. Lett. 32, 438 (1974).
   A. J. Buras, J. Ellis, M. K. Gaillard, and D. V. Nanopoulos, Nucl. Phys. B135, 66 (1978).
9. H. Fritzsch and P. Minkowski, Ann. Phys. 93, 193 (1975).
   H. Georgi and D. V. Nanopoulos, Nucl. Phys. B155, 52 (1979).
10. M. Gell-Mann, P. Ramond, and R. Slansky, unpublished.
11. F. Gürsey, P. Ramond, and P. Sikivie, Phys. Lett. 60B, 177 (1975).
    Y. Achiman and B. Stech, Phys. Lett. 77B, 389 (1978).
    For a review of E6 models and further literature see ref. 4.
12. J. Harvey, Nucl. Phys. B163, 254 (1980).
13. P. H. Frampton and T. W. Kephart, North Carolina preprint IFP 164-UNC.
14. J. Ellis, M. K. Gaillard, D. V. Nanopoulos, C. T. Sachrajda, Phys. Lett. 83B, 339 (1979).
15. For a review on axions see R. D. Peccei, Munich preprint MPI-PAE/PTh 45/81.

GROUND STATE METAMORPHOSIS FOR YANG-MILLS FIELDS ON A FINITE PERIODIC LATTICE

A. Gonzalez-Arroyo*, J. Jurkiewicz**, and C.P. Korthals-Altes+

Centre de Physique Théorique, Section 2
CNRS, Marseille, France

ABSTRACT

We study the weak coupling behaviour of the partition function of non-abelian gauge fields on a finite lattice. Periodic boundary conditions are imposed. Two different power laws in the coupling β^{-1} arise for the partition function, when the dimension d of space time is larger or smaller than a critical dimension d_c. For SU(2) $d_c = 4$ and we find at this dimension power behaviour corrected by $\log \beta$. The phenomenon is of practical importance in Monte Carlo simulations of the twisted action.

1. INTRODUCTION

Traditionally, large β (small kT) behaviour of continuous gauge groups[1,8] on a finite lattice means power law behaviour in β or kT.

In this note we want to report on unexpected behaviour of global quantities like the partition function or the total mean action : for non-abelian gauge theories on a finite periodic lattice we have two different power laws in β , depending on whether we are above or below a critical dimension d_c. d_c depends on the gauge group and equals four for SU(2). In this critical dimension both

*Universidad Autonoma de Madrid, Canto Blanco, Madrid, Spain
**Jagellonian University, Reymonta 4, Cracow, Poland
+Centre de Physique de Théorique, Section 2, CNRS, Luminy, Case 907 13288 Marseille Cedex 9, France

powerlaws coincide but are corrected by a logarithm in β.

The basic reason for the two powerlaw regimes is the presence of singular points on the zeromode manifold , apart from regular points. Small fluctuations around singular points do define a quartic term in the expansion of the action. Small fluctuations around regular points give a quadratic term. Hence a competition between $\beta^{-1/4}$ and $\beta^{-1/2}$ power behaviour, in the partition function or between $\frac{1}{4}$ kT and $\frac{1}{2}$ kT terms in the total action.

The string tension (usually defined through Wilson loops) is related to the ratio of twisted and untwisted partition functions [2,5]. So our results may be relevant for confinement. The arguments used in Section 2 show clearly that our result does not hold in a system with free boundary conditions ; this suggests that local quantities are not affected. In abelian theories the effect is always absent : the effect is specific to non-abelian theories and their defects.

The lay-out of this paper is as follows : in Section 2 we discuss the zeromode manifold of Yang Mills fields on a toroidal lattice ; in Section 3 the disease is diagnosed ; in Section 4 we discuss the relation between many-point and single-point lattices with periodic boundary conditions ; in Section 5, we discuss in detail the single point lattice and comments are reserved for Section 6.

2. THE ZEROMODES OF THE ACTION ON A HYPERTORUS

To obtain the large β behaviour of our gauge system knowledge of the configurations of link variables that minimize the action (zeromodes) is indispensable. They are well-known in the case of periodic boundary conditions[3] (p.b.c.'s) (with or without twist[2]), but it will turn out useful for the discussion in the next two sections to go through their enumeration once more.

Consider a hypercubic lattice in d dimensions of size $a_1 \times a_2 \times \ldots \times a_d \equiv V$ with p.b.c.'s in all directions $\mu = 1,\ldots,d$. Link variables $U_\mu(n)$ SU(N) are defined on all nearest neighbour links $\ell \equiv (n,\mu)$. Any lattice point n is given by a set of d integers $(n_1, n_2, \ldots, n_d)(1 \leq n_\mu \leq a_\mu)$. To count the zeromodes of the action we first fix the axial gauge :

$$\begin{aligned} &U_d(n) = 1 \quad \text{for all } n \text{ with } n_d \leq a_d - 1 \\ &U_{d-1}(n) = 1 \text{ for all } n \text{ with } n_d = 1,\ n_{d-1} \leq a_{d-1} - 1 \qquad (2.1) \\ &U_1(n) = 1 \quad \text{for all } n \text{ with } n_d = n_{d-1} = \ldots = n_2 = 1,\ n_1 \leq a_1 - 1 \end{aligned}$$

The set of links in (2.1) on which $U_\mu(n) = 1$ is called the gauge tree $G_T(d)$ and is represented for the case $d = 2$ in Fig. 1.

The gauge tree $G_T(d)$ has the following properties :

i) in the (d-1) dimensional subspace at $n_d=1$ it defines a gauge tree $G_T(d-1)$.

ii) from i) it follows that $G_T(d)$ visits all points on the lattice, consists of V-1 links and contains no closed loops.

iii) $G_T(d)$ leaves a constant gauge transformation undetermined. This is the only gauge freedom left.

Let the action be the usual Wilson action[1] for the gauge group SU(N) :

$$S = 1 - \frac{1}{2N} (\mathrm{Tr}\, U(P) + c.c.) \tag{2.2}$$

U(P) is the product of link variables $U(\ell)$ along the elementary plaquette P.

The two-dimensional case in Fig. 1 shows all the qualitative features of the d-dimensional case $(d > 2)$ (see Appendix A). The zeromodes in 2 (d) dimensions are those configurations with $U(\ell) = 1$ everywhere except when ℓ lies on one of the 1 (d-1)-dimensional "ladders" L_μ ($\mu = 1,\dots,d$). On each L_μ (consisting of links in the μ-direction with starting point $n = (n_1, n_2,\dots,n_\mu = a_\mu, \dots, n_d)$ we have a constant matrix $U_\mu^{(0)}$, such that :

$$U_\mu^{(o)} U_\nu^{(o)} = U_\nu^{(o)} U_\mu^{(o)} \quad \text{for all } \mu,\nu \text{ with } 1 \leq \mu < \nu \leq d \tag{2.3}$$

The actual construction of these zeromodes goes in two steps :

Step 1. Consider only those plaquettes attached to the gauge tree $G_T(d)$ to have zero action ("attached" means to have at least one link in common with $G_T(d)$). This fixes all links ℓ to have weight $U(\ell) = 1$, except for the links ℓ on L_μ ($\mu = 1,\dots,d$). On L_μ the only freedom left is the constant matrix $U_\mu^{(o)}$.

Step 2. Consider the $\frac{1}{2}d(d-1)$ 2-planes

$T_{\mu\nu}$ ($n_\lambda = 1$, $\lambda \neq \mu$, $\lambda \neq \nu$)

In these planes we have a pattern of gauge-fixed links as in Fig. 1 ; so all plaquettes in these planes have been used in Step 1, except the plaquette $P_{o\mu\nu}$. Fix the action on the $\frac{1}{2}d(d-1)$ $P_{o\mu\nu}$ to be zero, then (2.3) becomes true.

A comment is in order : for the zeromodes of a twisted action[3,5] we have the same construction as for the untwisted case above. Only in Step 2 the matrices $U^{(o)}$ will obey (instead of equ. (2.3)) :

$$U_\mu^{(o)} U_\nu^{(o)} = z_{\mu\nu} U_\nu^{(o)} U_\mu^{(o)} \tag{2.3'}$$

where the $z_{\mu\nu}$ is an element of the center Z(N) of SU(N) ; and $z_{\mu\nu}$ appears in the action S_p on all plaquettes $P_{\mu\nu}$ in the intersection of L_μ and L_ν (and <u>only</u> there) through :

$$S_p = 1 - \frac{1}{2N}\left[z_{\mu\nu}\,\mathrm{Tr}\,U(P_{\mu\nu}) + c.c.\right]$$

Note that zero action on the plaquettes specified in Step 1 and Step 2 implies zero action for all plaquettes.

Let us return to equ.(2.3). We call the zeromodes of the untwisted case "torons" and denote them by $\{U_\mu^{(o)}\}_{\mu=1}^{d}$. Locally a toron is a pure gauge : any contractible Wilson loop $W(C)$ gets the trivial value $W(C) \equiv \frac{1}{N}\,\mathrm{Tr}\,U^{(o)}(C) = 1$ because of equ.(2.3). On the other hand a Wilson loop C_μ that winds around the torus in the μ-direction will acquire a non-trivial value $W(C_\mu) = \frac{1}{N}\,\mathrm{Tr}\,U_\mu^{(o)}$. In other words : though a toron has zero action it cannot be written as a pure gauge with periodic gauge transformation $\Omega(n)$ except when all $U_\mu^{(o)} = 1$.

Step 1 in the construction of the torons suggests how to introduce a new set of variables. Choose the link $\ell_{o\mu}$ in L_μ (see Fig. 1) to be the link in the μ-direction with starting point $n = (\{n_\lambda = 1\},\ \lambda \neq \mu\ ,\ n_\mu = a_\mu)$. Then we define :

$$\begin{aligned} &U(\ell) \equiv \underline{U}(\ell) \quad \text{if} \quad \ell \notin L_\mu && \text{(a)}\\ &U(\ell) \equiv U_\mu \underline{U}(\ell) \quad \text{if} \quad \ell \in L_\mu,\ \ell \neq \ell_{o\mu} && \text{(b)}\\ &U(\ell_{o\mu}) \equiv U_\mu && \text{(c)} \end{aligned} \qquad (2.4)$$

for all $\mu = 1,\ldots,d$.

The variables $\underline{U}(\ell)$ are called Gaussian variables because if $\beta \to \infty$:

$$\underline{U}(\ell) = 1 + O(\beta^{-1/2}) \qquad (2.5)$$

The links on which the $\underline{U}(\ell)$ are defined are called Gaussian links and we have $\underline{N}(d)$ of them. The plaquettes attached to $G_T(d)$ are also $\underline{N}(d)$ in number (see Appendix A). This is useful to know, when we construct an upper bound for the partition function in Section 4b. From its definition it follows that

$$\begin{aligned} \underline{N}(d) &= \text{total number of links} - (\text{number of gauge fixed links}) - d\\ &= (d-1)(V-1) \end{aligned} \qquad (2.6)$$

The d variables U_μ must almost commute when $\beta \to \infty$:

$$[U_\mu U_\nu] \equiv U_\mu U_\nu U_\mu^{-1} U_\nu^{-1} = 1 + O(\beta^{-1/2}) \qquad (2.6')$$

as follows from equ.(2.4).

It is obvious that in a system with free boundary conditions the axial gauge leaves us with <u>only</u> Gaussian links.

3. THE TORON MANIFOLD

3a. The Regular and Singular Torons

The crucial observation is now : the toron manifold consists of two classes of torons $(U_1^{(o)},\ldots,U_d^{(o)})$:

i) regular torons : by definition they define a unique set of eigenvectors $\{\vec{v}_i\}_{i=1}^N$ up to phases

ii) singular torons : by definition there is degeneracy in the eigenvector system. Example : in SU(2) all singular torons are of the form $(C_1, C_2, \ldots, C_d)$ with $C_\mu \in Z(2)$.

The regular torons are characterized by the set of N eigenvectors $\{\vec{v}_i\}_{i=1}^N$ and the set of d diagonal matrices $(\Lambda^1, \ldots, \Lambda^d)$. The eigenvectors do transform under the N(N-1) off-diagonal constant gauge transformations in a non-trivial way (the N-1 diagonal constant gauge transformations only change the phases), so a regular toron carries N(N-1) "charges". Geometrically, we have N(N-1) + d(N-1) tangents to the toron manifold in a regular toron : N(N-1) from the charges, d(N-1) from the phases in the d diagonal matrices.

The singular torons are characterized by a set of p eigenspaces, with dimension V_i, $i = 1, \ldots p$, and at least one $V_i \geq 2$. Obviously $\sum_{i=1} V_i = N$. Now the singular torons are obviously neutral with respect to those off-diagonal gauge transformations that act entirely inside an eigenspace. We have $\sum_{i=1} V_i(V_i-1)$ of those gauge transformations. The remaining charges do transform in a non trivial way a singular toron. In SU(2) e.g. all singular torons are fully neutral.

3b. Fluctuations around Torons and Naive Power Counting

To simplify the discussion we first omit the fluctuations in the $\underline{N}(d)$ Gaussian variables, so we will look at a single point periodic lattice with $\underline{N}(d) = 0$ and discuss the corresponding partition function $Z_1(\beta)$.

The fluctuations around a regular toron are defining a positive quadratic form in the normal modes, i.e. the directions normal to the tangent plane. A normal mode is given by a change from $U^{(o)}_\mu$ to $U^{(o)}_\mu + d\,U^{(o)}_\mu$ where the eigenvectors change from $\vec{v}_i^{\,\mu}$ to $\vec{v}_i + d\vec{v}_i^{\,\mu}$. This gives d N(N-1) directions from which N(N-1) have to be subtracted since they lie in the tangent plane (see 3a). These fluctuations are of order $\beta^{-1/2}$ so we expect a power law

$$Z_1(\beta) \sim \beta^{-1/2\ (d-1)\ N(N-1)} \tag{3.1}$$

from the regular torons alone.

The opposite extreme of a regular toron is a purely neutral toron, commuting with all constant gauge transformations ; such a toron consists only of centergroup elements $\{C_\mu\}$. Then we have, expanding around $\{C_\mu\}$:

$$S = \frac{1}{2N} \mathrm{Tr} \left[d\,U^{(o)}_\mu, dU^{(o)}_\nu\right] \left[dU^{(o)}_\mu, dU^{(o)}_\nu\right]^+$$

Thus we have fluctuations of order $\beta^{-1/4}$ around a purely neutral toron and a power law behaviour

$$Z_1(\beta) \sim \beta^{-1/4\ d(N^2-1)} \qquad (3.2)$$

from the neutral torons alone.

It is evident that the power law (3.1) dominates for

$$d < d_c = \frac{2N}{N-1}$$

For $d > d_c$ it is the power law (3.2) that dominates.
In between these extremes we have singular torons characterized by set of p eigenspaces with dimension V_i that define a partly quadratic, partly quartic form. From what has been said in Section 3a, we deduce easily that such a toron gives a power law

$$Z_1(\beta) \sim \beta^{\left[1/2\ (d-1)\left\{N(N-1)-\sum_{i=1} V_i(V_i-1)\right\}\ -1/4\ d\sum_{i=1}(V_i^2-1)\right]} \qquad (3.3)$$

Using the fact that $N \leq \sum_{i=1} V_i^2 \leq N^2$ it is easy to see that (3.3) never dominates unless $V_i = 1$ $(i=1,\ldots,N)$ (regular toron) or $V_1 = N$, $V_i = 0$ $(i>1)$ (purely neutral toron). This naive power-counting is correct as we show in Section 5 for SU(2). For SU(N) $(N\geq 3)$ we will publish the results elsewhere [13].

Before we go to the verification of (3.1) and (3.2) we will discuss the role of the $\underline{N}(d)$ Gaussian variables $\underline{U}(\ell)$ present in a box of size $V = a_1 x\ a_2 x\ldots x a_d$. It will turn out that they do not interfere with (3.1), (3.2) and (3.3) (see Section 4).

4. RELATING THE ASYMPTOTIC BEHAVIOUR OF SINGLE POINT LATTICE TO THAT OF BIG BOX

The asymptotic behaviour of the partition function $Z_V(\beta)$ of a box with V points $(V = \prod_\mu a_\mu)$ can be expressed in terms of the asymptotic form of $Z_1(\beta)$. This we do by two inequalities :

$$\frac{C_\ell}{\beta^{[1/2\underline{N}(d)(N^2-1)]}} Z_{1a}(\beta) \leq Z_{Va}(\beta) \leq \frac{C_u}{\beta^{[1/2\underline{N}(d)(N^2-1)]}} Z_{1a}(\beta) \qquad (4.1)$$

$\underline{N}(d)$ is the number of Gaussian links (see Section 2) and the subscript a stands for asymptotic form.

In Section 4a, we will prove (4.1) in two dimensions, with $C_u = C_\ell$. In Section 4b, the upper bound in (4.1) is proven for any dimension. In Section 4c, the lower bound is shown.

4a. The Two Dimensional Case

Consider the two dimensional periodic lattice of size $a_1 x\ a_2 \equiv V$ (Fig. 1). We can express the partition function in terms of an integral over two link variables U_1 and U_2 as assigned in Fig. 3 :

$$Z_V(\beta) = \int dU_1\ dU_2\ C_V(\ [U_1 U_2]\ ,\ \mathbb{1}\ ) \qquad (4.2)$$

The symbol $[U_1 U_2]$ is a shorthand for :

$$[U_1 U_2] \equiv U_1 U_2 U_1^{-1} U_2^{-1}$$

and the chain $C_V(X_o, X_V^{-1})$ is defined as follows :

$$C_V(X_o, X_V^{-1}) \equiv \int dX_1 dX_2 \ldots dX_{V-1} \exp\left[-\beta\left\{S_p(X_o X_1^{-1}) + S_p(X_1 X_2^{-1}) + \ldots + S_p(X_{V-1} X_V^{-1})\right\}\right] \quad (4.3)$$

The action S_p in equ.(4.3) is the Wilson action (2.2) with two link-variables put to unity :

$$S_p(X_i X_j^{-1}) = 1 - \frac{1}{2N} (\mathrm{Tr}\, X_i X_j^{-1} + \mathrm{c.c.}) \quad (4.3')$$

The group measure is normalized to unity. See Appendix B for a derivation of equ.(4.2).

The chain in equ.(4.2) can be worked out using the character expansion of the Boltzmann factor :

$$\exp\left[-\beta S_p(U)\right] = \sum_\Lambda \chi_\Lambda(\mathbb{1}) \; \chi_\Lambda(U) \; W_\Lambda(\beta) \quad (4.4)$$

The character $\chi_\Lambda(U)$ is the trace of the irreducible representation $D_\Lambda(U)$, and Λ is the dominant weight.

In integrating out the chain (4.3), we use repeatedly :

$$\int dX_i \exp\left[-\beta\left\{S_p(X_{i-1} X_i^{-1}) + S_p(X_i X_{i+1}^{-1})\right\}\right] =$$

$$\chi_\Lambda(\mathbb{1}) \, \chi_\Lambda(X_{i-1} X_{i+1}^{-1}) \, W_\Lambda^2(\beta)$$

and obtain

$$C_V([U_1 U_2], \mathbb{1}) = \sum_\Lambda \chi_\Lambda(\mathbb{1}) \, \chi_\Lambda([U_1 U_2]) \, [W_\Lambda(\beta)]^V$$
$$\equiv \exp\left[-S_{12}([U_1 U_2], \beta, V)\right] \quad (4.4')$$

Therefore we have found that the partition function on a two dimensional periodic lattice of size $V = a_1 a_2$ equals the partition function on a single point lattice with periodic boundary conditions and with an effective action :

$$Z_V(\beta) = \int dU_1 dU_2 \exp\left[-S_{12}([U_1 U_2], \beta, V)\right] \quad (4.5)$$

Note that in contrast to the case with free boundary conditions our periodic model does not factorize[4] in a set of 1-dimensional spin-spin models. In our case we are left with a two-dimensional model with zero momentum modes only. All the other modes have been integrated out in (4.5), at the expense of having the effective action (4.4').

Had we taken instead of the Wilson action (2.2) the parade horse of the renormalization group, namely the heat kernel action[12] :

$$\exp -S_{H.K.}(\beta, U) \equiv N^{-1}(\beta) \sum_{\Lambda} \chi_{\Lambda}(\mathbb{1})\ \chi_{\Lambda}(U) \exp - \frac{1}{2\beta}(\Lambda+\delta)^2 \tag{4.5'}$$

(where δ equals half the sum of the positive roots in the Lie algebra of $SU(N)$ and $N(\beta)$ is a normalization such that the Boltzmann factor in (4.5') is normalized to unity at $U = \mathbb{1}$), then the action in (4.4') would not have changed in form, only β would have been replaced by β/V.

We are interested in the limit $\frac{\beta}{V} \to \infty$. This allows us to replace the $W_{\Lambda}(\beta)$ from (4.4) by the $W_{\Lambda}(\beta)$ of the heat kernel action (4.5'). In fact any action with the correct naïve continuum limit will permit this replacement in the large $\frac{\beta}{V}$ limit, because in this limit only large Λ contribute to (4.5) and $W_{\Lambda}(\beta)$ for large Λ gets its main contribution from the region around unity in group space. Thus we take :

$$W_{\Lambda}(\beta) = N(\beta)^{-1} \exp\left[-\frac{1}{2\beta}(\Lambda+\delta)^2\right]$$

In Appendix B we show that

$$N(\beta) \sim (\beta)^{\frac{N^2-1}{2}} \qquad \text{as} \qquad \beta \to \infty$$

As a consequence :

$$Z_{a}(\beta) \sim \frac{1}{\beta}\left[1/2(N^2-1)(V-1)\right] Z_{1a}(\beta) \tag{4.5''}$$

where $Z_{1a}(\beta)$ is the asymptotic part of

$$Z_1(\tfrac{\beta}{V}) = \int dU_1\, dU_2 \exp\left[-\frac{\beta}{V} S_{12}([U_1 U_2])\right]$$

The reader will recognize $V-1$ as the number of Gaussian links in $d = 2$. So in two dimensions our inequality (4.1) is an asymptotic equality.

4b. The Upper Bound in Any Dimension

To get the upper bound in equ.(4.1) we introduce a truncated action $\tilde{S}$: $\tilde{S}$ contains only the $N(d) + \frac{1}{2}d(d-1)$ plaquettes used in the construction of the zeromodes (see Step 1 and Step 2 underneath equ.(2.3)). Obviously we have $\tilde{S} \leq S$ and so

$$\tilde{Z}_V(\beta) \equiv \int \prod_{\ell} dU(\ell) \exp\left[-\beta\tilde{S}\right] \tag{4.6}$$

is an upper bound to $Z_V(\beta)$.

Since we threw away many plaquettes (in total $(\frac{1}{2}d-1)(d-1)(V-1)$) we can use the chain-methods from the preceding section to evaluate $\tilde{Z}_V(\beta)$. We kept enough plaquettes to keep the bound non-trivial. The result is :

$$Z(\beta) = [W_o]^{\left[\underline{N}(d) - \sum_{\mu<\nu}(a_\mu a_\nu - 1)\right]} \int \prod_{\mu=1}^{d} dU_\mu \prod_{\mu<\nu} C_{a_\mu a_\nu}([U_\mu U_\nu], \mathbb{1}) \tag{4.7}$$

Note that equ.(4.7) is valid for any β . It can be obtained through an easy induction on d. (See Appendix A). The bulk factor in front of the integral in (4.7) has factored out because of a simple property of chains :

$$\int dX_f \; C_V(X_o, X_f^{-1}) = [W_o(\beta)]^V \tag{4.8}$$

Note that the dependence on X_o disappears after integration over X_f. In the evaluation of $Z_V(\beta)$ we have used (4.8) to "cut the branches of the gauge tree" : X_f represents a link variable associated to only one plaquette, X_o is connected to X_f through a chain without branches and is the first branch point seen from X_f. (See Appendix A). The integral in (4.7) represents the boundary effects.

Three comments :

a) the upper bound in equ.(4.1) can be easily obtained by using (4.4') and (4.5') for the chains $C_{a_\mu a_\nu}([U_\mu U_\nu], \mathbb{1})$ in the integrand of (4.7),

b) if we have twisted boundary conditions $Z_{\mu\nu}$, all that changes in (4.7) is that $C_{a_\mu a_\nu}([U_\mu U_\nu], \mathbb{1})$ is replaced by $C_{a_\mu a_\nu}([U_\mu U_\nu], z_{\mu\nu})$

c) the mean action per plaquette from (4.7) is given by $-\frac{2}{d}\frac{W_o'(\beta)}{W_o(\beta)}$, and has the usual spin wave large β behaviour.

4c. The Lower Bound in Any Dimension

In contrast to the upper bound the lower bound will be only valid for large β . It is easily derived in terms of the Gaussian variables $\underline{U}(\ell)$ defined in Section 2 (equ.(2.4)). Let $\{U_\mu^{(o)}\}_{\mu=1}^{d}$ be a toron. Expand the action in the Gaussian variables $\underline{U}(\ell) = 1 + i\underline{A}(\ell)$ and keep the toron fixed. We will use an abbreviated notation :

$$\underline{A} \cdot X \equiv \sum_{\ell, a} \underline{A}^a(\ell)\, X^a(\ell)$$

The expansion defines a quadratic form :

$$S(\{U_\mu^o\}, \{\underline{U}(\ell)\}) = \underline{A} \cdot \mathcal{M}(\{U_\mu^o\}) \cdot \underline{A}' + O(\underline{A}^3) \tag{4.9}$$

The matrix $\mathcal{M}(\{U_\mu^o\})$ has only positive eigenvalues because

a) $S(\{U_\mu^o\}, \underline{U}(\ell) = 1)$ is a minimum of the action

b) the definition (2.4) prohibits the $\underline{A}(\ell)$ to be tangent to the toron manifold.

In the presence of a non-commuting set $\{U_\mu\}_{\mu=1}^{d}$ the expansion (4.9) becomes :

$$S(\{U_\mu\},\{\underline{U}(\ell)\}) = \sum_{\mu<\nu} N_{\mu\nu}\, S([U_\mu U_\nu]) + L(\{U_\mu\}).\,\underline{A} + \underline{A}.\mathfrak{M}(\{U_\mu\}).\,\underline{A}' + O(\underline{A}^3) \qquad (4.10)$$

The number $N_{\mu\nu}$ is the number of plaquettes contained in the intersection of the ladder L_μ and L_ν defined in Section 2. $L(\{U_\mu\})$ is only non-zero on these intersections. $S([U_\mu U_\nu])$ stands for

$$S([U_\mu U_\nu]) \equiv 1 - \frac{1}{2N}\left[\mathrm{Tr}\; U_\mu U_\nu U_\mu^{-1}\, U_\nu^{-1} + \mathrm{c.c.}\right]$$

The matrix $\mathfrak{M}(\{U_\mu\})$ can have eigenvalues equal to zero or negative. There is, however, a region R around the toron-manifold where $\mathfrak{M}(\{U_\mu\})$ is still positive definite. This region R depends only on the size of the system and the gauge group. Since we are interested in the large β behaviour for fixed size, fluctuations of order $\beta^{-1/4}$ or less away from the toron manifold are of interest to us and will be contained in the region R. Therefore we have :

$$Z_{Va}(\beta) \geq \int_R \prod_\mu dU_\mu \int d\underline{U}(\ell)\, \exp\left[-\beta\, S(\{U_\mu\}, \underline{U}(\ell))\right]$$

Use the expansion (4.10) and integrate out the $\underline{U}(\ell)$ variables :

$$Z_{Va}(\beta) \geq \beta^{\left[-1/2\,\underline{N}(d)(N^2-1)\right]} \int_R \prod_\mu dU_\mu \frac{1}{\det \mathfrak{M}(\{U_\mu\})^{1/2}}\,.$$

$$.\; \exp\left[-\beta\left\{\sum_{\mu<\nu} N_{\mu\nu}\, S([U_\mu U_\nu]) - \frac{1}{4} L(\{U_\mu\}).\mathfrak{M}(\{U_\mu\}).L'(\{U_\mu\})\right\}\right]$$

Use positivity of $\mathfrak{M}(\{U_\mu\})$ on R and the fact that the maximum eigenvalue of $\mathfrak{M}(\{U_\mu\})$ on R can be bound from above by Λ_m to get :

$$Z_{Va}(\beta) \geq [\beta\Lambda_m]^{-\underline{N}(d)\left(\frac{N^2-1}{2}\right)} \int_R \prod_\mu dU_\mu \exp\left[-\beta M \sum_{\mu<\nu} S([U_\mu U_\nu])\right] \qquad (4.11)$$

The remaining integral becomes asymptotically $Z_{1a}(M\beta)$ where M is the largest of the numbers $N_{\mu,\nu}$. Therefore

$$Z_{Va}(\beta) \geq [\beta\Lambda_m]^{-1/2\,\underline{N}(d)(N^2-1)}\; Z_{1a}(M\beta) \qquad (4.12)$$

and we have proven the lower bound in equ.(4.1).

5. BOUNDS FOR $Z_1(\beta)$ (PERIODIC ONE POINT LATTICE)

The bounds for SU(2) will be discussed below. They are of the form :

$$Z_{1\ell}(\beta) \leq Z_1(\beta) \leq Z_{1u}(\beta) \qquad \text{for any } \beta \qquad (5.1)$$

where the leading large β behaviour of $Z_{1\ell}$ and Z_{1u} will turn out to be identical up to a constant depending on d, the dimension. We use the parametrization :

$$U = \exp[\, i\psi\, \vec{n}.\vec{\sigma}\,]$$

where $\vec{\sigma}$ are the Pauli matrices ($\{\sigma_i \sigma_j\} = 2\delta_{ij}$) and $\vec{n}^2 = 1$. Then :

$$Z_1(\beta) = \frac{1}{(2\pi^2)^d} \int_0^{\pi} d\psi_1 \sin^2\psi_1 \;\; d\Omega_1 \ldots \;\; d\psi_d \sin^2\psi_d \, d\Omega_d$$
$$\times \exp\Big[-2\beta \sum_{\mu<\nu} \sin^2\psi_\mu \sin^2\psi_\nu \sin^2\vartheta_{\mu\nu}\Big] \qquad (5.2)$$

The angle $\vartheta_{\mu\nu}$ is defined by :

$$\cos\vartheta_{\mu\nu} = \vec{n}_\mu.\vec{n}_\nu$$

and

$$d\Omega_\mu = d\cos\vartheta_\mu \; d\varphi_\mu \qquad 0 \leq \vartheta_\mu \leq \pi \quad , \; 0 \leq \varphi_\mu \leq 2\pi$$

Symmetry arguments can be used to simplify equ.(5.2). Upon integration over Ω_μ ($\mu = 1,\ldots,d$) we get an integrand which is fully symmetric in variables $\{\psi_\mu\}$. Therefore :

$$Z_1(\beta) = \frac{4\pi}{(\pi^2)^d}\, d \int_0^{\pi/2} d\psi_1 \sin^2\psi_1 \int_0^{\psi_1} \prod_{\mu\geq 2} d\psi_\mu \sin^2\psi_\mu \, Y(\beta, \{\psi_\mu\}) \qquad (5.3)$$

where

$$Y(\beta, \{\psi_\mu\}) \equiv \int \prod_{\mu\geq 2} d\Omega_\mu \exp\Big[-2\beta \sum_{\mu<\nu} \sin^2\psi_\mu \sin^2\psi_\nu \sin^2\vartheta_{\mu\nu}\Big] \qquad (5.4)$$

Rotational invariance permitted us to do the Ω_1 integration. Equs.(5.3) and (5.4) will be the starting points for obtaining the bounds (5.1).

5a. The Upper Bound on $Z_1(\beta)$

We will only keep those plaquettes in S containing a link in the 1-direction :

$$S_u \equiv 2 \sin^2\psi_1 \sum_{\mu=2}^{d} \sin^2\psi_\mu \sin^2\vartheta_{\mu 1} \leq S \qquad (5.5)$$

Replacing S by S_u in (5.4) permits us to do the φ_μ integrations :

$$Y(\beta\, \{\psi_\mu\}) \leq (4\pi)^{d-1} \prod_{\mu=2}^{d} \int_0^1 dx_\mu \exp\Big[-2\beta \sin^2\psi_1 \sum_{\mu=2}^{d} \sin^2\psi_\mu\, x_\mu\Big] \qquad (5.6)$$

where $x_\mu = 1 - \cos\vartheta_{\mu 1}$. The x_μ integrations are trivial and using

$$\frac{2}{\pi}\psi_\mu \leq \sin\psi_\mu \leq \sin\psi_1 \quad \text{for} \quad \mu \geq 2$$

we find after doing the ψ_μ integrations ($\mu \geq 2$) :

$$Z_1(\beta) \leq d\,\pi^d \left(\frac{2}{\pi}\right)^4 \frac{1}{\beta^{d-1}}\, I(\beta) \tag{5.7}$$

where

$$I(\beta) = \int_0^{\pi/2} d\psi_1\, \psi_1^{\,3-d} \left[1 - \exp(-2\beta\psi_1^{\,4})\right]^{d-1} \tag{5.8}$$

Equ.(5.7) is valid for any β .

For large β we find :

$$\begin{aligned} Z_{1a}(\beta) &\leq \frac{2d}{4-d}\,\frac{1}{d-1} && d < 4 \\ Z_{1a}(\beta) &\leq \frac{2}{\beta^3}\log\beta && d = 4 \\ Z_{1a}(\beta) &\leq d\, 2^{-\frac{3}{4}d+4}\, C_u \frac{1}{\beta^{\frac{3}{4}d}} && d > 4 \end{aligned} \tag{5.9}$$

with $C_u \equiv \int_0^\infty dx\, x^{3-d}\,(1 - \exp - u^4)^{d-1}$

5b. The Lower Bound on $Z_1(\beta)$

Restrictions on the domain of the ϑ_μ variables lead to a lower bound. If $0 \leq \vartheta_\mu \leq \frac{\pi}{4}$ we can decouple the Ω_μ integrations since this restriction permits to write :

$$\sin^2\vartheta_{\mu\nu} \leq 2(\sin^2\vartheta_{\mu 1} + \sin^2\vartheta_{\nu 1})$$

Then we have

$$S \leq S_\ell \equiv \sum_{\mu=2}^{d} (1 - \cos\vartheta_{\mu 1})\, R \sin^2\psi_\mu \tag{5.10}$$

where $R \equiv 8\sum_{\mu=1}^{d}\sin^2\psi_\mu$. Again we can do the integrations on the $\vartheta_{\mu 1}$ variables resulting in :

$$Z_1(\beta) \geq d\,\frac{4\pi}{2d}\,\frac{1}{(\beta')^{d-1}} \int_0^{\pi/2} d\psi_1\, \sin^2\psi_1 \prod_{\mu\geq 2}\int_0^{\psi_1} d\psi_\mu \cdot R^{1-d}(1 - \exp - \beta' R\sin^2\psi_\mu)$$

where $\beta' = \beta(1 - \frac{1}{2}\sqrt{2})$.

The integration on the ψ_μ ($\mu \geq 2$) can be done by replacing R by $8 \sin^2 \psi_\mu$ in the exponent and by $8d \sin^2 \psi_1$ elsewhere. Finally we multiply the integrand by

$$\prod_{\mu=2}^{d} \frac{\sin^3 \psi_\mu}{\sin^2 \psi_1} \cos \psi_1 \leq 1$$

to obtain an integrand in the variables ψ_μ ($\mu \geq 2$) that can be explicitly integrated. As a result

$$Z_1(\beta) \geq \frac{C}{\beta^{d-1}} \int_0^{\pi/2} d\psi_1 \, \sin^2 \psi_1 {}^{-5+7} \left[\sin^4 \psi_1 - \frac{1}{4\beta'} \times (1 - \exp(-4\beta' \sin^4 \psi_1))\right]^{d-1}$$

Multiplication of the last integrand by $\cos \psi_1$ gives a simpler integral :

$$Z_1(\beta) \geq \frac{C}{\beta^{d-1}} G(4\beta') \tag{5.11}$$

where

$$C = \frac{4d}{\pi} \frac{1}{[2d\pi^2(1 - \frac{1}{2}\sqrt{2})]^{d-1}}$$

and

$$G(\beta) = \int_0^{\pi/2} dx \, x^{3-d} \left[1 - \frac{1 - \exp(- \quad x^4)}{x^4}\right]^{d-1}$$

Asymptotically (for large β) $G(\beta)$ behaves like $I(\beta)$ in (5.8), so from (5.11) follows the lower bound on $Z_{1a}(\beta)$. Combining both upper and lower bound we obtain, barring pathological behaviour :

$$\begin{aligned} Z_{1a}(\beta) &\sim \frac{1}{\beta^{d-1}} && d < 4 \\ &\sim \frac{1}{\beta^3} \log \beta && d = 4 \\ &\sim \frac{1}{\beta^{\frac{3}{4}d}} && d > 4 \end{aligned} \tag{5.12}$$

as predicted by naïve power-counting (see Section 3b).

6. COMMENTS

i) The bounds on $Z_1(\beta)$ still allows for indefinite oscillations as $\beta \to \infty$. However an exact calculation in $d = 2$ (see Appendix B) shows that these oscillations are absent. Positivity of the action imposes bounds on all derivatives of $Z_1(\beta)$ and they may rule out such behaviour. Apart from this, oscillatory behaviour would be hard to understand from a physical point of view.

ii) An alternative on the use of bounds in deriving the relationship between the asymptotic behaviour of $Z_V(\beta)$ and $Z_1(\beta)$ (see equ. (4.1)) is suggested by the outcome : it is only the zero-momentum modes in $Z_V(\beta)$ that do determine the change in power behaviour. All other modes do contribute to the usual $\beta^{-1/2}$ factors. Therefore in a systematic treatment[10] of the expansion around the zero-modes one should look at the role of zero-momentum excitations. In ref. 2, the zero-momentum contribution to the free energy was calculated in a continuum box with p.b.c.'s in $d = 4$; the integral did diverge for the SU(2) gauge group. This divergence is absent on the lattice and appears through $\log \beta$ as we find in equ.(5.12).

iii) The result (5.12) is relevant for Monte Carlo calculations[7] of the difference between mean twisted action $\langle S_t \rangle_t$ and the mean untwisted action $\langle S \rangle$:

$$\langle S_t \rangle_t - \langle S \rangle \cong \frac{3}{2}\beta^{-1} + \frac{\beta^{-1}}{\log \beta} \quad \text{for } d = 4 \text{ and } SU(2) \qquad (6.1)$$

The left hand-side is related to the string tension[2,3,5] in the infinite volume limit. Notice the term linear in $\frac{1}{2}\beta^{-1}$ which is absent in ref. 7b ; the factor 3 is just the difference between the dimensionality of the twisted zero-mode manifold (which equals 3, and has only regular points, see ref. 3) and the dimensionality of the untwisted zero-mode manifold (which equals 6 in a regular point). The logarithmic term comes from equ.(5.12), and makes the use of the traditional finite size scaling corrections somewhat questionable.

iv) Is our result (5.12) of any relevance to the continuum limit (i.e. first $V \to \infty$, then $\beta \to \infty$) ? Remember that the lower-bound in (4.1) was obtained through arguments leaning heavily on the finiteness of the volume (see below equ.(4.10)), thus restricting ourselves to $\beta \to \infty$ first. Thus (6.1) is valid when the correlation length is much bigger than the size of the box, outside the region where we see continuum physics. In any case, the phenomenon of ground state metamorphosis is independent of the action used, provided it has the correct naïve continuum limit.

v) The physics of the phenomenon is that the fluctuations of order $\beta^{-1/2}$ around a charged (i.e. regular) toron are overtaken by the fluctuations of order $\beta^{-1/4}$ around a neutral toron at $d = 4$, (SU(2)), where ground state metamorphosis takes place. In view of ref. 6 where configuration with coinciding eigenvalues are shown to represent magnetic monopoles[9], we would like to call the power law $\beta^{-3/4d}$ a magnetic power law, and $\beta^{-(d-1)}$ an electric power law. The appearance of a logarithm at the cross over point $(d = 4)$ is not unfamiliar.

vi) The phenomenon disappears on a lattice with free boundary conditions. The situation is reminiscent of that of the chiral anomaly : imposing a translationally invariant infrared cut-off (p.b.c.'s) introduces the change in power laws. If we renounce from translation invariance (free boundary conditions) we have only Gaussian links, therefore only fluctuations of order $\beta^{-1/2}$ and no change in power-laws.

APPENDIX A

First we show :

A.I. Every plaquette attached to the gauge tree $G_T(d)$ defines one Gaussian link and vice-versa.

Proof : In $d = 2$ the statement can be read of from Fig. 1 immediately : first fit in the plaquettes with 3 links in common with $G_T(2)$, then those with two links in common, etc. In the end we have fitted V-1 plaquettes (the crossed one in Fig. 1 is not attached to $G_T(2)$) which equals indeed the number of Gaussian links $\underline{N}(2)$. The proof proceeds by induction.

We want to prove A.I for a d-dimensional hypercube $(\prod_{\mu=1}^{d} a_\mu \equiv V)$ with the gauge tree $G_T(d)$ as defined in equ.(2.1.). Use property i) of gauge trees : for $n_d=1$ fixed, we have a d-1 dimensional hypercube with a gauge tree $G_T(d-1)$, consisting of $a_1a_2\ldots a_{d-1}-1$ links. We suppose the induction hypothesis to be valid for this (d-1)-dimensional hypercube : $\underline{N}(d-1)$ equals the number of plaquettes attached to $G_T(d-1)$ and lying entirely inside the d-1 dimensional subspace.

Attach first the $(a_1a_2\ldots a_{d-1}-1)$ plaquettes to $G_T(d-1)$ with one side in the negative d-direction. In the positive d-direction we can attach $(a_d-1)\{(d-1)a_1a_2\ldots a_{d-1}\}$ plaquettes to $G_T(d)$. This is in total a number of

$$P \equiv (d-1)\ (a_1a_2\ldots a_d-1) - (d-2)(a_1a_2\ldots a_{d-1}-1) \tag{A.1}$$

plaquettes attached to $G_T(d)$ with a link in the d-direction. Adding to A.1 $\underline{N}(d-1) = (d-2)(a_1a_2\ldots a_{d-1}-1)$ according to the induction hypothesis, gives indeed a total of $\underline{N}(d) = (d-1)(a_1a_2\ldots a_d-1)$ as we wanted to prove.

A.II. Proof of equ.(4.7).

For $d = 2$, equ.(4.7) reduces to equ.(4.2) so is correct in $d = 2$. For d dimensions we assume (4.7) to be true in the (d-1)-

dimensional subspace at $n_d = 1$ fixed :

$$Z_{d-1}(\beta) = \left[W_o\right]^{\left[N(d-1) - \sum_{\mu<\nu\leq d-1} (a_\mu a_\nu - 1)\right]} \times$$

$$\times \int \prod_{\mu\leq d-1} dU_\mu \prod_{\mu<\nu\leq d-1} \cdot C_{a_\mu a_\nu}\left(\left[U_\mu U_\nu\right], \mathbb{1}\right)$$

The quantity we want to calculate is $Z_d(\beta)$. All 2-planes with a component in the d-direction have a pattern of gauge fixed links as depicted in Figs. 1 and 2. According to Step 1 and Step 2 in Section 2, there are in all these 2-planes at least two free links, except in the (d-1) 2-planes $T_{d\mu}$ mentioned in Step 2. Since no gauge fixed link is present in any (d-1)-dimensional subspace at fixed $n_d > 1$, we can use chain methods in all $T_{d\mu}$ planes. In the $T_{d\mu}$ planes with free links we can use equ.(4.8) with X_f being the weight of the free links and X_o being a link variable at $n_d=1$. So all these 2-planes factor out in the bulk. Only the d-1 $T_{d\mu}$ planes in which all plaquettes are kept stay in the integrand. Using the result A.I for the bulk part we get equ.(4.7).

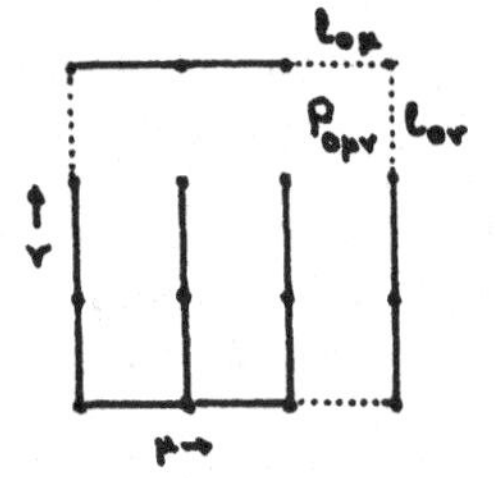

Figure 1.

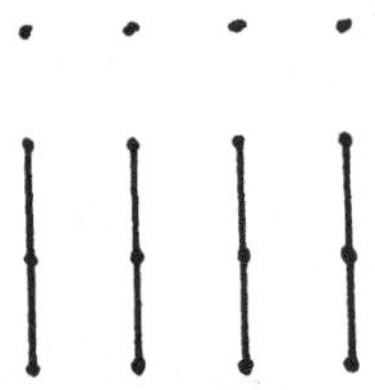

Figure 2.

APPENDIX B

The Two-Dimensional Case for All Groups SU(N)

i) First we will derive equ.(4.2) in the text. The easiest way to understand the result is to introduce "railway" gauge (see Fig. 3); it is then clear that :

$$Z_V(\beta) = \int dU_1 dU_2 dU_1' dU_2' \; C_{(a_1-1)a_2}(U_2, U_2'^{-1}) \; P(U_2' U_1 U_2^{-1} U_1'^{-1}) \times C_{a_2-1}(U_1', U_1^{-1}) \qquad (B.1)$$

In (B.1) P(X) stands for $\exp\left[-\beta\left\{1 - \frac{1}{2N}(\mathrm{Tr}\,X + \mathrm{c.c.})\right\}\right]$.

Use now $C(U,V^{-1}) = C(UX, X^{-1}V^{-1}) = C(YU, V^{-1}Y^{-1})$

and $\int dV\, C(U,V^{-1})\, C(V,W) = C(U,W)$

to write (B.1) as :

$$Z_V(\beta) = \int dU_1 dU_2 \; C_V(U_1 U_2 U_1^{-1} U_2^{-1}, \mathbb{1}) \tag{B.2}$$

Formula (B.2) (or equ.(4.2) in the main text) is the well-known result of decimation of variables[4], but now on a periodic lattice.

Equ.(B.2) can be further reduced to equ.(4.4') :

$$Z_V(\beta) = \int dU_1 dU_2 \sum_\Lambda \chi_\Lambda(\mathbb{1}) \; \chi_\Lambda(U_1 U_2 U_1^{-1} U_2^{-1}) [W_\Lambda(\beta)]^V \tag{B.3}$$

and orthonormality of matrix elements of irreducible representations :

$$\int dU \; D_{m_1 m_2}(U) \; \overline{D_{m_1' m_2'}(U)} = \frac{1}{\chi_\Lambda(\mathbb{1})} \delta_{m_1 m_1'} \delta_{m_2 m_2'}$$

can be used to do the integrations over U_1 and U_2 in (B.3). The result is :

$$Z_V(\beta) = \sum_\Lambda [W_\Lambda(\beta)]^V \tag{B.4}$$

ii) Asymptotic behaviour of (B.4) for the heat kernel action (SU(N)).

We argued in the text below equ.(4.5) that in the limit $\frac{\beta}{V} \to \infty$ the weight function $W_\Lambda(\beta)$ could be replaced by the asymptotic form of the weight function at the heat kernel :

$$W_\Lambda(\beta) = \frac{1}{N(\beta)} \exp\left[-\frac{1}{2\beta}(\Lambda+\delta)^2\right] \tag{B.5}$$

where Λ equals the dominant weight vector of the representation and where δ equals half the sum of the positive roots of the Lie-algebra of SU(N)[12c]. $N(\beta)$ is a normalization factor such that

$$\exp - S_{H.K.}(U,\beta) = \sum_\Lambda \chi_\Lambda(\mathbb{1}) \; \chi_\Lambda(U) \; W_\Lambda(\beta) \tag{B.6}$$

equals 1 when $U = \mathbb{1}$, and with $W_\Lambda(\beta)$ as in equ.(B.5).

We will now show

$$\begin{aligned} &\text{(a)} \quad N(\beta) \sim \beta^{1/2\,[N^2-1]} \quad \text{as} \quad \beta \to \infty \\ &\text{(b)} \quad Z_{1a}(\beta) \sim \beta^{-1/2 \; N(N-1)} \quad \text{as} \quad \beta \to \infty \end{aligned} \tag{I}$$

To this end we recall some basic features of the heat kernel action [12c].

The summation over Λ can be represented[12c] for SU(N) by a summation Σ' over N integers, called the electric variables $\{e_a\}_{a=1}^N$, with the constraint :

$$\sum_{a=1}^N e_a = 0, 1, \ldots, N-1 \tag{B.7}$$

In terms of the N angles $\{\vartheta_a\}_{a=1}^N$ appearing in the diagonal form of U we have :

$$\exp - S_{H.K}(U,\beta) = N^{-1}(\beta) \prod_{a<b} \frac{N/2}{(b-a)\sin\frac{1}{2}\vartheta_{a,b}} \frac{\partial}{\partial\vartheta_{a,b}} I(\vartheta,\beta) \tag{B.8}$$

The function $I(\vartheta,\beta)$ equals :

$$I(\vartheta,\beta) = \sum_{\{e_a\}}{}' \prod_{a<b} \exp\left[\frac{i}{N} e_{a,b}\vartheta_{a,b} - \frac{1}{2\beta N}(e_{a,b})^2\right] \tag{B.9}$$

and $\vartheta_{a,b} = \vartheta_a - \vartheta_b$

$e_{a,b} = e_a - e_b$

By introduction of an extra angular variable φ one undoes the constraint (B7) on the electric variables $\{e_a\}$:

$$\sum_{\{e_a\}}{}' = \sum_{\{e_a\}} \frac{1}{2\pi}\int_0^{2\pi} d\varphi \sum_{\ell=0}^{N-1} e^{i\varphi(E-\ell)} \quad , \text{ where } E = \sum_{a=1}^{N} e_a \tag{B.10}$$

We perform on (B.9) a duality transformation with the Poisson formula

$$\sum_{\ell} \delta(\ell-\ell') = \sum_m e^{i2\pi m\ell'}$$

and get [12c] an expression with summation over "magnetic" variables $\{m_a\}_{a=1}^{N}$, with the constraint

$$\sum_a m_a = 0 \tag{B.11}$$

This constraint comes about when integrating out the variable introduced in equ.(B.10). The resulting form for $I(\vartheta,\beta)$ is then :

$$I(\vartheta,\beta) = \sqrt{N}(2\pi\beta)^{\frac{N-1}{2}} \sum_{\{m_a\}}{}' \prod_{a<b} \exp-\frac{\beta}{2N}(\vartheta_{a,b} - 2\pi m_{a,b})^2 \tag{B.12}$$

and the heat kernel action can be written as :

$$\exp - S_{H.K.}(\beta U) = N^{-1}(\beta)\, C \sum_{\{m_a\}}{}' \prod_{a<b} \frac{(\vartheta_{a,b} - 2\pi m_{a,b})}{2\sin\frac{1}{2}(\vartheta_{a,b} - 2\pi m_{a,b})}$$

$$\times \exp - \frac{\beta}{2N}(\vartheta_{a,b} - 2\pi m_{a,b})^2 \tag{B.13}$$

with

$$C = \frac{\sqrt{N}}{\prod_{a<b}(b-a)} (2\pi\beta)^{\frac{N-1}{2}} \beta^{\frac{1}{2}N(N-1)} \tag{B.14}$$

and $m_{a,b} = m_a - m_b$.

The normalization $\exp - S_{H.K.}(U = \mathbb{1}, \beta) = 1$ means

$$N(\beta) \cong \frac{\sqrt{N}}{\prod_{a<b} (b-a)} (2\pi\beta)^{\frac{N-1}{2}} \beta^{\frac{1}{2} N(N-1)} \qquad \text{if } \beta \to \infty$$

This constitutes the first part a) of assertion I.

To show I(b) one has to realize that the leading behaviour of

$$\sum_{\Lambda} \exp\left[- \frac{1}{2\beta} (\Lambda + \delta)^2 \right]$$

comes from the function $I(\vartheta, \beta)$, equ.(B.12). Therefore

$$\sum_{\Lambda} \exp\left[- \frac{1}{2\beta} (\Lambda + \delta)^2\right] \cong \frac{1}{N!} \cdot \sqrt{N} \cdot (2\pi\beta)^{\frac{N-1}{2}}$$

and indeed we find

$$Z_{1a}(\beta) \sim \beta^{-\frac{1}{2} N(N-1)} \qquad \text{for } SU(N),\ d = 2$$

in accord with the prediction of naive power counting equ.(3.1).

The reader can easily convince himself that the Wilson action equ.(2.2) with gauge group SU(2) has as weights :

$$W_{\ell}(\beta) = \frac{2}{\beta} \cdot I_{2\ell+1}(\beta) \qquad \ell = 0, \frac{1}{2}, \ldots$$

The functions $I_n(\beta)$ are the modified Bessel functions[11].
Using the relation [11] :

$$\exp \frac{1}{2} (t + t^{-1}) z = \sum_{n=-\infty}^{+\infty} t^n I_n(z)$$

one can readily see that

$$Z_1(\beta) = \frac{2}{\beta} \sum_{\ell=0,\frac{1}{2},\ldots} I_{2\ell+1}(\beta)$$

(see equ.(B.4) with V = 1) gives

$$Z_1(\beta) = \frac{1}{\beta} (1 - e^{-\beta} I_o(\beta)) \cong \frac{1}{\beta} \quad \text{as} \quad \beta \to \infty$$

Thus the asymptotics of the Wilson action is the same as for the heat kernel action, as expected.

ACKNOWLEDGEMENTS

The authors are indebted to the Institute of Theoretical Physics in Utrecht for hospitality. They enjoyed fruitful discussions with

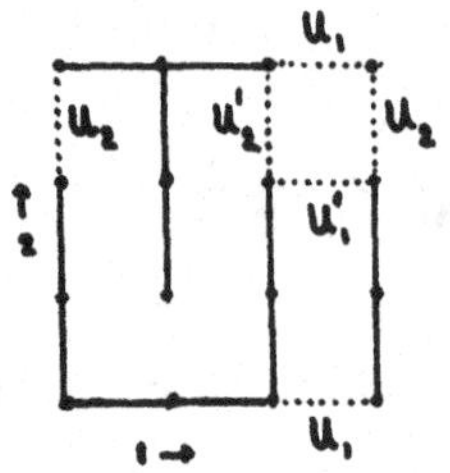

Figure 3: Railway gauge for periodic lattice (d=2).

Prof. G.'t Hooft. Special thanks are due to Dr. J. Groeneveld whose advice was essential for the genesis of this paper.

One of the authors (J.J.) acknowledges financial support from the Stichting Fundamental Onderzoek der Materie (F.O.M.).

REFERENCES

1. K.G. Wilson, Phys. Rev. D10, 2455 (1974).
2. G.'t Hooft, Nucl. Phys. B153, 141 (1979).
3. J. Groeneveld, J. Jurkiewicz and C.P. Korthals Altes, Physica Scripta 23, 1022 (1981).
   For a continuum generalisation, see G.'t Hooft, Comm. Math. Phys. 81, 267 (1981).
4. e.g. L.P. Kadanoff, Rev. Mod. Phys. 49, 267 (1977).
5. G. Mack in "Recent developments in gauge theories", G.'t Hooft et al. eds., Plenum Press, New York (1980).
   G. Munster, Nucl. Phys. B180 FS2 , 23 (1981).
6. G.'t Hooft, Caltech Preprint 68/819 (1981).
7. (a) J. Groeneveld et., Phys. Lett. 92B, 312 (1980).
   (b) G. Mack, E. Pietarinen, Phys. Lett. 94B, 397 (1980).
8. (a) B.E. Baaquie, Phys. Rev. D16, 2602 (1977).
   (b) V.F. Müller, W. Rühl, Ann. of Phys. 133, 240 (1981).
9. A.M. Polyakov, JETP Lett. 20, 194 (1974).
   G.'t Hooft, Nucl. Phys. B79, 276 (1974).
10. A.M. Polyakov, Nucl. Phys. B120, 429 (1977).
   G.'t Hooft, Phys. Rev. D14, 3432 (1976).
11. M. Abramovitch, I.A. Steegun, Hand book of Mathematical Functions.
12. (a) C.B. Lang, C. Rebbi, P. Salomonson, S. Skagerstam, CERN-TH 3021.
   (b) E. Onofri, P. Menotti, CERN-TH 3026.
   (c) A. Gonzalez-Arroyo, C.P. Korthals Altes, preprint Marseille CPT-81/P.1303, to be published in Nucl. Phys.
13. A. Gonzalez-Arroyo, J. Jurkiewicz, C.P. Korthals Altes, to be published.

LATTICE FERMIONS AND MONTE CARLO SIMULATIONS

P. Hasenfratz

CERN, Geneva, Switzerland

INTRODUCTION

QCD is asymptotically free. Short distance properties can be investigated by using perturbation theory. The rate by which the asymptotic behaviour is approached as the energy is increased is set by a scale parameter, by the so-called Λ parameter. Λ has the dimension of a mass. It is the basic external parameter of the theory, the "coupling" so to say. Its value is directly measurable in deep inelastic experiments.

Now, it is rather thrilling that having Λ (and the quark masses) QCD should predict all the hadron masses, widths and other spectroscopical data without any new parameters. Of course, these are non-perturbative problems - non-perturbative problems in a $d = 4$ relativistic quantum field theory.

The past in particle physics is rather bleak in this respect. In classical statistical physics however, the analogous problem - as it was discussed during the first week of this school - has a respectable history. Powerful non-perturbative methods have been developed, like the high temperature expansion, Monte Carlo simulation, renormalization group methods and so on. We would like to take advantage of this accumulated knowledge. The lattice not only provides for a gauge invariant regularization, but opens the way towards these non-perturbative techniques[1].

LATTICE REGULARIZATION

The continuous Euclidean space is replaced by discrete lattice points. Consider a cubic lattice. The lattice distance is denoted by "a", while a lattice point is characterized by $n_\mu = (n_1, n_2, n_3, n_4)$ integers. The Fourier transform of a function is defined by

$$\tilde{f}(p) = a^4 \sum_n e^{ip\cdot na} f(n) ,$$

$$f(n) = \frac{1}{(2\pi)^4} \int_{-\pi/a}^{\pi/a} d^4p \, e^{-ip\cdot na} \, \tilde{f}(p) .$$

The lattice provides for a cut-off in momentum space:

cut-off momentum = π/a.

When $a \to 0$, cut-off $\to \infty$, the regularization is removed. The lattice is only a regularization. There is no QCD and lattice QCD separately. Even the lattice enthusiasts do not want to live on a coarse grained lattice. Just, as people using dimensional regularization do not want to stay at 3.5 dimensions.

CONTINUUM LIMIT

$a \to 0$ must be taken. This limit is not trivial, however. Consider pure gauge theory for a moment. The model is described by two parameters: a and the bare coupling g. Assume, we succeeded in calculating a physical mass in this theory. On dimensional ground we have

$$m = \frac{1}{a} f(g).$$

In the $a \to 0$ limit g must also be turned in order to keep m finite. In the continuum limit m should be much smaller than the cut-off. In the language of statistical physics, the mass gap (dimensionless mass) goes to zero, i.e., the system approaches a continuous phase transition point:

$$\begin{Bmatrix} a \to 0 \\ g \to g^* \end{Bmatrix} .$$

g^* is the bare coupling of continuum QCD. Asymptotic freedom dictates $g^* = 0$.

The lattice distance a and the coupling g are tuned together. It is required by renormalizability that when a is small (the lattice is fine, the cut-off is large) physics should become independent of a. For instance, we have

$$\frac{d}{da} m_{phys} = 0 \ , \qquad \begin{Bmatrix} a \to 0 \\ g \to 0 \end{Bmatrix} ,$$

implying

$$m_{phys} = c \cdot \Lambda^{latt} ,$$

where

$$\Lambda^{latt}_{\begin{Bmatrix} a \to 0 \\ g \to 0 \end{Bmatrix}} = \frac{1}{a} e^{-\frac{1}{2\beta_0 g^2}} (\beta_0 g^2)^{-\frac{\beta_1}{2\beta_0^2}} , \text{ fixed} .$$

Every physical mass (dimensional quantity) can be expressed in terms of a simple, renormalization group invariant mass parameters. Λ^{latt} is defined in complete analogy to the Λ parameter of the usual continuum formulation.

Historically, first a simplified problem was investigated: pure gauge theory without dynamical quarks. Without discussing the results of this approximation in detail, let me remind you of two points only:

- the results obtained in pure gauge QCD are reasonable;
- the scope is narrow; most of the interesting questions involve dynamical quarks.

In trying to extend the theory (by including dynamical quarks) one meets theoretical and technical problems. The theoretical problems are centered around the question of chiral symmetry breaking while the technical problems are related to the fact that it is difficult to find an effective Monte Carlo procedure for this extended theory.

THEORETICAL PROBLEMS OF INTRODUCING FERMIONS ON A LATTICE

The naive way of putting fermions on the lattice follows the usual recipe. In the continuum action of a free Dirac particle the derivatives are replaced by differences resulting in the lattice action:

$$S = a^n \sum_n \left[\sum_\mu \frac{1}{2a} (\bar{\psi}(n)\gamma_\mu \psi(n+\hat{\mu}) - \bar{\psi}(n+\hat{\mu})\gamma_\mu \psi(n)) - m\bar{\psi}(n)\psi(n) \right].$$

Let us try to describe QED on the lattice. The gauge field should be introduced in a gauge invariant way

$$S_{QED} = a^n \sum_n \Big[\sum_\mu \frac{1}{2a}(\bar{\psi}(n)\gamma_\mu U_{n\mu}\psi(n+\hat{\mu}) - \bar{\psi}(n+\hat{\mu})\gamma_\mu U^+_{n\mu}\psi(n))$$

$$- m\bar{\psi}(n)\psi(n)\Big] + \frac{1}{g^2} \sum_{\substack{\text{plaquettes} \\ (\square_{n\ \ n+\hat{\mu}}^{n+\hat{\mu}+\hat{\nu}})}} (U_{n\mu}U_{n+\hat{\mu},\nu}U^+_{n+\hat{\nu},\mu}U^+_{n\nu} + cc),$$

where $U_{n\mu} = e^{igaA_{n\mu}} \in U(1)$. For m = 0 the model is invariant under the transformations

$$\psi_n \to e^{i\alpha}\psi_n \quad , \quad \bar{\psi}_n \to \bar{\psi}_n e^{-i\alpha}$$

and

$$\psi_n \to e^{i\alpha\gamma_5}\psi_n \quad , \quad \bar{\psi}_n \to \bar{\psi}_n e^{i\alpha\gamma_5}$$

The symmetry group is $U(1)_{vector} \otimes U(1)_{axial}$. This symmetry is exact for any value of the lattice constant, therefore it is there in the $a \to 0$ continuum limit also.

However, there is a general theorem due to Adler[2] claiming that although the classical theory is $U(1) \otimes U(1)$ invariant, there is no regularization which could respect both of these symmetries. In the quantum theory the chiral symmetry is necessarily explicitly broken. The axial vector current is not conserved, its divergence receives a non-zero contribution from the triangle graph[3]:

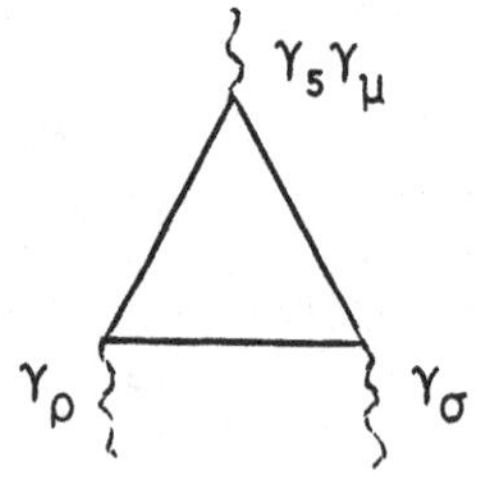

The construction above seemingly contradicts this theorem. The resolution of this paradox is the following. Though we wanted to describe the interaction of a single fermion with the electromagnetic field, actually our action describes 16 identical fermion species. Running around the triangle graph all of these fermions give a contribution to the axial anomaly, but - as it was shown by Karsten and Smit[4] - their contribution alternates in sign and adds up to zero.

It is easy to see that this action describes 16 fermion species. The free fermion part of the action

$$\sim(\bar{\psi}(n)\,\gamma_\mu\,\psi(n+\hat{\mu}) - \bar{\psi}(n+\hat{\mu})\,\gamma_\mu\,\psi(n))$$

gives the propagator in momentum space:

$$S(p) \sim \frac{1}{\sum_\mu \gamma_\mu \sin p_\mu}\ , \quad p_\mu \in (\overline{-\pi},\pi).$$

This propagator has 16 poles at the points p_μ = (0000), $(\pi,0,0,0)$, $(0,\pi,0,0),\ldots,(\pi,\pi,\pi,\pi)$.

Chiral symmetry implies species doubling on the lattice. If we want to describe a single fermion, the U(1) chiral symmetry must be explicitly broken. We should accept it, it cannot be otherwise due to Adler's theorem.

Similarly, for the general case with n_f fermions, we accept that the flavour singlet U(1) axial symmetry is explicitly broken since it is a phenomenon which is general, independent of the lattice and unavoidable. However, we would like to keep the SU(n_f) axial symmetry. That is the point where the solutions suggested until now are not satisfactory.

Let me briefly discuss Wilson's method of removing the degeneracy[5]. Wilson suggested to add a new term to the action with the following properties

- it gives large ($\sim$ cut-off) masses to the 15 unwanted fermions
- it goes to zero in the formal $a \to 0$ continuum limit.

After adding this new term the QCD action has the following form:

$$S = \sum_n \Big\{ -\bar{\psi}^{a,i}_\alpha(n)\psi^{a,i}_\alpha(n) + \sum_\mu K_i \bar{\psi}^{a,i}_\alpha(n)(1-\gamma_\mu)_{\alpha\beta} U^{ab}_{n\mu}\psi^{b,i}_\beta(n+\hat{\mu}) +$$

$$+ \sum_\mu K_i \bar{\psi}^{a,i}(n+\hat{\mu})(1+\gamma_\mu)_{\alpha\beta} U^{+ab}_{n\mu}\psi^{b,i}_\beta(n)\Big\} +$$

$$+ \frac{1}{g^2} \sum_{\text{plaquettes}} (\mathrm{Tr}\, U_p + cc),$$

where a,b; α,β; i are colour, Dirac and flavours indices respectively. $U_{n\mu} \in SU(3)$, the last term is the usual gauge field action.

By finding the coefficient of the quadratic term $\bar{\psi}(n)\psi(n)$, K_i can be related to the bare quark mass in the limit $(ma) \to 0$:

$$K_i = \frac{1}{2(m_i a) + 8} \qquad i = u,d,s,c,\ldots \quad (n_f \text{ flavours})$$

There is no species doubling in this prescription. Unfortunately, the new terms explicitly break chiral symmetry. We hope that $SU(n_f)_{axial}$ will be restored in the continuum limit, and even more, it will be broken spontaneously.

The amplitude of moving a quark by one lattice unit is proportional to K_i. Therefore the name: hopping parameter. If $g = 0$, $K = 1/8$ defines a massless fermion. For $g \neq 0$ there are mass corrections (there is no chiral symmetry which would prevent the occurrence of mass counterterms) and the value of K giving a massless fermion is not 1/8, but it receives perturbative and non-perturbative contributions. K_i should be renormalized. $K_i = K_i(g^2)$ is not known a priori.

TECHNICAL PROBLEMS AND METHODS

In a path integral formulation the fermion fields are represented by anticommuting c numbers, by Grassmann variables. There is no effective way of representing them on a computer. In every method the first step is to get rid of them.

The action is only quadratic in the fermion fields. In a concise notation it has the form:

$$S = \sum_{i,j} \bar{\psi}_i \Delta_{ij}(U)\psi_j + S_{gauge} \quad ,$$

where i,j represent all kinds of indices. One may integrate over the fermion fields in the vacuum functional or in any expectation value of fermion fields:

$$\int D\psi D\bar{\psi}\ e^{-S} = \det(\Delta(U))e^{-S_{gauge}}$$

$$\int D\psi D\bar{\psi}\ \bar{\psi}_i\psi_j\ e^{-S} = \Delta^{-1}(U)_{ij} \det \Delta(U)\ e^{-S_{gauge}}$$

The second example describes the propagation of a quark in the background field U. This background field is generated with a probability distribution governed by $S_{eff} = S_{gauge} - \mathrm{Tr}\,\ell n\,\Delta(U)$. The second term in S_{eff} represents the effect of virtual quark loops.

A. Direct numerical methods

One possibility is to study this effective gauge theory by direct numerical methods. As the problem of calculating $\mathrm{Tr}\,\ell n\,\Delta(U)$ can also be reduced to the calculation of $\Delta^{-1}(U)$ [6], one needs a fast algorithm for inverting $\Delta(U)$. The problem is especially acute in the effective action, since $\Delta^{-1}(U)$ [6], one needs a many times, in principle at every updating step.

There are different suggestions for this inversion problem[6-9]. The Gauss-Seidel iteration method (suggested by Petcher and Weingarten[7]) has the advantage of being free of statistical errors inherently present in other procedures. The matrix $\Delta(U)$ can be written as $\Delta(U) = 1 - KB(U)$. The equation

$$\chi = (1 - KB(U))^{-1}\phi$$

can be rearranged as

$$\chi = KB(U)\chi + \phi$$

and this form is ready for iteration.

The different methods suggested until now are common in being rather time consuming. There are interesting results in one and two dimensions. Marinari, Parisi and Rebbi analyzed the $d = 2$ Schwinger model recently[10] and the results are encouraging. But this problem is rather far from $d = 4$ QCD. The only result on a $d = 4$ non-Abelian theory is due to Petcher and Weingarten[7]. They studied an SU(2) gauge theory with two flavours on a 2^4 lattice. The size of the lattice reflects the difficulties involved. In order to measure the one-plaquette and the thermal-loop expectation values to a precision of ∿10%, approximately ∿50 hours of equivalent CDC computer time was needed.

B. Hopping parameter expansion combined with Monte Carlo

A rather different method is the so-called hopping parameter expansion combined with an ordinary Monte Carlo simulation over the gauge fields[11,12].

The hopping parameter expansion[5] is analogous to the high temperature expansion in statistical physics in many respects. The amplitude of moving a quark by one lattice unit is proportional to the hopping parameter K. An expansion in K is equivalent to an expansion in the length of quark paths in configuration space. In their propagation the quarks are constrained by the maximum order or the expansion, nevertheless they should still gather the essential information on the hadron's structure. This defines the conditions under which the expansion to a given order will be reliable. The size of the hadron in lattice units should not be too large, it has to be comparable to the regions covered by possible quark paths. Or, alternatively, the lattice distance a cannot be too small.

In Ref. 12 the method has been tried on the problem studied previously by Petcher and Weingarten[7]. For comparison, $K = 1/8$ was taken as in Ref. 7. For this value of K, the expansion seemed to be rapidly convergent and the results were reproduced easily, reducing the required computer time by orders of magnitude. Unfortunately, as we discussed earlier, K should be renormalized, and it is an open question, how the hopping parameter expansion works for physical K values.

The preliminary studies discussed above indicated that by using direct numerical methods the effect of virtual quark loops would be extremely difficult to take into account. There are arguments however, supporting the approximation where they are neglected (large N_c limit, Zweig rule, ...). The ambitious programme of calculating the meson and baryon spectrum seems to be feasible in this approximation.

In the hopping parameter expansion method the virtual quark loops can be included up to the given order rather easily. Of course, this method is plagued by the usual disease of expansions: the presence of systematical errors which are difficult to control.

In spite of the obvious difficulties spectacular progress might be expected in this field in the near future. Masses, widths and the like from basic principles ... an almost unbelievable prospect.

* * *

At the end of 1981 exciting results were published on this subject[13-16].

REFERENCES

1. K.G. Wilson, Phys. Rev. D10:2445 (1974);
   J.B. Kogut and L. Susskind, Phys. Rev. D11:395 (1975); for a recent summary and for further references see:
   P. Hasenfratz, EPS International Conference on High Energy Physics, Lisbon (1981), CERN preprint TH.3157 (1981) and 1981 International Symposium on Lepton and Photon Interactions at High Energies, Bonn, CERN preprint TH.3187 (1981).
2. S.L. Adler, Brandeis University Summer Institute in Theoretical Physics, Eds. S. Deser, M. Grisaru and H. Pendleton, MIT Press, Cambridge (MA) and London (1970).
3. S.L. Adler, Phys. Rev. 177:2426 (1969);
   J.S. Bell and R. Jackiw, Nuovo Cimento 60A:47 (1969).
4. L.H. Karsten and J. Smit, Nucl. Phys. B183:103 (1981).
5. K.G. Wilson, in New Phenomena in Subnuclear Physics, A. Zichichi Ed., Erice 1975, Plenum Press, New York (1977).
6. F. Fucito, E. Marinari, G. Parisi and C. Rebbi, Nucl. Phys. B180:[FS2] 369 (1981).
7. D.H. Weingarten and D.N. Petcher, Phys. Lett. B99:333 (1981).

8. D.J. Scalapino and R.L. Sugar, Phys. Rev. Lett. 46:519 (1981).
9. A. Duncan and M. Furman, Nucl. Phys. B190[FS3] 767 (1981).
10. E. Marinari, G. Parisi and C. Rebbi, Nucl. Phys. B190[FS3] 734 (1981).
11. C.B. Lang and H. Nicolai, Nucl. Phys. B200[FS4]:135 (1982).
12. A. Hasenfratz and P. Hasenfratz, Phys. Lett. 104B:489 (1981).
13. E. Marinari, G. Parisi and C. Rebbi, Brookhaven preprint BNL-30212 (1981).
14. D.H. Weingarten, Indiana University preprint IUHET-69 (1981) + Errata.
15. H. Hamber and G. Parisi, Brookhaven preprint BNL 30170 (1981), revised version.
16. A. Hasenfratz, P. Hasenfratz, Z. Kunszt and C.B. Lang, CERN preprint TH.3220 (1981).

INDEX